Science Fiction, Disruption and Tourism

THE FUTURE OF TOURISM

Series Editors: Ian Yeoman, *Victoria University of Wellington, New Zealand* and **Una McMahon-Beattie**, *Ulster University, Northern Ireland, UK*

Some would say that the only certainties are birth and death; everything else that happens in between is uncertain. Uncertainty stems from risk, a lack of understanding or a lack of familiarity. Whether it is political instability, autonomous transport, hypersonic travel or peak oil, the future of tourism is full of uncertainty but it can be explained or imagined through trend analysis, economic forecasting or scenario planning.

This new book series, *The Future of Tourism*, sets out to address the challenges and unexplained futures of tourism, events and hospitality. By addressing the big questions of change, examining new theories and frameworks or critical issues pertaining to research or industry, the series will stretch your understanding and generate dialogue about the future. By adopting a multidisciplinary perspective, be it through science fiction or computer-generated equilibrium modelling of tourism economies, the series will explain and structure the future – to help researchers, managers and students understand how futures could occur. The series welcomes proposals on emerging trends and critical issues across the tourism industry and research. All proposals must emphasise the future and be embedded in research.

All books in this series are externally peer-reviewed.

Full details of all the books in this series and of all our other publications can be found on http://www.channelviewpublications.com, or by writing to Channel View Publications, St Nicholas House, 31-34 High Street, Bristol BS1 2AW, UK.

THE FUTURE OF TOURISM: 6

Science Fiction, Disruption and Tourism

Edited by
**Ian Yeoman,
Una McMahon-Beattie and
Marianna Sigala**

CHANNEL VIEW PUBLICATIONS
Bristol • Jackson

DOI https://doi.org/10.21832/YEOMAN8670
Library of Congress Cataloging in Publication Data
A catalog record for this book is available from the Library of Congress.
Names: Yeoman, Ian, editor. | McMahon-Beattie, Una, editor. | Sigala, Marianna, editor.
Title: Science Fiction, Disruption and Tourism/Edited by Ian Yeoman, Una McMahon-Beattie and Marianna Sigala.
Description: Bristol, UK; Blue Ridge Summit, PA: Channel View Publications, 2022. | Series: The Future of Tourism: 6 | Includes bibliographical references and index. | Summary: "This book examines science fiction's theoretical and ontological backgrounds and how science fiction applies to the future of tourism. Focusing on disruption, sustainability and technology, it brings a new theoretical paradigm to the study of tourism in a post COVID-19 world and can be used to explore, frame and even form the future of tourism"— Provided by publisher. Identifiers: LCCN 2021036438 (print) | LCCN 2021036439 (ebook) | ISBN 9781845418663 (paperback) | ISBN 9781845418670 (hardback) | ISBN 9781845418687 (pdf) | ISBN 9781845418694 (epub)
Subjects: LCSH: Tourism—Planning. | Contents tourism. | Literary journeys. | Science fiction. Classification: LCC G155.A1 S354 2022 (print) | LCC G155.A1 (ebook) | DDC 306.4/819—dc23
LC record available at https://lccn.loc.gov/2021036438
LC ebook record available at https://lccn.loc.gov/2021036439

British Library Cataloguing in Publication Data
A catalogue entry for this book is available from the British Library.

ISBN-13: 978-1-84541-867-0 (hbk)
ISBN-13: 978-1-84541-866-3 (pbk)

Channel View Publications
UK: St Nicholas House, 31-34 High Street, Bristol, BS1 2AW, UK.
USA: Ingram, Jackson, TN, USA.

Website: www.channelviewpublications.com
Twitter: Channel_View
Facebook: https://www.facebook.com/channelviewpublications
Blog: www.channelviewpublications.wordpress.com

Copyright © 2022 Ian Yeoman, Una McMahon-Beattie, Marianna Sigala and the authors of individual chapters.

All rights reserved. No part of this work may be reproduced in any form or by any means without permission in writing from the publisher.

The policy of Multilingual Matters/Channel View Publications is to use papers that are natural, renewable and recyclable products, made from wood grown in sustainable forests. In the manufacturing process of our books, and to further support our policy, preference is given to printers that have FSC and PEFC Chain of Custody certification. The FSC and/or PEFC logos will appear on those books where full certification has been granted to the printer concerned.

Typeset by Nova Techset Private Limited, Bengaluru and Chennai, India.

Contents

	Figures and Table	vii
	Contributors	ix
	Foreword	xiii
	Introduction	
1	Science Fiction: A Theoretical Lens and Methodological Approach for Re-Imagining Tourism Futures *Ian Yeoman, Una McMahon-Beattie and Marianna Sigala*	3
	Part 1: Ontological Approaches	
2	Science Fiction and the Future of Tourism *Ian Yeoman and Una McMahon-Beattie*	19
3	The Future, the Devil and the Deep Blue Sea *Felicity Picken*	30
4	Prosuming Existential Authenticity in Dystopian Spaces *Peter Robinson*	42
5	Space Tourism – Science Fiction Becoming a Reality *Annette Toivonen*	56
	Part 2: Science Fiction	
6	A Life Without Limits: Design, Technology and Tourism Futures in *Westworld* *Leon Gurevitch*	73
7	Harry Potter and the Future of Tourism *Ina Reichenberger*	84
8	Wildlife Tourism in 2150: Uplifted Animals, Virtual and Augmented Reality and Everything In-between *Giovanna Bertella*	97
9	Tears in the Rain: Tourism in the World of *Blade Runner* and *Total Recall* *Peter Bolan*	109

10 Destination of the Dead: The Future for Tourism? 119
    Mairead McEntee, Ruairi McEntee, Una McMahon-Beattie and Ian Yeoman

Part 3: Disruption

11 Holidays with Inspector Maigret: Mixed Reality Adventures as Value Drivers in Future Tourism 133
    Stephan Bingemer

12 Digital Destinations and Avatar Tourists: A Futuristic Look at Virtual Reality Tourism and Its Real-World Impacts 145
    Daniel Guttentag

13 The 'Safety Bubble' and the Future of Enclave Tourism 161
    Cecilia de Bernardi

Part 4: Dystopia

14 The Coming of the Fugue and the Blind Tourist? 175
    Stuart Reid and Richard Ek

15 Technological Frontiers: From the Wild West Myth to the Dystopia of the Westworld's Post-human Theme Park 187
    Jane Lovell and Sam Hitchmough

16 The Future of Music Concerts and Tourism in Dystopian Times 200
    Daniel Wright

17 Exclusion Tourism: Sci-Fi Stalkers and Subjunctive Plays in Apocalyptic Destinations from Chernobyl to Plymouth, Montserrat 213
    Magdalena Banaszkiewicz and Jonathan Skinner

18 Hotel Anthropocene 234
    Martin Gren and Emily Höckert

Part 5: Concluding Thoughts

19 Developing a Theoretical Framework of Science Fiction and the Future of Tourism: A Cognitive Mapping Perspective 255
    Ian Yeoman, Una McMahon-Beattie and Marianna Sigala

Index 330

Figures and Table

Figures

Figure 5.1	Futures Map	59
Figure 10.1	Paint and old posters peel from the walls of a section behind the basketball arena, built for the 2004 Athens Olympics	127
Figure 10.2	The swimming facilities, while still in use, are slowly falling into disrepair	127
Figure 17.1	Montserrat Exclusion Zones from 1997 to 2011	222
Figure 17.2	Montserrat Hazard Level System Zones, 1 August 2014	223
Figure 17.3	Branded MVO and Hazard Alerts as of 21 December 2018	224
Figure 17.4	Monitoring the Soufriere Hills volcano from the MVO	224
Figure 17.5	Viewing the Soufriere Hills volcano from the MVO and the Montserrat Springs Hotel	225
Figure 17.6	Tour guide negotiating access to the exclusion zone	227
Figure 17.7	Guiding the exclusion zone – Plymouth	227
Figure 19.1	Science fiction and futures theories	259
Figure 19.2	The future, the devil and the deep blue sea	261
Figure 19.3	Prosuming existential authenticity in dystopian spaces	263
Figure 19.4	Space tourism – science fiction becoming a reality	265
Figure 19.5	Design, technology and tourist futures in *Westworld*	266
Figure 19.6	Harry Potter and the future of tourism	268
Figure 19.7	Wildlife tourism in 2150: uplifted animals, virtual and augmented reality and everything in-between	270
Figure 19.8	Tourism in the world of *Blade Runner* and *Total Recall*	272

Figure 19.9	Destination of the dead: the future for tourism?	273
Figure 19.10	Holidays with Inspector Maigret: mixed reality adventures as value drivers in future tourism	276
Figure 19.11	Digital destinations and avatar tourists: a futuristic look at virtual reality tourism and its real-world impacts	277
Figure 19.12	Safety bubble and the future of enclave tourism	279
Figure 19.13	The coming plague of the fugue and the blind tourist?	280
Figure 19.14	Technological frontiers: from the Wild West myth to the dystopia of the Westworld's post-human theme park	282
Figure 19.15	The future of music concerts and tourism in dystopian times	284
Figure 19.16	Exclusion tourism: sci-fi and subjunctive plays in apocalyptic destinations from Chernobyl to Plymouth, Montserrat	286
Figure 19.17	Hotel Anthropocene	288
Figure 19.18	Conceptualisation of science fiction and tourism	293
Figure 19.19	The essence of science fiction	294
Figure 19.20	Authenticity	297
Figure 19.21	Dystopia	301
Figure 19.22	Plurality of the future(s)	303
Figure 19.23	Disruption & transformation	305
Figure 19.24	Liminality	308
Figure 19.25	Tourism as portrayed as science fiction	310
Figure 19.26	Technological singularity has arrived: *Westworld*	314
Figure 19.27	Sustainability	316
Figure 19.28	The future of tourism: a science fiction theoretical framework	319

Table

| Table 17.1 | Exclusion Zone definitions | 222 |

Contributors

Editors

Ian Yeoman is an Associate Professor of Tourism Futures at Victoria University of Wellington and Visiting Professors at the European Tourism Futures Institute and Ulster University. Dr Yeoman is co-editor of the *Journal of Tourism Futures* and co-editor of Channel View's The Future of Tourism series. He is author and editor of over 20 books, including the forthcoming *Scenarios for Global Tourism: 2075*. Outside the future, Ian is New Zealand's number one Sunderland AFC fan.

Una McMahon-Beattie is Professor and Head of Department for Hospitality and Tourism Management at Ulster University (UK). Her research interests include tourism futures, tourism and event marketing and revenue management. Una is co-editor of Channel View's The Future of Tourism series and sits on the editorial board of the *Journal of Tourism Futures*. She is the author/editor of a number of books, including *The Future Past of Tourism: Historical Perspectives and Future Evolutions* (Channel View Publications, 2020).

Marianna Sigala is Professor of Marketing, University of Piraeus, Greece. She is an international authority in the field of technological advances and applications in tourism with numerous awarded publications, research projects, keynote presentations in international conferences. Professor Sigala is editor-in-chief of the *Journal of Hospitality and Tourism Management*, and co-editor of the *Journal of Service Theory & Practice*. In 2016, she was awarded the prestigious EuroCHRIE Presidents' Award for her lifetime contributions and achievements to tourism and hospitality education. In 2020, she become a member of the CAUTHE College of Fellows.

Chapter Authors

Magdalena Banaszkiewicz, PhD, is a cultural anthropologist and Assistant Professor in the Institute of Intercultural Studies at the Jagiellonian University in Krakow. Her research interests focus on the dissonances connected with tourism development. She is co-editor of *The Anthropology of Tourism in Central and Eastern Europe* (Lexington Books, 2018).

Giovanna Bertella is Associate Professor at the School of Business and Economics, UiT The Arctic University of Norway, Tromsø. Her research interests are: management, marketing, entrepreneurship/innovation, tourism and leisure studies (nature- and animal-based experiences, rural tourism, food tourism, events), food studies (veganism), futures studies.

Stephan Bingemer is a full-time professor for business administration and tourism at International School of Management Dortmund, Germany. He teaches at the Campus in Frankfurt am Main, Germany.

Peter Bolan, PhD, is a Senior Lecturer and the Director for International Travel and Tourism Management at the Ulster University Business School in Northern Ireland. His research interests and consultancy specialisms include film/screen tourism, digital tourism, golf tourism and food tourism. Peter also writes regularly for a number of business and hospitality trade publications.

Cecilia de Bernardi is a Postdoctoral Researcher at ETOUR (Mid-Sweden University). She is a doctor of Social Sciences in Tourism Research. Cecilia has previously worked with authenticity and Indigenous tourism, but she is currently doing research on policy and behaviour in the Mistra Sport and Outdoors programme.

Richard Ek is an Associate Professor in Human Geography, working at the Department of Service Management and Service Studies, Lund University, Sweden. He writes particularly on topics that juxtapose tourism theory and political philosophy, for instance the politico-ontological tension between the tourist and the citizen.

Martin Gren holds a PhD in Human Geography and is a Professor at the Linnaeus University (Sweden). His research is centred on the re-conceptualisation of the Earth in the Anthropocene and terrestrial politics in the planetary climate and ecological emergency.

Leon Gurevitch, PhD, is Associate Professor in the School of Design at Victoria University of Wellington, New Zealand where he publishes academic writing, usable software, data visualisation and photographics focused on the interface between science, technology and design.

Daniel Guttentag is an Assistant Professor in Hospitality and Tourism Management at the College of Charleston in South Carolina, USA. Daniel holds a PhD in Recreation and Leisure Studies and a Master's degree in Tourism Policy and Planning, both from the University of Waterloo, Canada.

Sam Hitchmough currently works as Director of Teaching in the History Department at the University of Bristol. He was Programme Director for American Studies at Canterbury Christ Church University. He is currently involved in a number of research projects: Red Power activism and Buffalo Bill's Wild West Shows in England.

Emily Höckert is a Postdoctoral Fellow in tourism studies at the University of Lapland in Finland. She is a member of the 'Intra-living in the Anthropocene' research group, which explores the current time of the Anthropocene by using feminist theories, new materialism and methodological innovations.

Jane Lovell, PhD, worked at the Royal Opera House Covent Garden and in tourism development at Canterbury City Council, where she staged events including international light shows. At Canterbury Christ Church University, Jane researches and publishes on authenticity and teaches Heritage and Creative Industry management and Creative Places.

Mairead McEntee is Associate Head of the Department of Hospitality and Tourism Management, Ulster University. Prior to entering academia, Mairead was an Associate Director in the Corporate Finance department of one of the UK's largest accountancy houses, specialising in the hospitality and tourism sector.

Ruairi McEntee is experienced in research in food chemistry and field chemical analysis, has long been fascinated with zombies and provided the technical zombie expertise required.

Felicity Picken is dedicated to understanding the diverse array of pleasurable encounters with oceanic space and their contribution towards living with the blue planet. In the 21st century and its reshaping of human–nature relations, oceans offer important spheres of human action, future action and critique of the limits of humanism.

Ina Reichenberger is a Hufflepuff and Senior Lecturer in Tourism Management at Victoria University of Wellington, New Zealand. Her research focuses on tourist behaviour, with a special interest in social interactions, co-creation and youth travel.

Stuart Reid is a Doctoral Candidate at the Department of Service Management and Service Studies, Lund University, Sweden. His research focus is the social construction of entrepreneuring in tourism lifestyle enterprise. He currently teaches in tourism innovation and strategic communication.

Peter Robinson is Head of the Centre for Tourism and Hospitality Management at Leeds Beckett University. He is a Fellow of The Tourism Management Institute and a Director of The Institute for Travel and Tourism. He is a member of The Tourism Society, a Principal Fellow of the Higher Education Academy and is an Executive Committee Member of The Association for Tourism in Higher Education. His research interests include Cold War tourism, visitor experiences, regeneration and operations management. He is also currently Chair of the Elvaston Castle and Gardens Trust.

Jonathan Skinner, PhD, is Reader in the Anthropology of Events in the Department of Events – School of Hospitality & Tourism Management, University of Surrey. His research interests focus on tourism regeneration and special events, often in conflicted destinations. He is author of *Before the Volcano: Reverberations of Identity on Montserrat* (Arawak Publications, 2004), and editor of *Writing the Dark Side of Travel* (Berghahn Books, 2012).

Annette Toivonen is a Senior Lecturer in Space Tourism in the Department of Tourism Business at Haaga-Helia University of Applied Sciences. She is currently finalising her PhD in sustainable 'New Space' tourism at the University of Lapland. She has also gained both an MSc and BA (Hons) in Tourism Research. Annette is the author of *Sustainable Space Tourism: An Introduction* (Channel View Publications, 2021).

Daniel Wright has completed a PhD in Tourism Development, specifically exploring the role of tourism in post-disaster situations. Research interests continue to explore the role of tourism as a tool for development and more recently exploring the future of tourism and related industries and sectors.

Foreword

Tourism's Unknown Yet Plausible Future

When I was an 11-year-old boy I was fascinated by three books by Jules Verne that I got from my father. I still remember the three blue paperbacks well: '*20.000 Mijlen Onder Zee*' (Twenty Thousand Leagues Under the Sea), '*Naar Het Middelpunt der Aarde*' (Journey to the Centre of the Earth) and '*Reis Om de Wereld in 80 Dagen*' (Around the World in Eighty Days). The books were distributed by ARAL gas stations in the Netherlands in 1973. Not hindered by expert knowledge, the stories and engravings captured my fantasy and helped me to imagine worlds that were unknown to me.

However, the creativity and imagination of any child that grows up is gradually restricted by what it hears and learns from its parents, friends, the community in which it grows up, and from its school teachers and professors. During their development a person gradually acquires a perspective on their surroundings, learns what works and what doesn't, and develops routines and habits (Buzan, 2009). However, it is questionable whether this so-called dominant thinking with its cognitive barriers is sufficient to address the challenges that tourism is faced with. Developments such as globalisation, informatisation and technologisation have made tourism, and tourism businesses and organisations in particular, so interdependent of demographic, economic, social, technological, ecological and political developments across the globe, that its future is surrounded with uncertainties.

To achieve a sustainable future of global tourism we need to acknowledge the limits of our mental capacities and face our individual uncertainties. We need to make sense of the world around us and how this relates to developments in tourism. We need to explore multiple futures and anticipate possible changes that they bring along.

This is the domain of scenario planning. At the European Tourism Futures Institute, we employ scenario planning as a moderated process in which multiple stakeholders share their knowledge, challenge each other's paradigms, develop a common understanding of the forces that drive their joint future and identify the forces that they perceive as most powerful and least predictable (Postma, 2013, 2015). This collaborative process results in multiple surprising and plausible narratives of the future. These future scenarios do not make truth claims but envision of how the developments and interdependencies of key uncertainties and driving forces could impact upon tourism and its relevant context and can therefore be referred to as science fiction (Bergman et al., 2010).

Science fiction is a contemporary approach which can be used to explore the future of tourism. The road that future tourism development will take is unpredictable. It is subject to a complex and dynamic web of interactions with political, technological, social, ecological and economic forces. Moreover, hyperconnectivity and interdependency within the web of interactions may facilitate wildcards, disruptions and innovations to have a tremendous impact that makes the long-term future of tourism rather uncertain. Within the borders of what we think is plausible, science fiction explores what could happen and the implications it may have for tourism in the long term.

This book introduces the reader to the theoretical and ontological backgrounds of science fiction and how this applies to the future of tourism. It also draws the reader into the unknown future of tourism – a future that may be disruptive, dystopian or utopian. May this book capture your fantasy, make you imagine futures you have never thought of and inspire you to address tourism's challenges of tomorrow in an innovative way to contribute to the industry's resilience.

Dr Albert Postma
Professor of Strategic Foresight and Scenario Planning
NHL Stenden University – European Tourism Futures Institute

References

Bergman, A., Karlsson, J.C. and Axelsson, J. (2010) Truth claims and explanatory claims – an ontological typology of futures studies. *Futures* 42 (8), 857–865. doi:https://doi.org/10.1016/j.futures.2010.02.003

Buzan, T. and Buzan, B. (2009) *The Mind Map Book: Unlock Your Creativity, Boost Your Memory, Change Your Life*. Harlow: BBC Active.

Postma, A. (2013) Anticipating the future of European tourism. In A. Postma, I. Yeoman and J. Oskam (eds) *The Future of European Tourism* (pp. 290–305). Leeuwarden, Netherlands: European Tourism Futures Institute.

Postma, A. (2015) Investigating scenario planning – a European tourism perspective. *Journal of Tourism Futures* 1 (1), 46–52.

Introduction

1 Science Fiction: A Theoretical Lens and Methodological Approach for Re-Imagining Tourism Futures

Ian Yeoman, Una McMahon-Beattie and Marianna Sigala

Chapter Highlights

- What is science fiction and why is the study of science fiction important in the context of the future of tourism?
- The aims, purpose and contribution of this edited collection.
- The future is more than COVID-19 as a global pandemic is no longer science fiction but reality.

Science Fiction and Reality

In ontological literature there are arguments about the concepts of time. Rovelli (2019) argues that *Presentism* is the idea that there is an objectively real 'present' which forms a three-dimensional continuum. The alternative is *Eternalism* which is the idea that present, past and future events are 'equally real'. Reality is formed by a four-dimensional continuum. The passage of time, or becoming, is not real, it is in some sense illusory. Rovelli (2019) argues that reality can also be judged as a temporal structure that describes 'becoming'. As such one might ask, 'what is real?' and 'when does something become real?'.

This is at the heart of when science fiction becomes reality. Throughout science fiction writing we see technological stories that eventually become true as science fiction has a history of influencing popular culture and inspiring engineers to turn ideas into reality (Chen, 2010). For example, Robert Heinlein mentioned a 'pocketphone' in *Assignment in Eternity*

(1953) and a portable phone fitting in a pocket in *Space Cadet* (1948). In real life, the radiotelephone (forerunner of the mobile phone) appeared in 1964 but was much bigger than a pocket (more like a shoebox). COVID-19 accelerated the use of Zoom and other video-conferencing platforms in our daily lives, but one of the earliest writing on video calling was 'telephot' described by Hugo Gernsback in *Modern Electrics* magazine (1911) and the 'phonotelephote' in Jules Verne's *Year 2889* (1899). Google's automatic language translator concept can be seen in the idea of a 'universal translator' in Murray Leinster's *First Contact* (1945). As Yeoman (2012) noted, *Star Trek* is a clear example of science fiction becoming reality, with the 'communicator' becoming the mobile phone, Dr McCoy's 'tricorder' becoming the MRI scanner, Spock's 'Personal Access Display Device' becoming the iPad and the 'replicator' becoming a 3D printer.

To some, science fiction is reality whereas to others it is just fiction. The perspective you adopt is influenced by your understanding of the role of the 'expert' and your cognitive understanding of a science domain (Hubert *et al.*, 2010). Take for example, Spielberg's film *Minority Report* (2002). Baudrillard (2006) reports that prior to the filming Spielberg invited 15 experts to a hotel in Santa Monica for a three-day 'think tank' convened by Peter Schwartz and Stewart Brand. He wanted to consult with the group to create a plausible 'future reality'. They created a '2054 bible' which was an 80-page guide to the technology and social trends featured in the film.

COVID-19: Science Fiction or Reality

COVID-19 has changed the world and to some it has come straight out of science fiction (Chakravorty, 2020), whereas to others it was predictable (Page *et al.*, 2006). Its unfolding in Wuhan and the associated conspiracy theories of an outbreak or leakage from a Chinese military science facility mirrors Dean Koontz's (1981) novel *The Eyes of Darkness* based upon a virus called Wuhan-400 with a death rate of 100%. Across the globe we see stories of hospitals on the verge of collapse, dissent from populations, economic deprivation, social isolation, violation of human rights and mental health issues; overall, a general dystopian outlook. What makes the COVID-19 crisis unique is not only its global scale, but its multi-dimensional scope and knock-on effects, a biological crisis that has been converted into a global economic, political, socio-psychological and even moral crisis (Sigala, 2020). Dilemmas posed by COVID-19 (such as opening up economies versus public health, and use of mobile tracing apps versus human rights of privacy and mobility) have challenged many foundational taken-for-granted assumptions and ethical values, making us think and re-think what the next normal might or should be. What previously seemed to be a science fiction utopia or dystopia, post pandemic seems to be an unavoidable future reality. This is what Margaret

Atwood (2004) predicted in her novel *Oryx and Crake*. As Chakravorty (2020) reminds us, the connection between science fiction and pandemics is not new.

It can be argued that COVID-19 is a representation of science fiction because its disruptions and implications have shaken foundational assumptions requiring every tourism actor to rethink and reset practices and mindsets to something that may or should not be related to the past. Doherty and Giordano (2020) have noted:

> Literary enthusiast Patrick Parrinder describes science fiction as a 'thinking machine' that provides an outlet to visualise *what could be,* and therefore allows both reflection on *what is,* and some idea for planning *what to do next*. This sheds some light upon the current interest in pandemic fiction. These movies engender deeper contemplation of the COVID-19 crisis because they combine the familiar with the novel. Balancing the two – both the factual and fictitious – is critical to instilling a sense of *ostranenie*: the unfamiliar presentation of a common thing that affords the viewer an enhanced perception of the familiar. Science fiction stitches truths about humanity into the fabric of its unfamiliar worlds: when we imagine ourselves in stories' novel scenarios, it provides good food for thought and the possibility to internalise applicable moral lessons.

As such, COVID-19 is a transformational agent and stressor (Sigala, 2020) demanding tourism actors to recalibrate priorities and values in order to redefine what a 'better' tourism future might be and to prioritise actions and plans on how to build back 'better'. Even if the need to change cannot be solely attributed to COVID-19, what the pandemic has done is to accelerate and magnify the urgent need to reset and reform tourism as it has highlighted and intensified long-discussed issues that were overdue for action (e.g. climate change, unstoppable economic growth, social values/ethics, equality etc.). Using science fiction as a 'thinking machine' or a means to find new forms of tourism, and realising that the future of the tourism is not the same as the past, requires new ways of conceptualising the future. Scenario planning is accepted as the main research methodology in futures studies; however, it is sometimes constrained by plausibility and the political reality of change (Yeoman & McMahon-Beattie, 2016). So, how can we reinvent the future and take a more radical approach to creating the future beyond rational thinking to account for innovation, disruption and the unexpected? How can we think outside, above and beyond the box?

Dator (1986, 2014; Dator & Yeoman, 2015) stressed that any future of value must have a degree of scepticism. Science fiction has attributes of soft falsification or truth distortion. The concept of scepticism is important as it challenges the notion of truth and expresses the question of what else is possible or indeed impossible (Sardar, 2010). This embraces the arguments surrounding conceptualisation, that is, the issue of knowing, the issue of being and our definitions of quality. According to Hales (1969: 2–3):

> Rapid change has affected human thought in another way. The future is no longer regarded as predestined – an existing landscape that will be revealed to us as we travel through it. It is now seen as the result of the decisions, discoveries and the efforts that we make today. The future does not exist, but a limitless number of possible futures can be created. From this mode of thought it is a natural step to the idea of establishing desirable goals towards which we can deliberately work, ranging from the conception of a desirable (but feasible) form of transport to of the scenario of for a desirable society.

What is Science Fiction?

At the beginnings of the 20th century, H.G. Wells anticipated that futures research would become a scientific discipline, with the development fuelled by the post-World War II uncertainty and the need to deal with complex technologies in an uncertain world (Bell *et al.*, 2013).

> Rapid change has affected human thought in another way. The future is no longer regarded as predestined – an existing landscape that will be revealed to us as we travel through it. It is now seen as the result of the decisions, discoveries and the efforts that we make today. The future does not exist, but a limitless number of possible futures can be created. From this mode of thought it is a natural step to the idea of establishing desirable goals towards which we can deliberately work, ranging from the conception of a desirable (but feasible) form of transport to of the scenario of for a desirable society.
>
> (Hales, 1969: 2–3)

Hales' lead editorial for the first issue of the journal *Futures* aspires to a utopian focus through science fiction on the creation of a better society, which echoes the history of Plato's (380 B.C.) *Republic* and More's *Utopia* (1561), in that each outline a particular vision for a better society. Bell *et al.* (2013) takes this proposition to explore the contributions and warnings of utopia and dystopia through science fiction as mechanisms for innovation, visions and business. He notes that science fiction acts as mechanisms for explanations of the future as prototypes. Prototypes act like mechanisms for interpretation or construction of fictional futures through stories of events or products (Johnson, 2011). While prototypes and scenarios are different concepts both have emancipatory powers in which the future imagined espouses a belief that the future could actually occur, thus being seen as an explanatory mechanism (Wyss & Duran, 2001). Science fiction tends to evoke a visualisation of the future, a better future for humankind with some sort of science representation through a fictional account hence 'science' and 'fiction' as science fiction (Ivory, 2011). However, one of the issues of an interpretation approach is that Jürgen Habermas often criticised hermeneutics as being unsuitable for understanding society as it was unable to account for questions of social reality thus alluding to fantasy (Mendelson, 1979). However, an interpretation approach is justified when signals are weak and lacking the attributes of prediction. A hermeneutic approach is to understand or interpret, that is, to show how something could occur. In the scenario planning and futures literature, this is portrayed as 'what if?', thus thinking about how to bring about reality even if the connections are weak; an interpretation of fact rather than stating the absolute.

Science Fiction Tourism

Landon (2009) states that science fiction tourism is based on the premise:

> ... of going somewhere else to see the sights whether through voyages extraordinaires, explicit time-travel narratives, or the implicit time travel of trips to the future. It should come as little surprise then that science fiction tourism is a concept that offers us new and valuable perspectives on the nature and function of science fiction.

This journey is notable, as we cannot separate science fiction and the ultimate mode of travel which is restricted to science fiction, that is time travel. What is time travel? Inevitably, it involves a discrepancy between one time and another. Any traveller departs and then arrives at his destination; the time elapsed from departure to arrival (positive, or perhaps zero) is the duration of the journey (Lewis, 1976). Time travel is the concept of movement between certain points in time, analogous to movement between different points in space by an object or a person, typically with the use of a hypothetical device known as a time machine. Time travel is a widely recognised concept in philosophy and fiction, particularly science fiction (Clute & Nicholl, 1999), H.G. Wells' (1895) *Time Machine* is the classic science fiction novel that made the phenomena popular.

Science fiction and tourism has often been associated with visitor attractions and exhibitions For example Coney Island or the Paris Exposition of 1889 both featured future worlds focused mainly on technological invention (Abbott, 2009). The attractions featured mechanical rides, disaster attractions and futuristic experiences. Kasson (1978: 61) described the Luna Park at Coney Island in 1903:

> Here visitors entered a spaceship in the middle of a large building for an imaginary ride to the moon. Peering out of portholes, they beheld a series of shifting images that gave the illusion of a flight into space, a sense reinforced by the rocking of the ship itself. After supposedly landing on the moon, passengers left the spaceship to explore its caverns and grottoes, where they met giants and midgets in moon-men costumes, the Man in the Moon upon his throne, and dancing moon maidens, who pressed bits of green cheese upon them as souvenirs of the lunar voyage.

In the contemporary era films and television series have featured visitor attractions such as *Westworld* (Crichton, Brynner, Benjamin, Brolin, & Lazarus, 2000) *Futureworld* (Heffron, 1976) and *Jurassic Park* (Spielberg, 1993).

Why this Book?

As the Scenario Planner at VisitScotland in 2006, Ian Yeoman facilitated a team to model and construct a set of scenarios which replicated the present COVID-19 pandemic reality (Page *et al.*, 2006). This is an example of science fiction coming true and a journey beginning. Science fiction was used to explore the possible and impossible, to construct futures based upon technologies which had not being invented, to think about the transformation of tourism, and to predict the end of tourism based upon a natural disaster. The process took rationality to its limits. However, as academic researchers we would normally view science fiction as nothing more than a piece of creative writing. It is not something based upon fact but imagination; it is not real but fantasy. So, what is the point some would say. As Doherty and Giordano (2020) argue science fiction is a 'thinking machine', it is about imagination and is right at the centre of scenario planning – the main research methodology used in futures studies. Thus, the purpose of this book is to understand the role of science fiction in tourism research and how it is used to portray and make us re-think the future of tourism. It explores if science fiction can be of benefit to tourism researchers in a rapidly changing world, as it provides them food for thought and a way of thinking, re-thinking and de-thinking of tourism futures. It helps set research agendas, directions and scope of research. In this vein, science fiction can be seen as a useful approach to foster and support transformation tourism research (Sigala, 2020).

Given the implications of COVID-19 and the overdue changes required in tourism, this book is more than just topical in nature and focus; it is also much needed to direct and foster tourism research that envisions beyond the past normal. As such, we fundamentally address the requirements for transformational tourism thinking and research through the contributions of the authors in this edited collection. Holistically, the combined contribution of the chapters is to understand and construct a theoretical position or framework between science fiction and the future of tourism. If one can find an underpinning theory, then we have the basis of using science fiction as a theoretical lens and methodological approach to explore, frame and even form the future of tourism. By focusing on a specific form of tourism or topic, every book chapter uses a practical example and evidence to discuss and explain the theoretical underpinnings and the methods that others can also use to vision and re-think tourism futures.

Structure of the Book

Each contributor was asked to classify their own contribution against the book sections of ontology, science fiction, disruption and dystopia based upon what the primary contextual contribution of their chapter was. As the selection was based upon a primary contribution, it must be noted that many chapters overlapped in the terms of classification given the interdisciplinary nature of the topic.

Part 1: Ontological Approaches

The purpose of this section is to provide an overview of theories, practices and use of science fiction in futures research. As such, the section will discuss science fiction as a paradigm in tourism research, in particular it will explore how contributors position the ontology of science fiction in relation to the future of tourism. The premise of this section is focus on explanatory claims with which the authors explain how such a future could be created. In Chapter 2 *Science Fiction and the Future of Tourism* Yeoman and McMahon-Beattie position the book by overviewing science fiction and its portrayal in tourism futures. The central feature of the chapter is an ontological typology of science fiction based upon the proposition that science fiction is used to explain how the future of tourism could occur, thus supporting the idea of explanatory claims (Bergman *et al.*, 2010). In Chapter 3, *The Future, the Devil and the Deep Blue Sea*, Picken asks us to view undersea worlds as a means to rethink the assumptions that underpin tourism. It explains why futures research is important for tourism scholarship. It outlines some of the challenges involved in researching tourism (and other) futures and illustrates these key points through research that is aimed at understanding recent and emerging undersea tourism attractions and the futures these open. Robinson in Chapter 4, *Prosuming Existential Authenticity in Dystopian Spaces*, argues there a real value in positioning the future of tourism as something which is outside the current realms of established research, taking the ontological position that utopia and dystopia are not opposites, he then explores the tourist experience in the 21st century and the disappointment of authenticity. Space tourism is a new sector of adventure tourism which has been lacking academic research with a futures perspective. In Chapter 5, *Space Tourism – Science Fiction Becoming a Reality*, Toivonen explores the development of this futuristic travel phenomena, including elements originating from science fiction. In addition to this ontological approach, the chapter investigates emerging weak signals and uses a postmodern paradigm to model the future of space tourism.

Part 2: Science Fiction

How does science fiction portray the future of tourism? Contributors explore the technologies, stories and issues raised in science fiction films, literature and other media and critically discuss the context from a tourism futures perspective. In Chapter 6, *Life Without Limits: Design, Technology and Tourism Futures in* Westworld, Gurevitch using a design theory perspective which intertwines media, tourism futures and design to explore the disruptive potential of technology to deliver experiences and the desire of tourists to feel free from the moral, social, economic and political constraints of their daily lives. In Chapter *7 Harry Potter and the Future of Tourism*, Reichenberger using the series as a metaphor for the concepts of prejudice, discrimination, power, authority and morality. Thus, Reichenberger addresses the significant impact of values and attitudes on tourist behaviour and highlights developments that will shed light on future trends in tourist behaviour, in turn shaping the tourism industry. In Chapter 8, *Wildlife Tourism in 2150: Uplifted Animals, Virtual and Augmented Reality and Everything In-between*, Bertella discusses the current research in both tourism and other disciplines in order to make a considered prediction about the future of wildlife tourism in 2250. Bertella examines the authenticity of the future wildlife tourism where technology has been used to enhance the tourism experience. In Chapter 9, *Tears in the Rain: Tourism in the World of* Blade Runner *and* Total Recall, Bolan addresses the worlds and their technology as depicted in the science fiction works by Philip K. Dick (1968, 2002) and explores their impact and influence on tourism. He examines the transformational impact of technology in tourism, from replicants to memory implants and self-driving cars to holograms. In Chapter 10, *Destination of the Dead: The Future for Tourism?*, McEntee and colleagues consider tourists a plague of zombies within the context of overtourism and sustainability. The chapter takes a novel look at tourism and its impact on the people and places that experience excessive numbers of tourists. Zombies are now a clear genre in popular culture, appearing in countless movies, TV programmes and comic books, all of which depict crowds of mindless bodies shuffling along aimlessly while leaving a trail of destruction in their wake. The same could arguably be said for some tourists, slowly walking along looking upwards at buildings with a selfie stick in hand while busy locals go about their daily business.

Part 3: Disruption

We live in a world of rapid change where predictive and rational approaches to understanding the future of tourism need to be created (Agar, 2015). Indeed, innovative and disruptive technologies are now a mainstream feature of tourism management (Yeoman, 2012). The use of science fiction involves thinking the impossible or scenario planning the 'what if?'. In this section authors explore big ideas, radical solutions and quantum leaps focusing on the role of disruption in tourism. In Chapter 11, *Holidays with Inspector Maigret: Mixed Reality Adventures as Value Drivers in Future Tourism*, Bingemer tells a story for readers to explore the future through the implementation of mixed reality. Furthermore, the design of such adventures are discussed. Guttentag, in Chapter 12, *Digital Destinations and Avatar Tourists: A Futuristic Look at Virtual Reality Tourism and Its Real-World Impacts,* begins with an overview of present-day virtual reality (VR) technology. As VR increasingly offers people the opportunity to experience tourism destinations and activities in a realistic virtual world, it seems inevitable that substitution for 'real' travel will begin to occur. This substitution phenomenon would have permanent and significant disruptive consequences on tourism destinations and firms across the globe, creating a qualitative shift in the tourism sector and the very notion of tourism itself. In 2020, your personal 'bubble' became a key word during the year because of COVID-19. Taking this concept, de Bernardi in Chapter 13 *The 'Safety Bubble' and the Future of Enclave Tourism* explores the dimensions of bubbles from a risk and safety perspective focusing on enclaves and their possible future(s).

Part 4: Dystopia

In this section authors explore the dark side of tourism and phenomena of dystopia. In Chapter 14, *The Coming of the Fugue and the Blind Tourist?* Reid and Ek present two future scenarios that will challenge established meanings of tourism. Building from the core story of the travel craze, Reid and Ek assert the effects of 'overtourism' lead to assorted policy restrictions (e.g. higher visa costs, seasonal restrictions and closures). These developments then drive the scenarios for alternative tourism futures, one stemming from new digital technology and one from new human practice. In Chapter 15, *Technological Frontiers: From the Wild West Myth to the Dystopia of the Westworld's Post-human Theme Park,* Lovell and Hitchmough explore a concept that science fiction frequently features, that is, 'Cyborgs' and they discuss the immersive technology employed within theme parks to counter reality. They note that the fascination with automatons is overlaid with an element of atavistic distrust, which the series *Westworld* explores in the context of a Western theme park. Theme parks are predicated on the idea of control and the chapter enquires what will happen when the aptly-named post-humans take over the tourist attraction. In Chapter 16, *The Future of Music Concerts and Tourism in Dystopian Times,* Wright runs a dystopian-utopian gauntlet in which he focuses on the benefit of concerts as a means of supporting people living in dystopian realities. The value of the chapter lies in the exploration of tourism in a futures context in times of conflict, human struggle and what are often referred to as the darker periods of human history. Wright draws on historical and current evidence to justify the ideas presented. Chapter 17, *Exclusion Tourism: Sci-Fi Stalkers and Subjunctive Plays in Apocalyptic Destinations from Chernobyl to Plymouth, Montserrat*, Banaszkiewicz and Skinner explore how science fiction texts have established a new form of tourist practice: 'exclusion tourism'. The chapter examines this in relation to venues that had been destroyed or abandoned but which, in their apocalyptic ruin, have subsequently emerged as conflicted tourist attractions associated with science fiction. This is achieved via an anthropological approach using participant observation with two apocalyptic destinations: the Chernobyl Exclusion Zone (Ukraine) and the Plymouth, Montserrat Exclusion Zone (Eastern Caribbean). In Chapter 18, *Hotel Anthropocene* Gren and Höckert use a narrative approach asking you to imagine a typical day in the life of tourist. The purpose of the chapter is to question the idea of environmental crises as something that we assume we will have to face in the future with the key concepts being identified.

Part 5: Concluding Thoughts

In Chapter 19, *Developing a Theoretical Framework of Science Fiction and the Future of Tourism: A Cognitive Mapping Perspective*, Yeoman, McMahon-Beattie and Sigala bring the contribution of the edited collection together based upon a theoretical framework for science fiction and the future of tourism. Based upon the contributions of the preceding 17 chapters the proposed framework is based upon an ontological and epistemological perspective. From an ontological perspective, science fiction has to be conceptualised to be better explained and understood (see Bergman's (2010) ontological typology developed on the idea of explanatory claims). Explanatory claims are based upon the characteristics that science fiction is about weak signals of the future, is an interpretation (way of thinking) of the future rather than being fact-based and is sceptical (untruthful) to many. The ontological perspective clearly links the 'what if' question in scenario planning (Chermack, 2004) and future studies (Kuosa, 2011). Yeoman, McMahon-Beattie and Sigala present seven concepts of knowledge at an epistemological level.

First, futures in science fiction writings are based upon an alternative universe or cosmic pluralism, thus the chapters represent an alternative to the present derived from his/her own perspectives and realities. Hence the concept *Plurality of the Future(s)* – the difference between the real world and science fiction. Second, given the popularity and bias towards dystopian literature in science fiction, the authors have proposed a new word to bridge dystopia and utopia, namely Dys*Topia* which reflects this bias. As the word *topia* is defined as a place of specified characters (Shklar, 1965), *dys* is a reflection of the contributing chapters but utopia must be recognised, especially in present circumstances of post recovery strategies and the reimagination of tourism with the focus on sustainability and a better world (Duong, 2020; Geraghty *et al.*, 2019; Hassan & Habib, 2020; Jones & Comfort, 2020; Mihăilescu, 2016; Strielkowski, 2020; UNWTO, 2020; Yeoman, 2020; Yeoman *et al.*, 2020a, 2020b; Sigala, 2020). Third, *Liminality* is the blurring of reality and the unreal, between fact and fictional or the passageway of in between. Extending this concept, the *Hyperreality of Authenticity* is a form of knowledge based upon how we see the world of tourism through science fiction (Winter, 2002). Fifth, *Disruption and Transformation* is everywhere in the science fiction literature. Science fiction represents the radical alternatives, the unthinkable, science which hasn't been invented yet and an opposite to the status quo. As science fiction does not represent reality, is not factual, and appears to be fictitious, many will treat it with *Scepticism*. Finally, the seventh concept is *Narrative* as all science fiction, like scenarios, are based around fictional stories. The authors identify three concepts upon which tourism is predominately portrayed in science fiction, namely *Technological Singularity has Arrived: Westworld, Sustainability*, and *COVID-19*. The first concept, *Technological Singularity has Arrived: Westworld* represents the strong connection between science fiction and futurism (Polak, 1955). In 2019, overtourism was the most popular word of the year and one of the most written about topics (Panayiotopoulos & Pisano, 2019). Overtourism and disaster therefore is the basis of the second concept of *Sustainability* from a dystopian perspective. The final concept is *COVID-19 and Pandemics*, which aptly conveys all our experiences since 2020. The concept has been added by the editors as this word will dominate academic research going forward, as many of its changes are here to stay and it represents a critical milestone of shaping tourism futures.

What the Book is Not

This is not a book about COVID-19 as this subject is well documented elsewhere. The central purpose of the book is to develop a theoretical framework about science fiction and the future tourism. COVID-19 undoubtedly illustrates when science fiction becomes reality. It has magnified the urgency and the vitality to re-think and reset 'better' tourism futures but we have taken a wider picture of science fiction and tourism so that the book is not dominated by one illustration of the present. Nevertheless, the book provides an interesting theoretical and methodological approach for those wishing to vision and investigate a tourism future in the post-pandemic period.

References

Abbott, S. (2009) Arthouse SF Film (Part IV Subgenres). In M. Bould, A.M. Butler, A. Roberts and S. Vint (eds) *The Routledge Companion to Science Fiction*. Abingdon: Routledge, Taylor & Francis Group.

Agar, N. (2015) *The Sceptical Optimist: Why Technology Isn't The Answer to Everything*. Oxford: Oxford University Press.

Atwood, M. (2004) *Oryx and Crake: A Novel*. New York: Anchor Books.

Baudrillard, J. (2006) Virtuality and events: The hell of power. *International Journal of Baudrillard Studies* 3 (2), 1–12.

Bell, F., Fletcher, G., Greenhill, A., Griffiths, M. and McLean, R. (2013) Science fiction prototypes: Visionary technology narratives between futures. *Futures* 50 (June), 10.

Bergman, A., Karlsson, J.C. and Axelsson, J. (2010) Truth claims and explanatory claims – an ontological typology of futures studies. *Futures* 42 (8), 857–865. doi:https://doi.org/10.1016/j.futures.2010.02.003

Chakravorty, M. (2020) Science fiction explores the interconnectedness revealed by the coronavirus pandemic. See https://theconversation.com/science-fiction-explores-the-interconnectedness-revealed-by-the-coronavirus-pandemic-139021

Chen, T.M. (2010) Science fiction becomes reality. *IEEE Network* 24 (4), 2–3. doi:10.1109/MNET.2010.5510910

Chermack, T.J. (2004) Improving decision-making with scenario planning. *Futures* 36 (3), 295–309. doi:https://doi.org/10.1016/S0016-3287(03)00156-3

Clute, J. and Nicholl, P. (1999) *Encyclopaedia of Science Fictions*. London: Orbit.

Dator, J. (1986) The futures of futures studies: A view from Hawaii. In L. Garita (ed.) *The Futures of Peace: Cultural Perspectives* (p. 519–527). San Jose: University of Costa Rica.

Dator, J. and Yeoman, I. (2015) Tourism in Hawaii 1776– 2076: Futurist Jim Dator talks with Ian Yeoman. *Journal of Tourism Futures* 1 (1), 36–45. http://dx.doi.org/10.1108/jtf-01-2015-0001.

Dick, P.K. (1968) *Blade Runner – Do Androids Dream of Electric Sheep*. New York, N.Y: Ballantyne Books.

Dick, P.K. (2002) *We Can Remember it for you Wholesale*. New York, N.Y: Citadel Press Books.

Doherty, J. and Giordano, J. (2020) What we may learn – and need – from pandemic fiction. *Philosophy, Ethics, and Humanities in Medicine* 15 (1), 4. doi:10.1186/s13010-020-00089-0

Duong, M. (2020) COVID-19 and after: Impact and ways forward. *East Asia Forum Quarterly* 12 (2), 31–33.

Featherstone, M. (2020) Whither globalization? An interview with Roland Robertson. *Theory, Culture & Society* 37 (7–8), 169-185. doi:10.1177/0263276420959429

Geraghty, L., Ziakas, V. and Lundberg, C. (2019) Guest editorial: Exploring the popular culture and tourism place making nexus. *Journal of Popular Culture* 52 (6), 1241–1249. doi:10.1111/jpcu.12867

Gernsback, H. (1911) Editorial. *Modern Electrics* 4 (1), 1–4.

Hales, J. (1969) Futures: Confidence from chaos. *Futures* 1 (1), 2–3. doi:10.1016/S0016-3287(69)80001-7

Hassan, S. and Habib, W. (2020) Aspects of sustainable tourism development and COVID-19 pandemic. *Preprints*. See file:///C:/Users/e78557/OneDrive%20-%20Ulster%20University/Downloads/preprints202008.0418.v1.pdf (accessed 18 January 2021)

Heinlein, R.A. (1948) *Space Cadet*. New York: Scribner's.

Heinlein, R.A. (1953) *Assignment in Eternity*. New York: Fantasy Press.

Hubert, B., Rosegrant, M., van Boekel, M. and Ortiz, R. (2010) The future of food: Scenarios for 2050. *Crop Science* 50, S33–S50.

Ivory, J. (Writer) (2011) *Howards End*. In Australia: Shock DVD.

Johnson, B.D. (2011) *Science Fiction Prototyping Designing the Future with Science Fiction*. San Rafael: Claypool Publishers.

Jones, P. and Comfort, D. (2020) A commentary on the COVID-19 crisis, sustainability and the service industries. *Journal of Public Affairs*. doi:10.1002/pa.2164

Koontz, D. (1981) *The Eyes of Darkness*. New York: Pocket Books.

Kubrick, S. (Writer) (1968) *2001: A Space Odyssey*. In S. Kubrick (Producer). USA: Metro-Goldwyn-Mayer.
Kuosa, T. (2011) Evolution of futures studies. *Futures* 43 (3), 327–336.
Landon, B. (2009) SF tourism. In M. Bould, A.M. Butler, A. Roberts and S. Vint (eds) *The Routledge Companion to Science Fiction* (p. 32–41). Abingdon: Routledge, Taylor & Francis Group.
Leinster, M. (1945) *First Contact*. New York: Astounding Science Fiction.
Lewis, D.K. (1976) The paradoxes of time travel. *American Philosophical Quarterly* 13 (2), 145–152.
Mendelson, J. (1979) The Habermas-Gadamer debate. *New German Critique* (18), 44–73. doi:10.2307/487850
Mihăilescu, C.-A. (2016) Theses on political reimagination. *Caietele Echinox* (30), 353–356.
More, T. (1561) *Utopia* (A. Robert, Trans. 3rd edn). Cambridge: Cambridge University Press.
Page, S., Yeoman, I., Munro, C., Connell, J. and Walker, L. (2006) A case study of best practice – Visit Scotland's prepared response to an influenza pandemic. *Tourism Management* 27 (3), 361–393. doi:https://doi.org/10.1016/j.tourman.2006.01.001
Panayiotopoulos, A. and Pisano, C. (2019) Overtourism dystopias and socialist utopias: Towards an urban armature for Dubrovnik. *Tourism Planning & Development* 16 (4), 393–410. doi:10.1080/21568316.2019.1569123
Plato (380 B.C.) *Republic* (R. Waterfield, Trans. Vol. 1). Oxford: Oxford University Press.
Polak, F. (1955) *The Image of the Future*. Amsterdam: Elsevier.
Rovelli, C. (2019) Neither presentism nor eternalism. *Foundations of Physics* 49 (12), 1325–1335. doi:10.1007/s10701-019-00312-9
Sardar, Z. (2010) The Namesake: Futures; futures studies; futurology; futuristic; foresight–What's in a name? *Futures* 42 (3), 177–184.
Shklar, J. (1965) The political theory of utopia: From melancholy to nostalgia. *Daedalus* 94 (2), 367–381.
Sigala, M. (2020) Tourism and COVID-19: Impacts and implications for advancing and resetting industry and research. *Journal of Business Research* 117, 312–321.
Spielberg, S. (Writer) (2002) *Minority Report*. In G.R. Molen, B. Curtis, W.F. Parkes and J. de Bont (Producer). USA: 20th Century Fox.
Strielkowski, W. (2020) International tourism and COVID-19: Recovery strategies for tourism organisations. *Preprints*. doi:10.20944/preprints202003.0445.v1
UNWTO. (2020) Policy brief: COVID-19 and transforming tourism. Availabe at: : https://webunwto.s3.eu-west-1.amazonaws.com/s3fs-public/2020-08/SG-Policy-Brief-on-COVID-and-Tourism.pdf (accessed 16 January 2021).
Verne, J. (1899) *In the Year 2889*. The Forum (February), p. 262.
Weil, J. (2011) Gaultier launches men's fragrance. *WWD* 202 (47), 6.
Winter, R. (2002) Truth or fiction: Problems of validity and authenticity in narratives of action research. *Educational Action Research* 10 (1), 143–154. doi:10.1080/09650790200200178
Wyss, G.D. and Duran, F.A. (2001) *OBEST: The Object-based Event Scenario Tree Methodology*. Alberquerque, New Mexico: Sandia National Laboratories.
Yeoman, I. (2012) *2050: Tomorrow's Tourism*. Bristol: Channel View Publications.
Yeoman, I., Fountain, J. and Meikle, S. (2020a) Could food and drink save the tourism industry? See https://www.newsroom.co.nz/ideasroom/2020/06/08/1223344/could-food-and-drink-save-the-tourism-industry (accessed 17 January 2021).
Yeoman, I. (2020b) Don't leave home – but then go see your country. See https://www.newsroom.co.nz/ideasroom/dont-leave-home-but-then-go-and-see-your-country (accessed 17 January 2021)
Yeoman, I., McMahon-Beattie, U., Findlay, K., Goh, S., Tieng, S. and Nhem, S. (2020) Future-proofing the success of food festivals through determining the drivers of change: A case study of Wellington on a Plate. *Tourism Analysis* 26 (2–3), 167–193.

Part 1
Ontological Approaches

2 Science Fiction and the Future of Tourism

Ian Yeoman and Una McMahon-Beattie

Chapter Highlights

- An overview of science fiction and how it portrays the future.
- Science fiction – myth or reality, truth or explanation?
- COVID-19 and robot prostitutes of Amsterdam.

Introduction

Science fiction, known as sci-fi or SF, is a broad genre that often contains speculations based on current science and technology. It contains element of fantasy, utopia, dystopia, structured with narratives, plots, stories in which its imaginary elements are sometimes conceivable other times not (Yeoman, 2012). Wright (2018) in the paper 'Cloning Animals for Tourism in the Year 2070' discusses the near animal extinction because of habitat change and destruction because of climate change and humankind. As a result, Wright presents three narratives illustrating scenarios in which animals could be cloned in the future for tourism purposes. In scenario one, animals are cloned to order for luxury dining experiences. In scenario two, animals are cloned for sport hunting and in scenario three, animals are cloned for education, conservation and zoos. Does this sound far-fetched? Do you view such a scenario with a degree of scepticism? But what if the scenario were true? Or, what if we could explain how such a scenario could occur? This chapter addresses the fundamental assumptions of those questions, overviewing what is science fiction and how it portrays the future of tourism.

What is Science Fiction?

The word 'science' acquired its modern meaning with the realisation that reliable knowledge is rooted in the evidence of the senses, carefully sifted by deductive reasoning and the experimental testing of generalisations. In the 17th century, writers began producing speculative fictions about new discoveries and technologies that the application of scientific method might bring about, the earliest examples being accommodated rather uncomfortably within existing genres of literature and narrative frameworks (Stableford, 2003). Gunn (2003) noted that:

> Science fiction started in the pulp magazines invented in 1896 by Frank A. Munsey. Mostly filled with adventure stories in a variety of locales and periods, they became more specialised beginning in 1915 with the introduction of Detective Story Monthly and then Western Story Magazine in 1919 and Love Stories in 1921. Hugo Gernsteck, an immigrant from Luxembourg had published popular science magazines with science fiction stories in them. In 1926 he mustered his resources (and his courage) and founded Amazing Stories. Soon competitors began to appear, fans and new writers were attracted and a genre was born. (Gunn, 2003: xvi)

Science fiction stories and science fiction writers had been around before, but what they wrote was not quite what we would know as science fiction and it was not even called science fiction. It was Gernsback in 1929 who came up with the name and it stuck (Gernsback & Westfahl, 1994). Before that, for example, Verne's adventure novels, were called 'voyages extraordinaires', and Wells' stories and novels were known as 'scientific romances'. Some critics have claimed that the direction in which Gernsback moved the new category was a blind alley and that it would have better existing as a kind of mainstream variant. However, it is difficult to imagine how it would otherwise have developed its own sense of identity, a body of enthusiastic and informed writers and readers, and shared assumptions that sometimes rigidified into conventions that made it such a success. The Gernsback tradition, modified by a succession of influential magazine editors from John W. Campbell at the *Astounding/Analog* to Michael Moorcock of *New Worlds*, shaped the way science fiction developed. It is a genre that allows writers to venture where they will and invites unsolicited thought and freedom.

What are Science Fiction Writings?

One of the exemplars of science writings is Greg Egan (Burnham, 2014) who publishes works that challenge readers with rigorous, deeply informed scientific speculation. He delves into mathematics, physics and other disciplines in his writings, putting him in the vanguard of hard science fiction renaissance of the 1990s. Egan used cutting-edge scientific theory to explore ethical questions. His two major novels, *Permutation City* (Egan, 1994) and *Schild's Ladder* (Egan, 2002), constitute a bold artistic statement that narratives of science are equal to those of poetry and drama, and that science holds a place in the human condition as exalted as religion or art. Egan's *Schlid's Ladder* is set 20,000 years in the future. Cass, a humanoid physicist from Earth, travels to the Mimosa orbital station and begins a series of experiments to test the extremities of the fictitious Sarumpaet rules, a set of fundamental equations in Quantum Graph Theory which hold that physical existence is a manifestation of complex constructions of mathematical graphs. The book is a series of 'what if's' of cognitive estrangements (Mendlesohn, 2003). Cognitive estrangement is inextricability linked to the encoded nature of science fiction, to style, lexical invention and embedding. Cognitive estrangement is the sense that something in the world described in the literature is incongruous with the readers experienced world. On a basic level this difference may be achieved of time, place and technological scenery. But if that is not done, the resultant fiction is didactic, educational and overly descriptive. As Mendlesohn (2003: 5) recalls:

> The technique common in early science fiction is commonly known as 'info-dump': a character lectures a capture audience about something they could be expected to know but which we do not. It is a very difficult thing to avoid and at the moment of conceptual breakthrough when the critical insight is won, and the world is revealed as bigger or different than one thought, it can be the only tool a writer has to convey information.

Indeed, even Egan uses this technique in *Schild's Ladder* when he allows one character a two-page lecture (pp. 88–90), 'mitigated by allowing the point of view protagonist to be slightly less familiar with the material than others in the audience, by couching this didacticism as a plea for "forbearance" on behalf of the new theory, and by allowing the point of view to assume that others in the audience are irritated' (Mendlesohn, 2003: 5).

Science Fiction Scholarship

The teaching of science fiction was started by its fans. Sam Moskowitz taught evening classes at the City College in New York in 1953 followed by courses at Colgate University, Eastern Mexico University and the College of Wooster (Stableford, 2003). From this, other courses proliferated, not only in English departments but in physics, chemistry, anthropology and others. As the study of science fiction developed so did scholarly activity. Journals included, *Extrapolation* in 1959, *Foundation* in 1972 and *Science Fiction Studies* in 1973. The Science Fiction Research Association was founded in 1970 to improve classroom teaching and to encourage.

The Future

The use of 'future' in the English language dates back to the 14th century. It derives from the Latin *futurus*, meaning 'about to be', which became assimilated to French as 'futuer'. Broadly speaking, future and its translation refer to the time that is to be or come hereafter. It is not clearly delineated in terms of time horizon; it may mean tomorrow, next year, the coming decade, the next 20, 30 or 50 years or even forthcoming centuries (Asselt *et al.*, 2010). Scientific research, which is positivist in nature, is grounded in empirical research that is objective and data rich. However, data about the future cannot be gathered from surveys or in other positivist ways as the future hasn't occurred. But here lies the conundrum, we live in a society, which is data driven and objective (Yeoman & Postma, 2014) thus the future and future studies are often dismissed based upon scepticism. This is a challenge for those working and researching in the futures industry!

Humankind did not always contemplate the future as a realm of action (Adams & Groves, 2007). In earlier times, the future was considered a sacred domain ruled by the Gods. Only in modern times did the idea arise that humans could influence or even shape the future. This view of the future as a realm of action encouraged interest in contemplating the future. The future can be engaged in a variety of ways from, for example, utopian/dystopian novels (Wells, 1902, 1977), science fiction (Yeoman & Mars, 2012) or business-style reports (https://www.foresightfactory.co/). Economists deploy econometrics to forecast the future economy. Climate models are used for a variety of purposes from the study of dynamic weather and climate systems to projections of future climate. Demographers use vital statistics that track births and deaths, combined with data such as marriage, divorce and migration to forecast populations. This approach to the future is predictive, numerate and singular. Here the future is strongly linked to the past connecting the past to the future (Yeoman & Mars, 2012). However, others hold the alternative belief that there are multiple futures as accuracy cannot be achieved with a singular, predictive approach to the future (Yeoman & McMahon-Beattie, 2005, 2014; Yeoman & Postma, 2014).

Science Fiction and the Future

At the beginnings of the 20th century, H.G. Wells anticipated that futures research would become a scientific discipline. This idea was fuelled by the post-World War II uncertainty and the need to deal with complex technologies in an uncertain world (Bell *et al.*, 2013). As (Hales, 1969: 2–3) noted:

> Rapid change has affected human thought in another way. The future is no longer regarded as predestined – an existing landscape that will be revealed to us as we travel through it. It is now seen as the result of the decisions, discoveries and the efforts that we make today. The future does not exist, but a limitless number of possible futures can be created. From this mode of thought it is a natural step to the idea of establishing desirable goals towards which we can deliberately work, ranging from the conception of a desirable (but feasible) form of transport to the scenario of a desirable society.

This lead editorial for the first issue of the journal *Futures* aspires a utopian focus through science fiction on the creation of a better society, which echoes the history Plato's Republic (1993) and More's Utopia (2016), in that each outline a particular vision for a better society. Bell *et al.* (2013) takes this proposition to explore the contributions and warnings of utopia and dystopia through science fiction as mechanisms for innovation, visions and business. Bell notes that science fiction acts as mechanisms for explanations of the future as prototypes. Prototypes act like mechanisms for interpretation or construction of fictional futures through stories of events or products (Johnson, 2011). While prototypes and scenarios are different concepts both have emancipatory powers in which the future imagined espouses a belief that the future could actually occur, thus an explanatory mechanism (Wyss & Duran, 2001). Science fiction tends to evoke a visualisation of the future, a better future for humankind with some sort of science representation through a fictional account hence 'science' and 'fiction' as science fiction (Forster *et al.*, 2011). However, one of the issues of an interpretation approach is that Jürgen Habermas often criticised hermeneutics as being unsuitable for understanding society as it was unable to account for questions of social reality thus alluding to fantasy (Mendelson, 1979). Nonetheless, an interpretation approach is justified when signals are weak and lacking the attributes of prediction. A hermeneutic approach is to understand or interpret, that is to show how something could occur. In the scenario planning and futures literature, this is portrayed as 'what if?', thus making us think about how to bring about reality, even if the connections are weak; an interpretation of fact rather stating the absolute.

Science Fiction and the Future – Truth or Explanation

Ontology is concerned with the study of being and assumptions are concerned with what constitutes reality, in other words, 'what is' (Scotland, 2012). Thus, researchers need to take a position regarding their perceptions of how things really are and how things really work. Yeoman and McMahon-Beattie (2016) build upon this type of epistemological approach to theory, but because of ontological hierarchy an ontological perspective is proposed (Bergman *et al.*, 2010; Bhaskar, 1979) focusing on the belief system of those that execute tourism futures research. One way to look at the future from an ontological perspective is Bergman's (2010) classification of the future which is based upon two dimensions of the future, namely truth claim and explanatory claim and from this a theoretical classification of futures research can be conceptualised. Bergman's classification sets out to create an ontologically grounded typology of future states or what Bergman calls forecasts. The rationale of the typology is that truthfulness is based upon what the author believes is going to happen in the future.

Bergman's approach to the future centres upon the arguments of Bhaskar (1979) which are cited in Bergman *et al.* (2010) as to what truth espouses to do, that is by asserting to forms of belief and relationships of knowing. Bhaskar (1993) asserts an ontological alethicism about the truth of things in themselves and their generative causes. It is no longer tied to explicit language but is implicit. This is a move that goes back to what ontology is fundamentally about, that is, 'being', 'becoming', 'existence' or 'reality'. Bergman *et al*'.s (2010) assertion is from an ontology perspective of 'justified belief', in which the authors simply claim the forecast is going to happen. Justified belief links to the philosophies of Theaetetus and Socrates which consider a range of theories of knowledge formulation, including, that true belief is 'given an account of meaning 'explained'' (Gettier, 1963). 'Explained' can act as a mechanism, a mechanism being a kind of effect or phenomenon the explanation produces. These mechanisms are irreducibly causal, are hierarchical and have structure (Hedstrom & Ylikoski, 2010). This links to Bergman's (2010) mechanism of 'explanatory claims' where the author of a statement explicitly indicates mechanisms as causes behind the events or states that they forecast. Applied within the context of science fiction, what are authors trying to convey? Does scientific fiction writing have a degree of truth and reality, thus linking it to probability and plausibility of the future (Ramírez & Selin, 2014). Or is the future to be explained? This proposition links to the 'what if' question often found in the scenario planning literature (Martelli, 2014; Ramírez & Selin, 2014; Robertson & Yeoman, 2014; Yeoman & McMahon-Beattie, 2018) which is the main research method to portray the future.

Science Fiction and Tourism Futures – Explanatory Claim

Yeoman and Mars' (2012) paper 'Robots, Men and Sex Tourism' portrays Amsterdam's red-light district in 2050. Here the android prostitutes are clean of sexual transmitted infections (STIs), are not smuggled in from Eastern Europe and forced into slavery and the city council has direct control over the android sex workers controlling prices, hours of operations and sexual services. This paper presents a futuristic scenario about sex tourism; it discusses the drivers of change and the implications for the future. The paper pushes plausibility to the limit as boundaries of science fiction and fact become ever increasingly blurred. As Parker *et al.* (2007: 247) point out science fiction 'involves systematically altering technological, social or biological conditions and then attempting to understand the possible consequences'. Science fiction has an explanatory framework, as stories point out a mechanism or structure of explanation in a narrative form; why else would Steven Spielberg have brought together 23 futurists to explain how technological futures might occur for the film *Minority Report* (Yeoman, 2012)? He did this in order to find linkage, causality and explanation for the science fiction features of the film. Science fiction has attributes of soft falsification or distorting the truth. It is all about the concept of scepticism which Sardar (2010) regards as important in challenging the notions of truth, as it expresses the questions of what else is possible. This is the argument of conceptualisation, the issue of knowing and the issues of being and quality. As the future is a conceptualisation of something that hasn't happened yet, evidence has only to be abstract rather than empirical. Thus, there will always be a degree of scepticism and falsification in the minds of decision-makers involved in futures research as there is a higher degree of conceptualisation. Hence, science fiction has low elements of truth in Bergman *et al'*.s (2010) framework.

According to Gilad (2004) scenarios throw off signposts, which are monitored via signals. Robertson and Yeoman (2014) argue that signposts are directions to truthfulness and are evidence of scenarios occurring now, while the signal is an indicator of what the future could be. Thus, these are the explanatory mechanisms and claims to the future. Platenkamp and Botterill's (2015) discussion about fallibilism focuses on explanation rather than truth. As the linkages are weak and fallibilistic in nature the proposition is not belief and truth but explaining how the scenario could occur. In the scenario planning literature, 'what if' questions are at the core of scenarios (van der Hiejden *et al.*, 2002). Asking scenario planning participants to suspend belief systems (hence truth) and focus on explanation in how something could occur links to explanatory mechanisms which are thin on causality and these have a tendency to be weak signals (Lukka, 2014). Thus, explanatory mechanisms and explanation are interpretations of the future.

Science Fiction and Tourism Futures – Truth and Reality

Pandemics were once confined to science fiction magazines and Hollywood. Richard Matheson's (1954) novel *I Am Legend* has inspired three very different adaptations. All they have in common is that a plague wiped out most of humanity and the survivors, save the main characters, have been turned into (essentially) vampires. *Outbreak* (Petersen, 1995) portrays a new viral haemorrhagic disease breaking out in a small American town. Scientists from the Centre for Disease Control and Prevention (CDCP) race against time to stop it from spreading. Unfortunately, they also must deal with a bloodthirsty army general who wants the virus for a bioweapon and is determined to prevent a cure. *12 Monkeys* (Gilliam, 1995) is a science fiction classic starring Bruce Willis as a time traveller going back to the 1990s to identify the reason behind a global pandemic that nearly wiped out humanity. *Contagion* (Soderbergh, 2011) is a Steven Soderbergh film where Gywneth Paltrow dies horribly from a deadly new virus. It is a smart thriller about how diseases spread, the difficulty in finding a cure and the way conspiracy theorists and incompetent or malicious authorities can make the situation worse. Does this feel familiar? This is where science fiction becomes reality. What these science fiction films portray are situations like the emergence of COVID-19, which is an infectious disease caused by severe acute respiratory syndrome. It was first identified in December 2019 in Wuhan, China, and has since spread globally, resulting in an ongoing pandemic. As we write this chapter, COVID-19 is changing the world. With the slow rollout of the vaccine, and with some countries still continuing to manage medical capacity through social distancing and closing borders, tourism has been and continues to be one of the hardest hit industries. In the early stages air travel had stopped, cruise liners were in port and mega-events such as the Olympics were potentially going to be cancelled (Gössling et al., 2020) and eventually did not run in their normal spectator fomat.

Richter (2003) highlighted that two of the corollaries of globalisation was rising – international travel and the emergence and re-emergence of infectious diseases. In 2005, VisitScotland, Scotland's National Tourism Organisation, was worried that a media frenzy had begun to develop globally with the spread of Avian Influenza (hereafter Avian Flu) from South East Asia to Northern Europe. The media interest was largely a result of the impact on chickens as a food source and the potential of Avian Flu to mutate and trigger a global flu pandemic. The organisation prepared a series of scenarios (Page et al., 2006). One of the scenarios was called *It's Here*. The scenario is based upon a situation in which the influenza virus is has spreading and has entered Scotland, like the present circumstances with COVID-19. People are very reluctant to travel, public gatherings have been cancelled by the Westminster government and tourism has been severely curtailed. Through the two-year period when the virus is active, tourism demand is oriented to isolated, rural destinations. The economic impacts according to the Moffat Model (Blake et al., 2006) show the following results:

- A 40–50% drop in domestic tourism and a 70–90% drop in international tourism and a decline in day trips signals a profound change in Scotland's tourism economy, with a £26 bn drop in GDP.
- Assuming multiple waves of the pandemic, the economy is subject to a number of shock waves over a 2-year period, posing ongoing problems for economic activity.
- The outcome could be a 10–38% drop in GDP, measured over a 10-year period to allow for the effects to be absorbed.
- Some sectors of the economy experience a 50% drop in output and considerable drops in employment and productivity, highlighting the interconnections between tourism and the wider economy.

Given the above, we are now in a situation with COVID-19 where science fiction has become a reality. We can understand and comprehend that reality and it is plausible, relevant and credible. COVID-19 is a dystopian future that has come true and it is a place that we do not like and, ideally, would have wished to have avoided. In order to plan for recovery, we must be proactive, like VisitScotland in 2005, in terms of an effective crisis management response and scenario planning.

Concluding Arguments

Thinking the impossible or making the impossible possible is one of the roles of science fiction in tourism futures research. Science fiction represents a degree of truthfulness; it could be the difference between myth and reality. Alternatively, it could be a mechanism to explain how something could come true, that is, the classic 'what if' question. Science fiction engages audiences, whether the focus is COVID-19 or robot prostitutes. It conjures up the imagination and creates interest and curiosity. With tourism research we often fail to argue the impossible as we like certainty, prediction and facts (Robertson & Yeoman, 2014; Yeoman & Postma, 2014). Science fiction is metaphorical in nature, a different and new world that engenders innovation and alternative ideas. Dator (2014) notes that one of the roles of futures studies is to be transformational and think the impossible. This is certainly possible with the genre of science fiction.

References

Adams, B. and Groves, C. (2007) *Future Matters: Action, Knowledge, Ethics*. Leiden/Boston: Brill.
Asselt, M., Klooster, S., Notten, P. and Smits, L. (2010) *Foresight in Action: Developing Policy-oriented Scenarios*. London: Routledge.
Bell, F., Fletcher, G., Greenhill, A., Griffiths, M. and McLean, R. (2013) Science fiction prototypes: Visionary technology narratives between futures. *Futures* 50 (June), 10.
Bergman, A., Karlsson, J.C. and Axelsson, J. (2010) Truth claims and explanatory claims – An ontological typology of futures studies. *Futures* 42 (8), 9.
Bhaskar, R. (1979) *The Possibility of Naturalism: A Philosophical Critique of the Contemporary Human Sciences*. Hemel Hempstead: Harvester Press.
Bhaskar, R. (1993) *Dialectic: The Pulse of Freedom*. London: Verso.
Blake, A., Durbarry, R., Eugenio-Martin, J.L., Gooroochurn, N., Hay, B., Lennon, J., Sinclair, M.T., Sugiyarto, G. and Yeoman, I. (2006) Integrating forecasting and CGE models: The case of tourism in Scotland. *Tourism Management* 27 (2), 292–305.
Burnham, K. (2014) *Greg Egan*. Urbana, Illinois: University of Illinois Press.
Dator, J. (2014) Four images of the future. SET: Research Information for Teachers 1, 61–63. See http://www.nzcer.org.nz/mzcerpress/set/articles/four-images-future (accessed 18 May 2020).
Egan, G. (1994) *Permutation City*. Sydney: Millennium Orion Publishing Group.
Egan, G. (2002) *Schild's Ladder*. New York: Gollancz.
Forster, E.M., Merchant, I., Ivory, J., Hopkins, A., Redgrave, V. and Thompson, E. (2011) *Howards End*. In Australia: Distributed by Shock DVD.
Gernsback, H. and Westfahl, G. (1994) How to write 'science' stories: The editor of 'Scientific Detective Monthly' tells how to and how not to write them. *Science Fiction Studies* 21 (2), 268–272.
Gettier, E.L. (1963) Is justified true belief knowledge? *Analysis* 23 (6), 121–123.

Gilad, B. (2004) *Early Warning*. New York, USA: AMACOM.
Gilliam, T. (Writer) (1995) *12 Monkeys*. United States: Universal City Studios.
Gössling, S., Scott, D. and Hall, C.M. (2020) Pandemics, tourism and global change: A rapid assessment of COVID-19. *Journal of Sustainable Tourism* 1–20. doi:10.1080/0 9669582.2020.1758708
Gunn, J. (2003) Foreword. In E. James and F. Mendlesohn (eds) *The Cambridge Companion to Science Fiction* (pp. xv–xviii). Cambridge: Cambridge University Press.
Hales, J. (1969) Futures – Confidence from chaos. *Futures* 1 (1), 2–3. doi:10.1016/ S0016-3287(69)80001-7
Hedstrom, P. and Ylikoski, P. (2010) Causal mechanisms in the social sciences. *Annual Review of Sociology* 36 (August), 49–67.
Johnson, B.D. (2011) Science fiction prototyping designing the future with science fiction. *Synthesis Lectures on Computer Science* 3 (1), 1–190. Morgan and Claypool Publishers. https://doi.org/10.2200/S00336ED1V01Y201102CSL003
Lukka, K. (2014) Exploring the possibilities for causal explanation in interpretive research. *Accounting, Organizations and Society* 39 (8), 559–566.
Martelli, A. (2014) *Models of Scenario Building and Planning*. Milan / London: Brocconi University Press/Palgrave MacMillan.
Matheson, R. (1954) *I am Legend*. New York: Gold Medal Books.
Mendelson, J. (1979) The Habermas-Gadamer debate. *New German Critique* 18, 44–73. doi:10.2307/487850
Mendlesohn, F. (2003) Introduction: Reading science fiction. In E. James and F. Mendlesohn (eds) *The Cambridge Companion to Science Fiction* (pp. 1–12). Cambridge: Cambridge University Press.
More, T. (2016) *Utopia* (edited by George M. Logan; translated by Robert Adams, 3rd edn). Cambridge: Cambridge University Press.
Page, S., Yeoman, I., Munro, C., Connell, J. and Walker, L. (2006) A case study of best practice – VisitScotland's prepared response to an influenza pandemic. *Tourism Management* 27 (3), 361–393. doi:https://doi.org/10.1016/j.tourman.2006.01.001.
Parker, M., Fournier, V. and Reredy, P. (2007) *The Dictionary of Alternatives. Utopianism & Organization*. London: Zed Books.
Petersen, W. (Writer) (1995) *Outbreak*. California: Warner Bros.
Platenkamp, V. (2015) Academic insulae the search for a paradigm in and for tourism studies. *Tourism Analysis: An Interdisciplinary Journal* 20 (5), 561–571.
Plato (1993) *Republic*; translated by Robin Waterfield. Oxford: Oxford University Press.
Ramírez, R. and Selin, C. (2014) Plausibility and probability in scenario planning. *Foresight* 16 (1), 54–74. doi:10.1108/FS-08-2012-0061
Richter, L.K. (2003) International tourism and its global public health consequences. *Journal of Travel Research* 41 (4), 340–347. doi:10.1177/0047287503041004002
Robertson, M. and Yeoman, I. (2014) Signals and signposts of the future: Literary festival consumption in 2050. *Tourism Recreation Research* 39 (3), 321–342. doi:10.1080/02 508281.2014.11087004
Sardar, Z. (2010) The Namesake: Futures; futures studies; futurology; futuristic; foresight -What's in a name? *Futures* 42 (3), 177–184. doi:10.1016/j.futures.2009.11.001
Scotland, J. (2012) Exploring the philosophical underpinnings of research: Relating ontology and epistemology to the methodology and methods of the scientific, interpretive, and critical research paradigms. *English Language Teaching* 5 (9), 9–16.
Soderbergh, S. (Writer) (2011) *Contagion*. California: Warner Bros.
Stableford, B. (2003) Science fiction before the genre. In E. James and F. Mendlesohn (eds) *The Cambridge Companion to Science Fiction* (pp. 15–31). Cambridge: Cambridge University Press.
van der Heijden, K., Bradfield, R., Burt, G., Cairns, G. and Wright, G. (2002) *The Sixth Sense, Accelerating Organizational Learning with Scenarios*. Chichester: John Wiley & Sons.

Wells, H.G. (1902) The discovery of the future. *Nature* 65 (1684), 326–331.
Wells, H.G. (1977) *The Time Machine; The War of the Worlds*. A critical edition/ed by F.D. McConnell. New York: Oxford University Press.
Wright, D. (2018) Cloning animals for tourism in the year 2070. *Futures* 95, 58–75.
Wyss, G.D. and Duran, F.A. (2001) *OBEST: The Object-based Event Scenario Tree Methodology*. Albuquerque, NM: Sandia National Laboratories.
Yeoman, I. (2012) *2050: Tomorrow's Tourism*. Bristol: Channel View Publications.
Yeoman, I. and Mars, M. (2012) Robots, men and sex tourism. *Futures* 44 (4), 365–371. doi:http://dx.doi.org/10.1016/j.futures.2011.11.004
Yeoman, I. and McMahon-Beattie, U. (2018) Teaching the future: Learning strategies and student challenges. *Journal of Tourism Futures* 4 (2), 163–167. doi:10.1108/JTF-12-2016-0054.
Yeoman, I. and McMahon-Beattie, U. (2005) Developing a scenario planning process using a blank piece of paper. *Tourism and Hospitality Research* 5 (3), 273–285.
Yeoman, I. and McMahon-Beattie, U. (2014) New Zealand tourism: Which direction would it take? *Tourism Recreation Research* 39 (3), 415–435. doi:10.1080/02508281.2014.11087009
Yeoman, I. and McMahon-Beattie, U. (2016) An ontological classification of tourism futures. In M. Scerri and L.K. Hui (eds) *CAUTHE 2016: The Changing Landscape of Tourism and Hospitality: The Impact of Emerging Markets and Emerging Destinations*. Sydney: Blue Mountains International Hotel Management School.
Yeoman, I. and Postma, A. (2014) Developing an ontological framework for tourism futures. *Tourism Recreation Research* 39 (3), 299–304. doi:10.1080/02508281.2014.11087002

3 The Future, the Devil and the Deep Blue Sea

Felicity Picken

Chapter Highlights

- Undersea tourism is inherently oriented towards the future.
- The 21st century is noted to be the most oceanic century of our time.
- Tourism is a key driver of the desire and opportunity to visit undersea places.
- Undersea worlds invite a rethinking of assumptions that underpin tourism.

Introduction

Tourism is known for its transformative capacities. As an activity with an unremitting global reach and an important role in shaping the many desires and experiences of novelty and difference, the tourist has become a 21st-century metaphor for contemporary ways of being in the world (Bauman, 1996). In creating and recreating toured versions of the world that fashion and sustain tourist demand, the tourism industry relentlessly gestures towards the future.

Oceanic worlds are also futuristic places, particularly in the increasingly accessible spaces beneath the waves. The ocean depths, like its counterparts in outer space, have an enduring relationship to the diverse imaginative faculties of our kind. It is only in recent history that scientific knowledge has laid any claims of proof in knowing this place and even now lays claim to knowing only ten percent (Ellis, 2005; Rozwadowski, 2005). The futurism of oceanic worlds has resulted from their inaccessibility to much of our daily lives. The life-depleting properties of oceanic nature have given the oceans an alien status in the land-based humans' world (Helmreich, 2009). Any close relationship between these worlds has mainly existed in a future time. In some ways this time has arrived or is in the process of arriving. The 21st century has been recognised as an oceanic century (Mentz, 2009; Petersen, 2010; Picken & Ferguson, 2014; Steinberg, 1999). Portending a planetary future that is 'more oceanic'

(Deloughrey, 2017: 34), the blue planet is regarded as the most omnipresent icon of our time (Sachs, 1999).

While ocean surfaces are quite familiar today, the increasing visibility and accessibility of oceanic depths is more novel. Mythology; arts marine; photography; film; aquaria; marine parks; submersibles; satellites and technologies of immersion combine in the current day to describe a long history of challenging oceanic frontiers. These have included trials and fatal errors, scientific discovery, technological innovation, risk and labour to access and know the world 'underneath' (Deloughrey, 2017). While these are not yet familiar places, they are now more accessible than they have ever been and a significant gateway for this accessibility is tourism. The natures of both tourism and the undersea worlds that are being revealed at oceanic depths support the importance of futures research. Encouraged by the imaginary, the temptations of the possible and the allure of the unknown, there is an intensity of innovative activities unfolding through undersea worlds and it is this action of unfolding that is most challenging for research which must confront ways to address, and keep step with, the very near future.

Whether near or far, the *future* is a risky game that is prone to speculation and encourages the deployment of the imagination to fill out the gaps that exist around knowledge. Neither of these sits particularly well within the academy of scientific knowledge production. On epistemological grounds, *future tellers* tend to sit outside these walls and speculators are strongly discouraged. Futures-oriented research is bedevilled by a wide array of unique obstacles and many questions that are addressed through a working project, of interdisciplinary scope and indeterminant parameters (Kuosa, 2011). For tourism research, these broader implications are present in questions such as whether it is more ethical to describe or to prescribe 'futures' and whether our stance is influenced by dystopian or utopian visions (Coleman & Tutton, 2017). There are many questions of method and what kind of experiments are permissible or useful in the field (Kuosa, 2011; Sardar, 2010). Related to this are questions of how to describe, explain and/or analyse phenomena that are unfolding, the evidence of which is still being produced and that has no benefit of hindsight (Picken, 2015c). Ontologically, questions include whether an understanding of future or near-future realities oblige a reconsideration of assumptions that underpin the research? As described by Sardar (2010), in his review of the field of futures studies, an enduring guiding principal is that it cannot be assumed that a future that is different will subscribe to today's assumptions. He further explains that this would not be desirable anyway, since part of the enduring appeal of future-oriented research is the possibility of challenging 'inherited consciousness and views' (2010: 179).

Focusing on this final question, of futures research and its capacity to shift assumptions, this chapter will draw upon innovations in undersea tourism and how the opening up of this new cultural territory has the capacity to first reveal an assumption – of land-based tourism thinking – and then to render this somewhat imprecise. The common, and in this case literal, sense of 'diving into' futures research is explored as a set of challenges (and opportunities) that are also characteristic of the open and uncertain nature of futures themselves.

Tourism and Oceans as Futures

The value of future-oriented research in places where tourism and the oceans come together is highlighted by the futures-oriented nature of both tourism and oceans. The metaphor of the tourist for contemporary life recognises the powerful role of tourism in the ever-ubiquitous evidence of a globalising world. Tourism enrols people and places into systems of mobility, developing a global consciousness among the people and assigning places to a well-developed semiology of signs, to a world that is increasingly designed to be visited (Bærenholdt *et al.*, 2004; Bauman, 1998; Franklin, 2004; MacCannell, 2001; Urry, 2000; Van den Abbeele, 1980). As a metaphor for contemporary life, tourism is both a comment and critique of new and emerging hyper-mobile consumerist lives and of lives that are made dynamic by this same system. Facilitating touristic kinds of travel and experience involves both the practical logistics of making the experience possible and also the symbolic practices that create a desire to be part of the experience before this. While not often explicitly framed as a future-oriented activity and industry, modern tourism remains a key inventor of desirable futures for people who are primed, through their lives in a mundane 'everyday', to be persuaded to travel away from routine and towards extraordinary realities (Franklin, 2004). With a history of production that is characterised by the constant pursuit of change, of extending and diversifying across spaces (Bailey, 1989: 108), tourism often operates at the cutting edge of time, perched on the verge of delivering new imaginaries and experiences.

The blue planet is a new object that emerged as a sobering and wonderful visual reality in the late 1960s through the images beamed back to Earth by various space missions at the time (Helmreich, 2011b). Earth as the blue planet, that is, as distinctive by virtue of its blue, oceanic nature was to become an icon of the 21st century. At the same time, the idea of the oceanic world as an 'undiscovered country' has persisted since the mid-19th century (Rozwadowski, 2005: 37). Despite the vast successes of science and technology, the oceanic world continues to constitute an 'historical, linguistic and experiential gap in our collective understanding' (Mentz, 2009: 998). Regarded, even by those who know it well, as an 'alien' place (Helmreich, 2009), it is this unfamiliarity, lack of knowing and relative inaccessibility that casts the ocean as a future place. A future place in this context is one that is poised to become better known and more accessible. As Lavery (2015) notes, the opacity of the ocean is reflected in fictive imaginaries that furnish the invisible spaces. These invisible spaces perform both geologic and social underworlds (2015: 26) where there is a 'long history of making things disappear' (2015: 28) and of activities that transgress and evade the eyes of the law and of history (2015: 26). What we have inherited instead are 'fleeting glimpses' (Ellis, 2005), of lives that are carried on beyond much of our everyday activity (Beatley, 2014). These same qualities, that put oceanic worlds in opposition to and inversion of the orderly world of land (Mentz, 2009: 1001), qualify them as suitable tourist places. The very elements that combine to form the mystique of oceans, their inaccessibility and their fertility in futures research, also amalgamate some of the most important elements of a promising tourist destination. In characterising liminality, novelty, adventure, excitement, awe and wonder, undersea is already a compelling tourist space.

The oceanic future is reasoned against the problem-raising futures of global climate change and the oceans' leading role within this. The unfolding geological era of the Anthropocene, with its watchfulness for any sign of change in Earth's systems, strongly supports the oceans' raised status in the world of humans today. The oceanic future is also reasoned against the problem-solving advancements in science and technology that open, and continue to, greater expanses of the Earth's oceanic estate for exploration and exploitation in the expanse of 'blue-green capitalism' (Helmreich, 2009: 26). The recent framing of these global aspirations is the 'blue economy' and the special role this accords the tourism and leisure industry in growing this.

> Many tourists nowadays seek a unique and customised experience rather than the more traditional type of 'sun and-sea' package holiday. These changes on the demand side require reaction and adaptation by operators and destinations. The sector should develop new products promoting attractiveness and accessibility of coastal and marine archaeology, maritime heritage, underwater tourism among other innovative activities.
> (European Commission, 2014)

This portends the diminishment of the extreme part of undersea tourism activities or, more specifically, that undersea environments will be less 'extreme'. While the extreme experiences found at oceanic depths through various free and assisted dive expeditions attend physical, psychological and emotional transgressions of thresholds, of previously held limits and of journeying across frontiers (Picken, 2016, 2018), each year, longer durations undersea, advancements in SCUBA, fins and skins as well as refashioned surface to air respiratory aids invite more people to experience undersea through tourist resorts all over the world. In encouraging and facilitating immersive experiences this way, tourism contributes towards the experimental end of both oceanic and touristic practice. Significant, and growing, numbers of tourist bodies furnish our experiential knowledge of immersion undersea. As such, dive tourists and tourism are sites of innovative tourism futures.

Ocean Depths as Tourist Playground

In the slow defeat of oceanic inaccessibility is a socio-scientific twinning that, while not new to tourism and leisure, opens a sizeable territory for the mobilisation of significant amounts of global capital, labour, science, technology and creativity. With increasing scale and intensity, blue tourism and leisure ride the wave of technological advancement to refashion our ability and desire to be undersea (Picken, 2015c: 293).

In the 1800s, both scientists and tourists began to populate oceanic spaces (Rozwadowski, 2005) but even before this, marine tourism was one of the earliest forms of pleasure travel, including any touristic activity that is associated with oceans and seas. Today, this includes established tourist attractions and activities like aquaria and marine parks, beach resorts, cruising and sailing, snorkelling and diving as well as novel and unconventional tourist attractions and activities, particularly within the oceans' depths, as a 'space race kind of urgency' develops to capture 'emerging undersea leisure markets' (Jackson & della Dora, 2009). Equally, technologies that open up ever greater and longer access to undersea follows its own relentless advancement meaning this is a dynamic and changeable place for tourism activity.

Both established and new forms of oceanic-based tourism conform to the idea of consuming oceanic worlds and this is promoted through the value of experiencing or encountering oceans for purposes that meet socially derived desires (Cater, 2010). The term 'consuming oceanic worlds' is distinguishable from 'consuming oceans', which is more exactly linked to resource extraction as with fish, oil, minerals and energy. Instead, consuming oceanic worlds is a practice that more often aims to keep these worlds intact (and alive) to be experienced through film and media, through captivity, on ocean surfaces or through bodily, vehicular or pod facilitated immersion. Attuned to the tertiary and quaternary 'experience economies' (Pine & Gilmore, 1999), consuming oceanic words follows a desire for an oceanic presence that is accomplished through the creative, media, digital, cultural, leisure, tourism and entertainment industries. Each of these, by stages, have come to occupy an increasingly significant place in producing our relations with oceanic spaces. Incorporating science and technology, myth and imaginary in machinations designed to elicit interest, desire and a heightened presence of oceanic places, tourism itself is a formidable presence in this space. Today, as if confirming the end of the modern, maritime driven producer society, 'the old piers [are] turned into heritage sites' (Lambert *et al.*, 2006: 483), just as Portugal reconstitutes its oceanic identity from one that was past, but is now the future (Beatley, 2014: 108) and Dubai engages in a scale of terraforming oceanic islands and innovating marine spectacle that is unprecedented (Jackson & della Dora, 2009). Off-shore, increasing and diverse forms of underwater leisure are performing significant catalysts for drawing more people into closer relations with oceanic worlds. The 'cultural purchase' of undersea leisure and tourism reinforces the growing status of 'stylistic self-expression', 'freewheeling hedonism' and 'self-conscious display' through undersea experience (Osgerby, 2006: 89).

In the 21st century a picture emerges where the sea is increasingly presented as a space for current and future leisure, tourism and recreational consumption (Mentz, 2009; Steinberg, 1999). Early seaside resort development is characterised by the limit of land, 'at the end of the line' and the 'edge of territories' that also signified 'the fading edge of culture' (Pons *et al.*, 2009: 157). These edges have now been compromised through more widespread, intense and diverse relations with oceans. Significant numbers of people now encounter these worlds for their alter-sensuality with an emphasis on the pleasurable, exciting, thrilling, imaginative and aesthetic dimensions of oceanic worlds (Jackson & della Dora, 2009; Lambert *et al.*, 2006; Picken, 2015a; Rozwadowski, 2005). Closer encounters come from skin-to-ocean relations, of immersion with or without a breathing apparatus, skin and fins. By contrast, lower intensity encounters and greater digital distribution of undersea worlds are created through underwater photography and filming. Expanding upon existing dive tourism (Musa & Dimmock, 2013), and niches within this like thermal vent tourism (Ellis, 2005), is greater awareness of undersea cultural heritage and investments in undersea attractions like art and sculpture museums (Picken, 2015a, 2015b). These are encouraged by prophecies of transformation through submersible technologies like undersea resorts and living spaces.

Prototypes of underwater pods and submerged hotels are promising an early, albeit exclusive, taste for spending extended periods of time beneath the sea (see for example Deep Ocean Technology http://www.deep-ocean-technology.com/). New materials and technological design aids mean that yesterday's failed experiments are making way for multiple, transnational enterprises in the pursuit of opening new cultural territories undersea. These are being developed through expanded knowledge and experiences of the life and conditions found beneath the sea following serious efforts to introduce permanent habitats on the ocean floor (Seedhouse, 2011). All these attempts to seed a built environment in a world that is socially and spatially disorganised, require actions that proceed 'without reference to an extant world' (Jackson & della Dora, 2009: 2092, emphasis added). Like aquaria on land, air and space craft in the atmospheres undersea settlements are (to be) active, rather than inert, life support systems, like 'skins' that perform the scuba suit, made up of matter and processes that are sensitive to maintaining optimal living conditions within oceanic environments. These are sophisticated, 'advanced life support systems' that constitute 'protective architectures' that offer 'counter-measures' to the impossibility of living undersea (Pell & Vermeulen, 2012). Even as many of the technologies involved in these kinds of ventures are 'speculative' or in the design phase, they still act as powerful 'attractants of capital, investment and curiosity' (Jackson & della Dora, 2009: 2089).

Implications of Ocean Depths for Futures-oriented Tourism Research

That the oceans have not occupied an important position among many of the contributing, 'parent' disciplines of tourism research, is an explanation, but not an excuse for tourism research to continue to avoid a more in-depth engagement with this space. All of these disciplines have begun to discover how the vastly different conditions of undersea worlds offer the opportunity to take a different view of human relations with the natural world. The extending of our ecological space into the oceans brings extensions to ontological space: in rethinking what it is to live with the blue planet. The case for crafting alternative views of human-nature relations rests upon one of the most obvious, and mundane, facts: that the vastly different nature of oceanic worlds (like light, pressure and density) produce an altered sense of perception (like orientation, sight, movement and sound) and that these conspire to perform alternative ways of being, doing and thinking. Lavery's (2015) proposition that oceans increasingly provide an avenue to 'think with the emergent' is particularly useful here, as is her suggestion that this can be illustrated through the method of 'sampling recent moments of undersea emergence' and of documenting the 'potential and imagined futures' of ocean spaces (2015: 25).

Tourism research has also begun to furnish some of these ways of doing and being in describing the experiences of SCUBA and free diving (Merchant, 2011a; Merchant, 2011b; Picken 2018; Picken & Ferguson, 2014) and through undersea art instillations and sculpture museums (Picken, 2015c). The free diver is a site of medical transformation, extending and the limits and altering the behaviour of the body as it falls and rises through the water column (Picken, 2018). The SCUBA diver is a cyborg (Picken & Ferguson, 2014), the result of coupling of technologies and bodies that sense and perform differently undersea, inviting ways to reconcile this experience back on land (Merchant, 2011a, 2011b). These, and other, experiences that are co-productions of body and technology are becoming more naturalised in tourism and tourist practices (Mars, 2018; Yeoman & Mars, 2012; Yeoman & McMahon-Beattie, 2015). New tourist attractions that are emplaced undersea construct new post-human aesthetics as with the undersea museums of Jason de Caires Taylor (see https://www.underwatersculpture.com/) and in the 'Sinking World' undersea photographic exhibitions of Andreas Franke (see http://thesinkingworld.com/). These cast artistic representation of humanity into the oceanic depths and invite their transformation into coral reef substrata in the undersea museums, or through the adaptations wrought by the 'oceanic patina' developed upon sunken images. At the level of the tourist, diving has been described as an activity enabled through the process of becoming cyborg (Picken & Ferguson, 2014), a coupling of technologies and body that is transmutes and extends vital survival codes including an expanded respiratory system. In undersea worlds, the tourism industry must concern itself with both the survivability and the desirability of engaging with this space. This includes the novel requirement of an appreciable 'aesthetics of life support', of 'waterproofing' and 'tethering' (Pell, 2014; Pell & Vermeulen, 2012) that leads, ultimately, towards the creation of new possibilities for interacting undersea and an enthusiasm for advancing the 'bio-techno-aquatic adaptations' (Pell, 2014: 102) that are required to achieve these. Furthermore, these bio-technological 'cyborgian' realities are increasingly naturalised through new tourism practices across a diverse range of domains (Yeoman & McMahon-Beattie, 2015).

Current, planned and envisioned 'toured oceans', especially undersea, are taking place during a time when regard for the oceanic worlds is beginning to shift from the margins or between the places that matter (Steinberg, 1999) towards occupying a more central place on the planet. Regard for the oceans as an enduring resource is increasingly met with regard for the oceans as a suite of vital ecologies and a growing sentiment that oceanic worlds are inherently beneficial, not least in offering valuable experiences for people to enjoy through tourism and at leisure. Tourism shares oceanic places with these existing and emerging paradigms and industries including the ubiquitous presence of science and technology. These relations are most evident in tourists' role in supporting environmental, 'eco' forms of marine tourism, as casual forms of 'surveillance' of oceanic worlds and in shared projects of discovery as tourists have increasingly played the role of citizen ocean scientists (Cigliano *et al.*, 2015; Marshall *et al.*, 2012; Newman *et al.*, 2012)

Finally, when considering the implications of a deeper engagement of tourism with the oceanic depths, it also worth noting that the futures of oceanic depths share similarities with 'space', the other extreme atmosphere on the planet. Alongside the ontological and methodological challenges and opportunities for both futures and tourism research within oceanic depths, there are also synergies to be considered between the conditions and challenges of space tourism. While these altered conditions of outer (atmospheric) space and inner (deep ocean) space are penetrated more deeply, they both still occupy human extremes and can be said to be in varying states of emergence (Freeland, 2005; Picken, 2016); highly dependent on science and technology; are derived in part as collateral activities that have evolved 'sideways' from their origins in defence or military innovation; and they both rely on the 'power and importance of imagination' in the development of their respective trajectories (Cole, 2011). While Conversely, the imagination, including a strong presence of science fiction, has held a central role in our understanding of oceanic worlds (Rozwadowski, 2005). As the inaccessibility of oceanic depths continues to erode, these places are increasingly harnessed in the exploration of how they 'constrain and allow forms of connectivity, meet and exceed human scales and interests, inspire and refuse imaginative engagement' (Lavery, 2015: 25). Tourism is a key contributor towards these new forms of connectivity. 'Marine mysteries' (Ellis, 2005) continue to abound and so remain significant contributors to forming relations with oceanic worlds. In the digital 21st century, this is increasingly stimulated by 'technoscientific-fuelled imaginings of the sea' (Helmreich, 2009: 6). In this way tourism operates in a space where not only accessibility but also imaginaries of the oceanic worlds are inseparable from highly scientised and technological framings.

Concluding Arguments

This chapter has reviewed some of the ways that oceans are becoming implicated in tourism futures and how, within oceans, tourism is an increasingly significant activity and field of research. The 21st century is unique in witnessing the intensification of awareness of, and activity with, oceanic worlds, including 'plastic oceans'; ocean acidification and sea-level rise; endangered species and endangering industries. Somewhat paradoxically, this is countered by a growing desire and ability to experience and be within oceanic environments for pleasure. The 21st century is one in which we have 'extended the ecological limits of our species' (Petersen, 2010) and, developed new oceanic imaginaries (Deloughrey, 2017). Investigations into the avenues through which oceanic worlds are becoming a much larger presence include not only those that relate to the science and technologies of oceans but also those that relate to the sociality of oceans. It is in this latter sphere that tourism emerges as a key activity through which widespread awareness, relatedness and intimacy with oceanic environments occurs. The industry, and its innovations, engage increasing numbers of people in an increasingly diverse way both on land, on the sea and beneath it.

Of importance to futures-oriented research are new experiences of depth, through longer periods of immersion, to deeper depths through submersible vehicles and through the promise of undersea hotels and living experiences. These are being recognised as future-oriented not only for the changes they bring to the dimensionality of human experience, but for the changes they inspire in the way people, and researchers, think about the world and their place within it. Ocean depths are not only becoming a more diverse playground for the tourists of the future but are increasingly recognised as a critical device, challenging what is revealed as a land-centric rendering of human life on Earth.

While science and technology appear to determine the rate and quality of new undersea experiences, the observation made of outer space settlements also applies in this case, that physical survival does not ensure social survival (Pass, 2006: 1). At the individual and aggregate level, tourism and leisure are instrumental in driving transformations in the way people relate to oceanic environments. Tourism, following a well-rehearsed formula on land of creating a desire to explore oceanic places and then facilitating this exploration, is positioned to be a key industry in the future development of oceanic spaces and of people's attitude, sentiment and experiences with these.

References

Bauman, Z. (1996) From pilgrim to tourist–or a short history of identity. In S. Hall and P. du Guy (eds) *Questions of Cultural Identity*. London: Sage.
Bauman, Z. (1998) *Globalisation: The Human Consequences*. New York: Columbia University Press.
Bærenholdt, J., Haldrup, M., Larsen, J. and Urry, J. (2004) *Performing Tourist Places*. Aldershot: Ashgate.
Bailey, P. (1989) Leisure, culture and the historian: Reviewing the first generation of leisure historiography in Britain. *Leisure Studies* 8 (2), 107–127.
Beatley, T. (2014) *Blue Urbanism: Exploring Connections Between Cities and Oceans*. Washington: Island Press.
Cater, C. (2010) Any closer and you'd be lunch! Interspecies interactions as nature tourism at marine aquaria. *Journal of Ecotourism* 9 (2), 133–148.
Cole, S. (2011) Space tourism: prospects, positioning, and planning. *Journal of Tourism Futures* 1 (2). See https://www.emeraldinsight.com/doi/full/10.1108/JTF-12-2014-0014 (accessed 14 January 2019).
Coleman, R. and Tutton, R. (2017) Introduction to special issue of sociological review on 'Futures in question: Theories, methods, practice'. *The Sociological Review* 65 (3), 440–447.
Deloughrey, E. (2017) Submarine futures of the Anthropocene. *Comparative Literature* 69 (1), 32–44.
Cigliano, J.A., Meyer, R., Ballard, H.L., Freitag, A., Phillips, T. B. and Wasser, A. (2015) Making marine and coastal citizen science matter. *Ocean & Coastal Management* 115, 77–87.
Ellis, R. (2005) *Singing Whales and Flying Squid: The Discovery of Marine Life*. Guildford Connecticut: The Lyons Press.
European Commission (2014) A European Strategy for more Growth and Jobs in Coastal and Maritime Tourism, Communication from the Commission to the European Parliament, the Council, the European Economic and Social Committee and the Committee of the Regions. See https://ec.europa.eu/maritimeaffairs/publications/european-strategy-more-growth-and-jobs-coastal-and-maritime-tourism_en (accessed 14 March 2018).
Franklin, A. (2004) Tourism as an ordering: Towards a new ontology of tourism. *Tourist Studies* 4 (2), 277–301.
Freeland, S. (2005) Up, Up and … back: The emergence of space tourism and its impact on the international law of outer space. *Chicago Journal of International Law* 6 (1), 1–22. See http://chicagounbound.uchicago.edu/cjil/vol6/iss1/4 (accessed 14 March 2018).
Helmreich, S. (2009) *Alien Ocean: Anthropological Voyages in Microbial Seas*. Berkley: University of California Press.
Helmreich, S. (2011a) Nature/culture/seawater. *American Anthropologist* 113 (1), 132–144.
Helmreich, S. (2011b) From spaceship earth to Google ocean: Planetary icons, indexes, and infrastructures. *Social Research* 78 (4), 1211–1242.
Helmreich, S. (2015) Old waves, new waves: Changing objects in physical oceanography. In J. Gillis and F. Torma (eds) *Fluid Frontiers: New Currents in Marine and Maritime Environmental History* (pp. 76–88). Cambridge: White Horse.

Jackson, M. and della Dora, V. (2009) 'Dreams so big only the sea can hold them': Man-made islands as anxious spaces, cultural icons, and travelling visions. *Environment and Planning A* 41 (9), 2086–2104.
Kuosa, T. (2011) Evolution of futures studies. *Futures* 43 (3), 327–336.
Lambert, D., Martins, L. and Ogborn, M. (2006) Currents, visions and voyages: Historical geographies of the sea. *Journal of Historical Geography* 32 (3), 479–493.
Lavery, C. (2015) Indian Ocean depths: Cables, cucumbers consortiums. In M. Jones (ed.) *The Johannesburg Salon – Volume Ten. The Johannesburg Workshop in Theory and Criticism*. See https://jwtc.org.za/resources/docs/salon-volume-10/JWTC_vol_10_salon.pdf (accessed 1 August 2018).
MacCannell, D. (2001) Tourist agency. *Tourist Studies* 1 (1), 23–37.
Mars, M. (2018) Sex robots: The future of desire. *Journal of Tourism Futures* 4 (1), 111–112.
Marshall, N., Kleine, D. and Dean, A. (2012) CoralWatch: Education, monitoring, and sustainability through citizen science. *Frontiers in Ecology and the Environment* 10 (6), 332–334.
Mentz, S. (2009) *At the Bottom of Shakespeare's Ocean*. London: Continuum.
Merchant, S. (2011a) Negotiating underwater space: The sensorium, the body and the practice of scuba-diving. *Tourist Studies* 11 (3), 216–234.
Merchant, S. (2011b) The body and the senses: Visual methods, videography and the submarine sensorium. *Body & Society* 17 (1), 53–72.
Musa, G. and Dimmock, K. (eds) (2013) *Scuba Diving Tourism*. New York: Routledge.
Newman, G., Wiggins, A., Crall, A., Graham, E., Newman, S. and Crowston, K. (2012) The future of citizen science: Emerging technologies and shifting paradigms. *Frontiers in Ecology and the Environment* 10 (6), 298–304.
Osgerby, B. (2006) Rapture of the deep: Leisure, lifestyle and the lure of sixties scuba. In D. Bell and J. Hollows (eds) *Hostoricizing Lifestyle: Mediating Taste, Consumption and Identity from the 1900s to 1970s*. Aldershot: Ashgate.
Pass, J. (2006) Astrosociology as the missing perspective. *Astropolitics* 4 (1), 85–99.
Pell, S.J. (2014) Aquabatics: A post-turbulent performance in water. *Performance Research* 19 (5), 98–108.
Pell, S.J. and Mueller, F. (2013) Designing for depth: Underwater play. *In Proceedings of the 9th Australasian Conference on Interactive Entertainment: Matters of Life and Death*, September, ACM. See https://dl.acm.org/doi/10.1145/2513002.2513036 (accessed 1 August 2018).
Pell, S.J. and Vermeulen, A. (2012) Space science is alive with art. In *Proceedings of European Space Agency (ESA) & International Society for Gravitational Physiology (ISGP) Life in Space for Life on Earth Symposium* – Joint 12th European Space Life Sciences Meeting, 33rd ISGP, Aberdeen, UK
Peterson, K. (2010) Future promises for contemporary social and cultural geographies of the sea. *Geography Compass* 4 (9), 1260–1272.
Picken, F. (2015a) Real things, tourist things and drawing a line in the ocean. In G. Johannesson, C. Ren and R. Van Der Duim (eds) *Tourism Encounters and Controversies: Ontological Politics of Tourism Development*. Surrey: Ashgate.
Picken, F. (2015b) Making heritage of modernity: Provoking Atlantis as a catalyst for change. *Journal of Heritage and Tourism* 11 (1), 58–70.
Picken, F. (2015c) Accounting the blue planet in tourism: Undersea and the opportunity for inclusive approaches to knowledge production. In *Proceedings of 25th CAUTHE 2015: Rising Tides and Sea Changes: Adaptation and Innovation in Tourism and Hospitality*, Gold Coast, Queensland, February, pp. 293–302.
Picken, F. (2016) Extreme tourism. In L. Lowry (ed.) *The SAGE International Encyclopedia of Travel and Tourism*. London: Sage.
Picken, F. (2018) Free-falling the water column: Raptures and ruptures of the deep. In H. Saul and E. Waterton (eds) *Affective Geographies of Transformation, Exploration and Adventure*. London: Routledge.

Picken, F. and Ferguson, T. (2014) Diving with Donna Haraway and the promise of a blue planet. *Environment and Planning D: Society and Space* 32 (2), 329–341.

Pine, J. and Gilmore, J. (1999) *The Experience Economy: Work is Theatre and Every Business a Stage*. Cambridge, MA: Havard Business School Press.

Pons, P.O., Crang, M. and Travlou, P. (2009) Corrupted seas: The Mediterranean in the Age of Mobilities. In P.O. Pons, M. Crang and P. Travlou (eds) *Cultures of Mass Tourism: Doing the Mediterranean in the Age of Banal Mobilities*. Farnham: Ashgate.

Rozwadowski, H. (2005) *Fathoming the Ocean: The Discovery and Exploration of the Deep Sea*. Cambridge, MA: Harvard University Press.

Sachs, W. (1999) *Planet Dialectics: Explorations in Environment and Development*. London: Zed Books.

Sardar, Z. (2010) The namesake: Futures; future studies; futurology; futuristic; foresight – What's in a name? *Futures* 42, 177–184.

Seedhouse, E. (2011) *Ocean Outpost: The Future of Humans Living Underwater*. Chichester: Springer.

Steinberg, P. (1999) navigating to multiple horizons: Toward a geography of ocean-space. *Professional Geographer* 51 (3), 366–375.

Urry, J. (2000) *Societies Beyond Mobility*. London: Routledge.

Van Den Abbeele, G. (1980) Sightseers: The tourist as theorist. *Diacritics* 10, 2–14.

Yeoman, I. and Mars, M. (2012) Robots, men and sex tourism. *Futures* 44 (4), 365–371.

Yeoman, I. and McMahon-Beattie, U. (2014) Trends: Cyborg games. *Journal of Tourism Futures* 1 (1), 78–81.

4 Prosuming Existential Authenticity in Dystopian Spaces

Peter Robinson

Chapter Highlights

- An exploration of prosumption and co-creation in the formation of subcultural experiences.
- A reflection on the notion of authenticity and heritage in a post-tourist context.
- An evaluation of urban exploration as an opportunity to rediscover the lost art of exploration.
- Identification of a new market for dystopic urban experiences.

Introduction

As tourists both consume and produce places, they become prosumers, mediating new places to visit and creating new attractions that have the potential to offer new urban experiences which may be perceived as being more authentic. This leads to disruptive forces which change the way in which some individuals engage with spaces – in the case of this chapter city spaces – raising questions about the way in which emergent subcultures may change tourist experiences. The chapter considers, through the lens of heritage tourism, what these phenomena mean in the context of changing individual experiences, the prosumption of spaces and the emergence of dystopic tourism destinations.

Dystopia, defined by Clute and Nicholls (1993: 360) as 'an image of future societies, pointing fearfully at the way the world is supposedly going in order to provide urgent propaganda for a change in direction' is generally regarded as the opposite of Utopia (Gordin *et al.*, 2010) – a world of perfection, an ideal society which may be so far from reality that the idea is based on pure escapism (Booker, 1994). There is an uneasy paradox between the two, and any delineation is equally clear and unclear.

The Disappointment of Authenticity

In 1976 MacCannell suggested that authenticity was the Holy Grail of Tourism. He argued that travel was primarily motivated by a search for authentic places and authentic experiences. However, the very notion of authenticity is highly contested, and the idea that a tourist will discover any truly authentic artefact, place or space is problematic. In a society so highly influenced by hyper-consumerism, globalisation and the experience economy it is more likely that historic authenticity becomes even harder to reach. At this level of experience, it is likely that touristic endeavours result in disappointment – a fascinating emotion and one which is rarely discussed in literature; that sense of a product, a service, an experience or an aspiration just not quite delivering what was hoped for. There is no great tragedy or sadness in disappointment, just a sense that we have been let down, failed by organisations and institutions that promise so much yet don't deliver. Disappointment is that frustrating feeling of helplessness and dissatisfaction, of missed opportunities and regret. Our response to disappointment is to try again to find a replacement experience which doesn't disappoint, yet disappointment is an internalisation of the comparison between expectation and reality – we are to blame for our own disappointment. We believed the advertising campaigns, other people's recommendations and the reality we convinced ourselves of beforehand. Recent research around online review sites, in particular Tripadvisor (Oriade & Robinson, 2018), clearly shows a key reason for asking questions on such fora relates to minimising the risk of disappointment (Oriade & Robinson, 2019).

Prosumption and Co-creation

As tourism evolved, post-Fordist structures of consumption exemplified the shifting nature of tourist consumption from package holidays to more tailored experiences. With greater consumer choice and social mobility, tourism relies not upon the traditional ideas of production and consumption, but on the notion of co-creation and prosumption (Ritzer *et al.*, 2012). Increasingly, tourists rely on information and advice from hosts, local people, the Tourist Information Centre within the destination and people we have never met online on review sites before we make the journey. Most people will, after some form of travel experience, share their thoughts, feelings, photographs and recommendations with friends, family and, increasingly, online.

It is the emergence of Web 2.0 that has empowered consumers to produce these recommendations and marketing materials on behalf of destination management organisations, attractions, hotels and other tourism related businesses (Ritzer *et al.*, 2012). The fact that consumers can be exploited through the use of incentives to complete online reviews, to leave feedback on Tripadvisor and to post images to Twitter means that they are working for free for those businesses, replacing the role of marketers, consuming, producing and reproducing the destination for a potentially global audience. The application of prosumption reflects a more active and innovative participation in the process of production (Prebensen & Foss, 2011). As postmodern theory is opposed to such binaries as production and consumption it is more appropriate to adopt prosumption as a model for understanding and explaining consumer behaviour, which also reflects the emergence of the post tourist.

Post-Tourists

Post-tourists are those consumers who openly embrace, but with some irony, the increasingly inauthentic, commercialised and simulated experiences offered by the tourism industry (Smith *et al.*, 2010). They are also fascinated with the ordinary lives of others in an extraordinary environment (Featherstone, 2007). In this case, it then matters what we define as extraordinary. In 1970 the role of a coal miner in the UK was not extraordinary. Fifty years later the world of the coal mine has become, to younger generations, most extraordinary. This explains why, in the UK alone, there are numerous coal-mining related attractions. In a world of instant communication, instant answers to questions using Google and same-day ordering, anything which remains hidden from view might be defined as extraordinary. Prentice (1993), Harris (1989) and Rudd and Davis (1998) observe that because manufacturing is unseen, and does not take place in public view, it is not fully understood, which has also created a particularly nostalgic view of industrial occupations. They argue this has driven demand for industrial heritage tourist sites. While younger generations view manufacturing with curiosity as they have never experienced factory work, older generations view industrial heritage with a sense of nostalgia.

These forgotten worlds are a powerful attraction. Prescott (2011) suggests the hidden medical world is just as enthralling. She discusses specifically the fascination with derelict hospitals and asylums among the urban exploration (UE) community (discussed later) and suggest that such sites provide an opportunity to access places associated with life and death. It is the mortuary and the maternity ward which are most commonly photographed. Such sites are associated with memorialisation, commodification and industrialisation. As birth and death are retracted from the public realm, rationalisation, particularly of the latter, is often explored through dark tourism, while the stories associated with the lived experiences of others are explored through the wider heritage industry. Podoshen *et al.* (2014) describe this dystopic dark tourism as a mutation in social desire that is ethical in its attention to reality, that exceeds social orthodoxies and is experimental in its encounter with fringe elements that portend social crisis and contradiction. Where better to experience this than in an abandoned mortuary, in places of tragedy or where murders have occurred – dystopic places which were never conceived as tourist destinations.

Places of Heritage (not History)

The heritage industry is a key proponent of modern tourism. Sites are promoted on the basis of historical significance, the narrativised stories of those who lived there and a thematic approach to storytelling. Within postmodern societies, individuals search for meaning and stimulation through events and images thus establishing that the relationship between history and identity is essential to reinforce national culture (Steiner & Reisinger, 2006). Consequently, the presentation of history and heritage to tourists relies upon a social process where competing interests argue for their interpretation of history, which then become legitimised through the nature of their ownership and interpretation (Bruner, 1994).

Dann (1996) argued that growth in heritage tourism had been a response to social dissatisfaction and concerns about the future. He noted that although the heritage tourism sector had tended to provide grand and glorified bourgeois representations of the past (which reflect arcadian ideas of Utopia), during the late 20th century there was a shift toward also presenting the lives of ordinary people in a heritage context. This has been achieved in many cases by the opening up of servant's quarters, coal mines, factories and other historic sites which subsequently interpret a past that is seen as worse than contemporary society, rather than better. However, these spaces still fail to convey the real unpleasantness (or dystopia) of life in centuries past. Indeed, Strain's 'illusion of demediating mediation', discusses 'the spell of commodity cast over a deluded or self-deluding public by a manipulative "culture industry" the purveyor of conformity, boredom' and reemphasises the postmodern simulacra (Baudrillard, 1988: 152) which is so often embraced by post-tourists.

These simulacra are further reinforced through complex hermeneutics influenced by tourist photography which, more often than not, reinforce particular sites and particular narratives as having a greater dominance than others. Thus, all the images we share online further contribute to the marketing collateral for the places we visit. Reviews on Tripadvisor, pictures on Instagram and discussions on Facebook all influence other people to do what we have done, or to not do it at all. We contribute to the contrived scenes and pseudo-events which others (tourists) may believe to be authentic or accurately reconstructed as authentic, or that others accept as less-reliable mis/interpretations of the authentic (post-tourists). It is likely that in the course of touristic endeavours that many are perpetually disappointed by their inability to get beyond commoditised interpretations of authenticity (Boorstin, 1964; Wang, 1999). Yet in reality it is likely that we seldom expect or even seek to find the real authentic experience, and thus we become post-tourists as we accept the fakery offered by the tourism industry as part of an economic transaction.

Indeed, the experience of living history museum visitors bears testament to this and reinforces the extent to which visitors are increasingly adopting a post-tourist nuance to their experiences. As consumers we seek the positive nostalgic aspects of history. We enjoy purchasing pseudo-authentic bread made with modern ingredients, themselves produced on a commercial farm somewhere some distance away with an uninteresting and unrelated provenance, from the pseudo-authentic bakers which has to comply with food safety legislation. We use public toilets which have been installed for our convenience, with lavatorial furniture which post-dates the museum set by some 150 years. We don't in reality want to experience 19th-century life with all its illness, disease, high mortality rates and lack of sanitation, healthcare and limited social mobility that would complete the authentic experience. Even if we accept that these elements of historic life are missing, we are there with other people who arrived in cars, from home or a holiday accommodation and who are all joining each other in engaging with a historic narrative played out through actors, interpretation panels, audio guides and story books that present an official and narrativised version of the place we are visiting. Of course, this is not to demean the value of such sites for entertainment, for education and for social benefit, merely it illustrates the disappointment at failing to find an authentic reality. For the post-tourist this doesn't necessarily matter anyway as the environment becomes a place for entertainment.

In this context visitor experiences are seen as playful, as a mix of authentic and inauthentic experience, and we recognise the inauthentic nature of the tourist experience and even embrace it (Smith *et al.*, 2010). Such experiences offer the post-tourist the opportunity for escapism. Any interpretation of historic reality offers a filtered understanding predetermined by individual interests and enthusiasm. Individual choices of items from the collections store and previous decisions about which historic archives to keep or which artefacts are significant. These collections of historic evidence contribute to a collective nostalgia.

The Holy Grail of Existential Authenticity

Wang (1999) proposes that existential authenticity provides a better lens through which to view tourist experiences as any other interpretation of authenticity makes it impractical to discover truth and reason. As tourists need to be bodily involved in the world (Crouch & Lubren, 2003; Pons, 2003) they experience places in relation to that place and are refigured in the process of their encounter. As Heidegger observed, our explorations of the world co-define both our heritage and our destiny. As only these experiences can be considered truly authentic this may be the version of authenticity which we are all seeking.

The feelings associated with these existential experiences are activated by the liminal process of the activities, which, away from home, are strengthened as there are none of the usual constraints placed on the tourist by their everyday lives (Wang, 1999). It is possible, therefore, to challenge the constraints that contemporary society and the modern tourism industry places on the traveller when we start to question and reinterpret the formalised spaces, structures and signs that dictate the way a space should be used.

When we start to question the legitimacy of no entry signs and the paradox of access for all (a mantra of many heritage organisations) the idea of space hacking starts to provide a platform for further discussion, while the challenge this presents to accepted rules helps to explain the growing interest in urban subcultures. Graffiti, parkour and, more recently UE provide strong evidence of a desire among small groups of likeminded people to challenge and question the nature of urban spaces and to reshape the subsequent engagement with, and exploration of these areas.

The Lost Art of Exploration

That today everything is made to undergo such processes of aesthetic synchronisation and idealisation had, for Baudrillard (1988: 18), produced conditions under which reality has effectively disappeared, becoming subject to the aestheticising processes of mass advertising and designer capitalism; 'there is nothing left to see'. Such statements have also been made when it comes to finding new tourist destinations.

The popular philosopher Alain DeBotton (2002) suggested that it is an essential desire of humanity to discover meaning in the world through exploration. The explorers of the latter half of the last millennium were essentially tourists or travellers (travel deriving from the French word *travaille*, 'to work'), painting and documenting what they found, and promoting sites which then developed into tourist routes and tourist destinations. Consequently, the places to see which were commonly regarded as being places worth visiting were already embedded within a semiotic of national significance before the birth of mass tourism.

In more recent decades mass tourism has made it easier and safer to visit and explore places, while authentic exploration still retains some sense of fear and risk taking (Holder, 2005). The trivialisation of 'adventure' in the 20th century is in part a result of the safety offered to tourists by accompanied and organised package tours which deprive travellers of the opportunity to use their initiative to experience real adventure. As a consequence, it takes more time and effort to create travel risks than to avoid them (Boorstin, 2012). By the end of the 1950s in Britain there was little left to explore and discover (Burden & Kohl, 2006; Boorstin, 1964). As DeBotton (2002) observed, exploration by its very nature is of little purpose when its results remain hidden, and so the popularisation of sites that have been discovered is inevitable. The more we explore, the less there is left to discover (Strain, 2003).

Exploring the City

Twenty years ago, Law (1993) observed the repackaging of industrial cities into tourism products – a spectacle of architecture, design, image and consumption. These gentrified urban spaces are considered by some to be impersonal and deeply contradictory (Wearing *et al.*, 2010). In post-industrial cities the regeneration of urban environments has led to the creation of new spaces for tourists which emerge from specific 'state-directed industries' (Hutton, 2006). Despite the benefits of tourism, Richards (2011) observed that the gentrification and serial reproduction of culture brings about issues of cultural commodification, authenticity and dissonance. Central to the modern urban tourism experience is the co-location of retail, leisure and hospitality facilities which enable visitors to celebrate individual identity in homogenous shops as they hunt for the next Instagram photo while purchasing designer clothes and other identikit products which celebrate individuality. These newly created spaces replace the old and present a new way to engage with and understand the many different and contested histories of the area. However, there remain dominant stories which are seen as having greater validity and interest when it comes to creating spaces designed to attract people. Such urban storytelling (Matthey, 2014) creates homogenised spaces with homogenised interpretations of history and instead merely reinforces the 'collective governmentality of citizens', while creating spaces which are not far removed from the 'society of spectacle' theorised by Debord (1984). These areas which seek to embrace contemporary architecture, modern transport solutions and pedestrianised leisure spaces reflect something of the 1950s' vision of Utopia. Such areas are often, by definition, and as a result history, geography and investment, areas which thrive on multiculturalism. Regenerated zones become Debord's 'situationist city' (1984) existing only in the parts as they are experienced leading to the creation of subjective versions of the city based on the way individuals engage with these spaces – thus no two individuals will experience the same city. Regeneration in particular also directs people to visit and spend their money in specific areas, sometimes imperceptibly changing the notion of the city space. As a result, areas on the fringes become forgotten. Shops move to new commercial areas and the economic drivers of the urban space shift from manufacturing and production to tertiary and quaternary industries. The creation of new spaces and the dereliction of these less popular areas both create new opportunities. Those presented by the latter offer a glimpse of a more dystopic future.

Urban Exploration and Dystopian Places

Urban explorers seek out these sites, often identified as temporary, obsolete, abandoned or derelict spaces (TOADS) (Dodge, 2006; Garrett, 2010; Paiva & Manuagh, 2008). UE is defined as the activity of 'seeking out, visiting and documenting interesting human-made spaces, most typically abandoned buildings' (Ninjalicious, 2005: 4) and has grown to become a recognised leisure activity over the past decade (High & Lewis, 2007; Mott & Roberts, 2014). UE sites are nearly always off-limits to the general public (Ninjalicious, 2005), and those who explore these places are most frequently motivated by both the historic and artistic opportunities (Garrett, 2010, 2012). Most urban explorers are keen photographers (Godwin, 2010).

Web 2.0 has enabled wider interest in UE to flourish (Garrett, 2014). Today there are numerous websites dedicated to UE in most major developed cities in the world (Garrett, 2012). These sites feature images which can be easily translated into retail items – most commonly as artistically styled photographic coffee table books (Strangleman, 2013). The publication of images and media attention (both about UE and the use of UE images of notable buildings) has further increased general awareness and participation. However, many who are interested will likely prioritise safety and legality over their desire to access and document these hidden dystopic city spaces, yet the emergence and popularisation of the activity is a key issue for considering future urban tourism opportunities.

UE and tourism share common backgrounds where the former has stayed true to the notion of exploration (complete with associated risks), while tourism has enabled larger numbers of people to visit en-masse the places which have been explored, discovered and made safe (Holder, 2005). Dodge (2006) suggested that UE is imbued with the thrill of accessing unauthorised spaces and further enhanced by the desire to find and experience an alternative aestheticism of space.

UE may range from legally permitted touristic endeavours (such as Heritage Open Days and authorised tours with groups such as the 21st Century Society (Craggs *et al.*, 2013)) to illegal activities and trespass. It includes factory tours, visits to workplaces, behind the scenes tours, visiting ruins, bridge climbing, tunnel games, elevator surfing, skateboarding, train-hopping, parkour/freerunning, exploring construction sites, drains and tunnels, buildering, urban climbing, roofing and sneaking into movies and concerts.

Ninjalicious (2005) and Dodge (2006) both suggest that activities which could be seen as UE or UE related may illustrate similar consumption and production behaviours to tourism. Often it is the perfect example of prosumption (produced by the same community that consume the activity and produced for a wider audience through photography). While there may be similarities with tourism, the places which become the subjects for UE are dystopic. Craggs *et al.* (2013) posit that UE, like any official tour, is a way for people to engage with their interest in architecture and heritage. The official tour would be recognised as tourism, while visiting somewhere without permission would be considered trespass, or, in this instance, UE. In this situation only the 'permission' element of the visit differentiates one from the other, especially for significant buildings which may occasionally be open for limited public access prior to restoration or demolition.

Similarly, Lynch (1990) distinguished that where 'a ruin' is seen as 'pleasant' and 'worthy of reverence', 'an abandoned place' is associated with 'entropy', 'dereliction' and 'death'. The difference is one of semantics and accessibility. Ruins – well maintained, with interpretation, picnic sites and managed access are preserved and managed for tourism; derelict places represent those which should not be visited, as they are not 'marked out' for visitation. They are beyond the usual frames of reference for tourists but are the chosen sites for UE.

Garrett (2012) suggested that UE is almost a form of anti-tourism, a rebuttal of the commonly accepted experiences of heritage as provided by the mainstream tourism industry, based upon 'smokestack nostalgia' (Cowie & Heathcott, 2003) creating an inauthentic shared nostalgia and experiences of histories which limit the potential for critical investigation of industrial decline and the people it affected. Heritage tourism tends to trade on a collective nostalgia based upon a reconstructed social past or idealised version of history (Kibby, 2000) rather than a more interpretive approach. UE has increasingly been associated with 'ruin porn' which subjugates such sites to being fetishised, seen voyeuristically, stripped of any social or cultural connections (Strangleman, 2013), and often associated with 'darker' connotations and meanings, but this could also be seen as just another example of urban storytelling.

Critics of UE suggest that photography overrides any real interest in the place, the people or the reality of industrial life. This leads to the production of over-aestheticised images of the 'industrial sublime' (High & Lewis, 2007) responsible for creating a false nostalgia which overlooks the real lives of those who inhabited the spaces. Such a view may be a reaction to the proliferation of photographic collections rather than a critically valuable insight into the attraction of these sites. This fascination with the photography of the dystopic and abandoned draws parallels with the hermeneutic Circle of Representation (Albers & James, 1988; Hall, 1997) which explores the continued cycle of the production of images which, over time, creating an illusory system of markers and sites to be visited (Urry, 1990, 2002). This model is seen as a driver for the demarcation of sites as being worth visiting – in other words, becoming tourist attraction. Over time, then, it is possible for some of these places to become attractions.

Detroit, for example: A place which looks like a city but within which persists a 'death-in-life existence' (Wilkins, 2010). It is a city in decay, which has become popular with urban explorers and photographers. It raises questions about the economic and governmental system that produced it and questions what could have been. Some areas are rebuilt in failed attempts at regeneration, others are vacuously derelict and have consequently become fascinating to tourists and journalists. This is a consequence of the city consuming itself, as out of town retail draws consumers away from the city centre, only to be replaced by subsequent new out of town developments which draw people to new sites of consumerism. Those areas left behind are neglected and fall into disrepair (Debord, 1984). They reflect a dystopic otherness which is just as interesting and engaging as heritage attraction. Echoing to a history of a vibrant motor industry and Motown music culture, there is a sense of reality within the decay and dereliction.

More Culture, Less Subculture

Pinder (2009) notes that UE has been important in recent arts, cultural and writing practice through projects that seek to engage with city spaces and their potentialities beyond galleries and other formal arts institutions. He suggests that these are linked to the earlier politicised spatial practices of the situationists and to the visionary and literary traditions of urban wandering. Bennett (2011) has already suggested that UE is in fact a middle-class activity, as have High and Lewis (2007) who suggest that such exploration transforms places into 'the exotic' (a notion also explored by DeBotton, 2002). They talk about UE as dark tourism for the middle classes. Country house explorers such as Binney (1984) were seeking to raise awareness of the plight of the buildings they explored, and therefore share common ground with Bennett's (2011) view on middle aged explorers viewing UE as a form of custodianship. Craggs *et al.* (2013) refer to the tours of abandoned buildings that are organised by the Twentieth Century Society as another example of legitimising UE.

TOADS offer something of a paradox. They represent the antithesis of the signposted and official tourist sites which come with a range of tourism management constraints and social controls including schedules, itineraries, queuing, finances and accessibility (Dann, 1996; Wang, 1999). Despite being off-limits they offer the freedom to explore the former life of buildings and spaces without limits. They challenge the fantasy and illusion of tourism, while the production of photography draws remarkably similar parallels to the role of photography within tourism (Urry & Larsen, 2011).

The sites are talked about in this chapter as places of dystopia because of the dark, unnatural, unfamiliar senses they evoke. Edensor (2005) observes that these places provide access to an authentic urban experience that can create powerful bodily sensations as they surprise, confound, scare and amaze. Giblett's (1996) psycho-analytical critique discusses the attraction and repulsion of dark, dank, smelly, slimy attributes of liminal spaces which are emphasised by those who explore these places.

In the 21st century most of our travel decisions are largely informed by a combination of official marketing materials, informal media sources and information and reviews provided by other people who share images and reviews online. As other disruptive influences impact on the tourism sector, the relationship between host communities, visitors and tourism NGOs will continue to evolve.

As tourism has made it easier for tourists to visit places, so explorers want to discover new places in a world where there is little left to explore and less left to discover. Abandoned buildings, views from illegally accessed rooftops and stories of adventure in the cities abandoned office blocks will become the narrative of touristic experiences beyond the geopolitically enforced rules of visitor attractions and urbanscape environments. In many ways UE is just part a of a continuum of experience and practice which reflects a common interest in visiting buildings. 'Being within' and 'experiencing first hand' is common to all visiting practices regardless of the method and legitimacy of access, but unlike the emergence of post-tourists, urban explorers are seeking an alternative form of existential authenticity created by the individual and prosumed by a small but growing community. What this may mean for tourism in the coming decades is ripe for further investigation.

While Wang (1999) refers to broader tourism experiences, this intrapersonal authenticity involves 'self-making' or 'self-identity' and is an implicit dimension for tourism motivation (Crouch & Lubren, 2003). From a constructivist perspective, most tourists search for symbolic rather than objective (original) authenticity. For those seeking objective authenticity, UE offers an intriguing opportunity. The value of viewing UE from this embodiment perspective will enable researchers to focus on groups that have been marginalised in previous research and academic practice.

Concluding Arguments

Exploration isn't for everyone. There is unlikely to be an explorative turn in tourist activities or tourism studies. However, there are explorers and as the world becomes ever more accessible, there will be less to explore. Those individuals will seek out new places to in a world of aesthetic synchronisation and idealisation. UE offers this opportunity, but those who travel the world to visit TOADS will be hard for policymakers and marketers to influence as such tourists make their own fun (Steiner & Reisinger, 2006). However, these same individuals are also prosumers and may play an interesting role in the future development and demarcation of attractions and significant places.

In an industry increasingly focussed on entertaining visitors (which has resulted in a rapidly growing concern about overtourism), it is likely that there will be some growth in the number of experiences presented as dystopic available commercially. However, while zombie attractions, film sets and virtual reality may all offer experiences capable of conjuring up a sense of dystopia, any facility which has been designed in this way merely reflect the continued inauthenticity of the tourism industry. The true explorers will move on from these official sites to identify new urban experiences.

It is, therefore, the desolate and derelict representational aesthetics and the secret and transgressive nature of UE which makes it such a fascinating opportunity for further investigation. The opening up of hidden spaces to a wider population through photography offers society a fascinating insight into lost worlds. The development of archive images, of oral and visual histories, of new forms of interpretation and the creation of new histories may be consequences which engage a far greater audience. The hidden dystopic places that few people explore may then come to elucidate new meanings for the wider tourism sector as we reflect on and react to some of the criticisms of the ways the modern tourism industry addresses the problems it has created in a desire for utopic authenticity.

References

Albers, P.C. and James, W.R. (1988) Travel photography: A methodological approach. *Annals of Tourism Research* 15 (1), 134–158.
Baudrillard, J. (1988) *The Consumer Society: Myths and Structures*. London: Sage.
Binney, M. (1984) *Our Vanishing Heritage*. Kent: Arlington Books Publishers Ltd.
Booker, M.K. (1994) *The Dystopian Impulse in Modern Literature: Fiction as Social Criticism*. Westport: Greenwood Press.
Boorstin, D.J. (1964) *The Image: A Guide to Pseudo-events in America*. New York: Atheneum.
Boorstin, D.J. (2012) *The Image: A Guide to Pseudo-events in America*. New York: Atheneum.
Burden, R. and Kohl, S. (eds) (2006) *Landscape and Englishness (Spatial Practices)*. Rodopi B.V.: Amsterdam.
Bruner, E.M. (1994) Abraham Lincoln as authentic reproduction: A critique of postmodernism. *American Anthropologist* 96, 397–415.
Clute, J. and Nicholls, P. (1993) *The Encyclopaedia of Science Fiction*. New York: St. Martins Press
Cowie, J. and Heathcott, J. (eds) (2003) *Beyond the Ruins: The Meanings of Deindustrialization*. Ithaca: Cornell University Press.
Craggs, R., Geohegan, H. and Neate, R. (2013) architectural enthusiasm: Visiting buildings with the twentieth century society. *Environment and Planning D: Society and Space* 31 (5), 879–896.
Crouch, D. and Lubren, N. (2003) *Visual Culture and Tourism*. Oxford: Berg Publishers.
Dann, G.M.S. (1996) The people of tourist brochures. In T. Selwyn (ed.) *The Tourist Image: Myths and Myth Making in Tourism*. New York, NY: John Wiley & Sons.
Debord, G. (1984) *The Spectacle of Society*. Michigan: Black and Red
DeBotton, A. (2002) *The Art of Travel*. London: Penguin
Dodge, M. (2006) *Exposing the Secret City: Urban Exploration as Space Hacking*. University of Manchester seminar. See http://personalpages.manchester.ac.uk/staff/m.dodge/cv_files/geography_seminar_space_hacking.pdf (accessed 18 April 2018).

Edensor, T. (2005) *Industrial Ruins: Space, Aesthetics, and Materiality*. Oxford: Berg Publishers.
Featherstone, N. (2007) *Consumer Culture and Postmodernism* (2nd edn). London: Sage.
Garrett, B.L. (2010) Urban explorers: Quests for myth, mystery and meaning. *Geography Compass* 4 (10), 1448–1461.
Garrett, B.L. (2012) *The Adventurers Club Monday 28th May 2012*. See URL http://www.placehacking.co.uk/ (accessed 17 June 2012).
Garrett, B.L. (2014) Undertaking recreational trespass: Urban exploration and infiltration. *Transactions of the Institute of British Geographers* 39 (1), 1–13. See doi:10.1111/tran.12001 (accessed 18 May 2020).
Giblett, R. (1996) *Postmodern Wetlands*. Edinburgh: Edinburgh University Press
Goodwin, R. (2010) London Evening Standard 15th March 2010, Secrets of an UrbEx mission. Available at: http://www.standard.co.uk/lifestyle/secrets-of-an-urbex-mission-6709707.html (accessed 16 August 2021).
Gordin, M., Tilley, H. and Prakash G. (2010) *Utopia/dystopia: Conditions of Historical Possibility*. Princeton: Princeton University Press.
Harris, J. (1998) *No Voice from the Hall*. London: John Murray.
Hall, S. (ed.) (1997) *Representation: Cultural Representations and Signifying Practices*. London: Sage Publications and Open University Press.
High, S. and Lewis, D.W. (2007) *Corporate Wasteland: The Landscape and Memory of Deindustrialization*. Ithaca: Cornell University Press.
Holder, A. (2005) *Tourism Studies and the Social Sciences*. Oxford: Routledge.
Hutton, T.A. (2006) Spatiality, built form and creative industry development in the inner city. *Environment and Planning A* 38, 819–1841.
Kibby, M. (2000) Tourists on the mother road and the information superhighway. In M. Robinson, P. Long, N. Evans, N.R. Sharpley and J. Swarbrooke (eds) *Reflections on International Tourism: Expressions of Culture, Identity, and Meaning in Tourism* (pp. 139–149). Newcastle: University of Northumbria.
Law, C. (1993) *Urban Tourism: Attracting Visitors to Large Cities*. London: Mansell Publishing
Lynch, K. (1990) *Wasting Away*. San Francisco: Sierra Club Books.
MacCannell, D. (1976) *The Tourist: A New Theory of the Leisure Class*. New York, NY: Schocken Books.
Matthey, L. (2014) *Building Up Stories: Sur l'action urbanistique à l'heure de la société du spectacle intégré*. Geneva: A-Type.
Ninjalicious (2005) *Access all Areas: A User's Guide to the Art of Urban Exploration*. Ontario: Infilpress.
Oriade, A. and Robinson, P. (2019) Prosuming tourist information: Asking questions on Tripadvisor. *International Journal of Tourism Research* 21 (1), 134–143.
Paiva, T. and Manuagh, G. (2008) *Night Visions: The Art of Urban Exploration*. San Francisco: Chronicle Books.
Pinder, D. (2009) Urban exploration. In D. Gregory, R. Johnston, G. Pratt, M.J. Watts and S. Whatmore. *The Dictionary of Human Geography*. Oxford: Wiley Blackwell.
Podoshen, J.S., Venkatesh, V. and Jin, Z. (2014) Theoretical reflections on dystopian consumer culture: Black metal. *Marketing Theory* 14 (2), 207–227.
Pons, P. (2003) Being on holiday. Tourist dwelling, bodies and place. *Tourist Studies* 3 (1), 47–66.
Prebensen, N.K. and Foss, L. (2011) Coping and co-creating in tourist experiences. *International Journal of Tourism Research* 13 (1), 54–67.
Prentice, R. (1993) *Tourism and Heritage Attractions*. London: Routledge.
Prescott, H. (2011) Reclaiming ruins: Childbirth, ruination, and urban exploration photography of the ruined maternity ward. *Women's Studies Quarterly* 39 (3/4), 13–132.
Rudd, M.A. and Davis, J.A. (1998) Industrial heritage tourism at the Bingham Canyon Copper Mine. *Journal of Travel Research* 36 (3), 85–89.

Richards, G. (2001) *Cultural Attractions and European Tourism*. Wallingford: CABI.
Ritzer, G., Dean, P. and Jurgenson, N. (2012) The coming of age of the prosumer. *American Behavioral Scientist* 56 (4), 379–398.
Smith, M., Macleod, N. and Hart Robertson, M. (2010) *Key Concepts in Tourism Studies*. London: Sage.
Strangleman, T. (2013) 'Smokestack nostalgia,' 'ruin porn' or working-class obituary: The role and meaning of deindustrial representation. *International Labor and Working-Class History* 84, 23–37.
Strain, E. (2003) *Public Places, Private Journeys. Ethnography, Eentertainment and the Tourist Gaze*. London: Rutgers University Press.
Steiner, C. and Reisinger, Y. (2006) Understanding existential authenticity. *Annals of Tourism Research* 33 (2), 299–318.
Urry, J. (1990) *The Tourist Gaze: Leisure and Travel in Contemporary Society*. London: Sage.
Urry, J. (2002) *The Tourist Gaze* (2nd edn). London: Sage.
Urry, J. and Larsen, J. (2011) *The Tourist Gaze 3.0*. London: Sage.
Wang, N. (1999) Rethinking authenticity in tourism experience. *Annals of Tourism Research* 26 (2), 349–370.
Wearing, S., Stevenson, D. and Young, T. (2010) *Tourist Cultures: Identity, Place and the Traveller*. London: Sage.
Wilkins, G. (2010) *Distributed Urbanism: Cities After Google Earth*. Oxford: Routledge.

5 Space Tourism – Science Fiction Becoming a Reality

Annette Toivonen

Chapter Highlights

- The chapter explores the development of future space tourism.
- The current industry trend is to create space tourism accessible to the masses.
- Space tourism is at variance with current sustainable travel trends in tourism more generally.
- The space tourism industry needs to act regarding environmental protection.

The Era of Space Tourism

For many generations, space exploration simply represented a wishful dream, which finally became reality in the 1950s *Space Race* between the United States and the former Soviet Union. Inspired by this development, Hollywood producers started to create space-related science fiction movies that enhanced audiences' imagination and desire to visit space. 2001: A *Space Odyssey* (1968), for example, constituted one of the first space travel films, introducing human-like intelligent computers.

It was commonly imagined that, within just a few decades, such a form of travel would be made available to the public. Some airlines, such as Pan American, even opened reservation lists for trips to the moon. However, it was three decades before even government-funded space flights occurred, when the first segment of the international space station (ISS) was launched into orbit in 1999 (Anderson, 2005). Nevertheless, in early 2018, a real-world, science fiction-like attempt at conquering space occurred, in the form of shooting a Tesla car with a human-like passenger on a rocket into space (SpaceX, 2018).

Typologically, space tourism consists of terrestrial space tourism, atmospheric space tourism, suborbital space tourism and beyond earth orbit astrotourism. The Kennedy Space Center in Florida, US and

Baikonur Cosmodrome in Kazakhstan comprise some of the current global spaceports on earth, from which space vehicles can be launched into orbit (Carter et al., 2015).

Only a handful of people have thus far visited space as paying tourists in orbital spaceflight, compared to about 550 professionals, mostly US astronauts. In order to do so, one must reach the altitude of space according to the Federation Aeronautique Internationale definition of the boundary of space being over 100 km (FAI, 2018). Dennis Tito was the first paying space tourist customer, travelling on a Russian Soyuz rocket in 2001. His $20 million trip involved a five-day stay at the ISS and included six months of astronaut training and hours of physical exercise (Wall, 2011).

There are currently many ongoing projects, including high altitude ballooning, aimed at starting to offer regular space tourism experiences to paying passengers. In December 2018, a Virgin Galactic-crewed vehicle, built for commercial passenger service, reached suborbital space with a team member in a passenger seat (Virgin Galactic, 2018). In July 2021, Virgin Galactic and Blue Origin officially entranced into the market of private spaceflight; both demonstrating successful suborbital flights, alongside their billionaire owners as well as the oldest and youngest passenger ever have flown in space (Blue Origin, 2021).

Rapid developments in technology during the 21st century have accelerated predictions for the future space tourism industry. According to Ashford (2002), space tourism is likely to generate significant public enthusiasm for space exploration as soon as it becomes safe and affordable. Suborbital space travel, whereby passengers are transported above the earth to experience weightlessness and star sighting, is a new adventure tourism experience offered by private space tourism companies. The wildest imaginations include visions of the moon becoming akin to a gas station, in which its dark spots (widely considered to be ice) are converted into fuel, so that travellers can continue to Mars (Tett, 2018).

The ultimate goal of such ventures is to develop a space tourism that is accessible to the masses; however, the high cost of a journey has limited the pioneering stage to wealthy, elite travellers. Future space tourists' missions are likely to include vacation packages for a range of lifestyles and budgets. Pioneering space tourists must undergo health checks to satisfy tour operators, safety authorities and insurance companies. Both compulsory and optional training programmes about surviving an emergency and coping with space sickness will also need to be developed (Ashford, 2002).

The 'best deal' options with the lowest amounts of travel, training and mission time have formerly been 'zero gravity flights', on which tourists experience the feeling of weightlessness, and 'edge of space' flights, where the curvature of the earth can be witnessed. The latest, however more expensive, suborbital flights will offer both weightlessness and a view of earth. Future translunar cruises could include travel to the Moon or space hotels (relatively) nearby in lunar orbit (Anderson, 2005).

Development of the space industry has been constrained by entrenched thinking, but the monopoly of large government space agencies is becoming increasingly untenable. Indeed, the market has come to recognise that clear opportunities to expand space travel to the wider public will exist in the future. An actively growing space tourism movement can now be found, benefitting in the last decade from governmental initiatives such as NASA's Space Launch to prepare the technology for a reusable launch vehicle. Numerous companies competed for the $10 million XPRIZE, awarded in 2004 to SpaceShipOne, the first NGO to demonstrate a reusable and piloted suborbital vehicle capable of carrying three people up to 100 km (Ansari XPRIZE, 2018).

Space tourism is forecast to become a lucrative business, with a significant potential for expansion (Global Information, 2018). Swiss finance firm UBS analyses space tourism industry to be worth $4 billion by 2030 (UBS, 2021). Plans by the US to reform their oversight of space development and regulations for commercial space flight launch and re-entry operations are also being developed, possibly motivated by the fact that the country may soon be eclipsed by China in space exploration. In June 2019, NASA declared a directive that further opens the International Space Station for commercialisation and space tourism with the goal of developing a robust economy in low-Earth orbit (NASA, 2019).

NASA's (1998) feasibility study presented some positive outcomes of space tourism, acknowledging its revolutionary political, social and economic potential. One of the main challenges in the development of space tourism has been finding economic funding. In the last decade, Silicon Valley-based private sector billionaires such as Elon Musk (Space X), Jeff Bezos (Blue Origin), as well as Richard Branson (Virgin Galactic) in the UK, entered the space race by establishing their own private space tourism companies that target the public. However, critics have questioned the synthesis of influential private sector commerce and publicly-funded infrastructure, as it raises concerns regarding policy and the concentration of power (Tett, 2018).

The next section focuses on the potential for commercial space tourism to become a key aspect of the contemporary adventure tourism industry, and – with reference to different futuristic theories and interviews – analyses some important issues regarding the industry's attempts to become sustainable in the future.

Futures Mapping as Future's Methodology

International tourism research has traditionally reflected on previous tourism research in order to envision its future. The forecasting of new travel phenomena has tended to be based on the history, current trends and perspectives of researchers (Ryan, 2018). Futurist thinking helps anticipate the future, how changes occur and how they are predicted. In the beginning, when a change emerges, few people know of its novelty. Eventually, the novelty affects the everyday lives of most people and everyone is familiar with it. At its foundation, anticipating a change involves stable factors, megatrends, trends, weak signals and wild cards (Hiltunen, 2013).

Future road-mapping represents an extended look at the future, composed of the collective knowledge and imagination of visionaries, and can be used as a planning tool to predict developments and decision-making (Matos & Afsarmanesh, 2004). One of the current megatrends in the tourism industry is sustainability (Mojic & Susic, 2014). Becken (2015), for example, suggests that environmental assessments should always be undertaken prior to any development of tourism activities and projects. Given that tourism was originally framed as co-existing with its environment, research concentrating on new resources and survival planning is capable of facilitating new, environmentally based discussions in space tourism discourse.

There are existing futures research road-mapping frames such as 'Futures Map', which handles road-mapping as one of the aspects of futures mapping (Kuusi *et al.*, 2015: 4). In the Futures Map (see Figure 5.1), the scenario is a specified path connecting the present state to at least one picture of the future. The map identifies possible futures with a 'planning horizon' and a 'mapping horizon'. The Futures Map has different

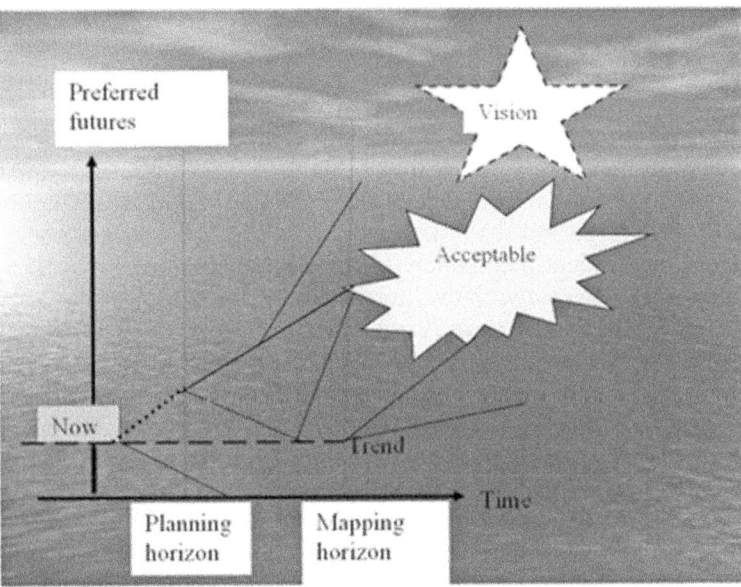

Figure 5.1 Futures Map (Kuusi *et al.*, 2015)

quality criteria to improve its pragmatic validity. Criteria 1 and 2 identify most relevant or important possible futures, criteria 3 and 4 cover casually-relevant facts by their identified futures, and criteria 5 and 6 concentrate on the needs of key customers (Kuusi *et al.*, 2015: 6).

For this chapter, a new roadmap for sustainable space tourism was constructed by the methods of grounded theory (Glaser & Strauss, 1967). The data were attained via five in-depth, 45-minute interviews. The participants included senior researchers in futures fields (Interview 1) and space technology (Interview 2); a Finnish Green Party politician (Interview 3): a space tourism entrepreneur whose current focus is virtual reality (Interview 4) and an internationally renowned space engineering professor (Interview 5). The interviews were conducted in Helsinki, because Finland become a new space nation in 2017. The participants' visions of the future of space tourism were treated anonymously. The number of interviews was decided to be comprehensive after similar observations began to occur.

The interview questions were developed regarding understandings of reflective practice and attention to ethics and values, which have stimulated critical theoretical developments, especially in sustainable tourism planning. The questions could be placed under three themed categories: space tourism, sustainability and the future. The interviews were transcribed word-for-word, to ensure reliability of the data for analysis.

The first part of the analysis, open coding, involved coding the interview line-by-line. The aim was to compare similarities across the respondents' visions for the future. The emerging concepts then were moved from describing such Future Visioneering to broader concepts such as Searching for Knowledge. A tentative core variable for these dimensions was named Making Space Human.

A delimitation of categories was achieved through selective coding; the primary categories related in some ways to the core variable. The compartment of concepts and categories established the patterns named by eight categories. In the last stage of the analysis, the sorting of analytical memos took place and the linkages between the variables and the eight categories were identified. Following the grounded theory principles, the findings were reflected with the literature.

Based on the data analyses, eight themed categories were formed. The categories were named: space legislation, economic impacts, alternative energy sources, circular economy, contemporary trends, health space tourism, space colonies, and virtual travel and robotisation. All these categories were linked to the core variable, Making Space Human, by describing some current elements of the human society that are considered to be important also in the space environment.

The findings were placed on the Futures Map (Kuusi *et al.*, 2015). The 'planning horizon' reflects historical trends and aims for more enhanced future and 'the mapping horizon' opens more visions for a longer time scale. The themes placed under planning horizon were economic impacts, legislation, alternative energy sources and circular economy (named as group 1). These represented the historical or current ways of living of the developed world. Contemporary trends, health space tourism, space colonies and virtual travel and robotisation (named as group 2) are acceptable trends or exist in visioneering minds so they were set under the mapping horizon. Planning horizon was established as a subcategory for Searching for Knowledge and mapping horizon for Future Visioneering, both under the umbrella core Making Space Human.

Planning Horizon

Economic impacts

Duval and Hall (2015: 451) claim that a concern exists regarding the planning and policy implementation of future space tourism, especially with respect to: (a) the impacts of future agreements developed by politicians and the private sector, and (b) the importance of mutual accords geared toward leaving some parts of space completely untouched by tourism. The data supported this statement and also highlighted the significance of economic modelling and global legislation.

The new era in space exploration has brought about a completely new commercial market sector. In the past, considerable expertise, scientific research and economic capital was required to become a space nation, without any guarantee of direct rewards.

> Today a complete rocket can be purchased privately for a cost of approximately $100 million from the previous national owner. Private launch providers selling space inside such rockets – and customers, including small satellite companies, can already be found. The cost of the space inside a rocket is shared; hence, even academic institutions with relatively small budgets have been able to test small measuring satellites in a real space environment. (Interview 2)

For example, Finland became a space nation in 2017 after Aalto 1 was launched to measure electrons hitting the upper atmosphere (Aalto University, 2018).

> Numerous possibilities exist to use space economically. For example, it would be more efficient to run solar panels in space than on the ground because UV light energy is more abundant in space. Plans have already been devised to locate solar panel stations in orbit and beam down energy using microwaves. (Interview 5)

However, unsolved issues with asteroids and other debris have hitherto prevented such activity from taking place. Aside from space tourists simply experiencing the space environment, other actors than the tourism industry may initiate mass space travel in the future. For instance, asteroids are full of precious metals such as diamonds and platinum, and thus space travel may become work related (Ross, 2002).

> The development of synergies or by-products could be facilitated, including adventure products and services that do not yet exist. Indeed, space tourism could pioneer the creation of such services, which may be provided by other industries in the future. (Interview 4)

At present, considerable foresight and capital are required to forecast the economic potential of space tourism. With the assistance of governmental planning, companies should mutually plan, invest and share operational facilities to recuse the economic costs involved. Eventually, the expense will help determine whether space tourism can become a mass product or whether it will be practised only by the elite.

Space legislation

As space becomes commercialised, mutual global legislation will be required to supervise future business and trading. The Outer Space Treaty (1967) corresponds to maritime law and constitutes the only legislation in existence for common space responsibilities. Otherwise, space is equivalent to the former 'Wild West' with a 'first come first own' attitude and approach to rights.

> For example, a future hotel or mining colony located on the Moon will raise questions as to whether the Moon should be regarded as a separate state as well as which country's (or countries') legislation would be legally binding there. If the development of a mutual global space law proves too challenging, space should be treated similarly to Antarctica, which is divided into sectors belonging to different nations. (Interview 2)

Space is no longer only entered by countries with space programmes, but also various private companies working as subcontractors or even commercial contractors such as satellite operators. The military has diverse interests in space and is constantly developing technologies from which corporate space tourism companies may benefit (Insinna, 2018). However, there is currently a risk that a 'no mans' land' attitude will arise, with the potential to cause ownership wars and jeopardise such synergistic operations.

> Once the private space tourism industry has developed infrastructure that can run independently, government sponsorship and funding is likely to cease. Without the support of a governmental legislative framework and preventative space law, private companies will face new problems related to owning and immaterial rights. (Interview 1)

In addition to ownership, a further major legislative concern comprises responsibility for space debris. Even a small particle of debris from a former satellite or similar may prevent a safe landing back to earth. This type of safety threat could seriously harm the entire space tourism industry.

Some existing technologies may help remove debris from orbit without requiring a propellant. An electric sail (1–2 km in length) can be attached to a small satellite and, once the decision is taken for it to leave orbit, the wire is unspooled from the satellite and the solar wind pushes it into outer space. (Interview 2)

Preventative space legislation should make such debris collectors compulsory in the future to ensure safe access to space and protection of the space environment in general.

Alternative energy sources

According to Fawkes (2006), space tourism and sustainable development are connected at 'resources and survival' levels. It is today a major business risk not to include sustainability in a business model; for example, the aviation industry already practises voluntary carbon offsets paid both by the industry and the customer (Broderick, 2009). Space tourism will contribute to higher levels of emissions, although research into alternative fuels to lower their impact, including rubber-base propellants, is ongoing (Carter *et al.*, 2015).

Purposeful integration with the disciplines of transportation and engineering is required in order to develop new renewable energy technologies and efficient transport systems that are free of fossil fuels (Becken, 2015). In particular, the data highlighted alternative energy sources and the importance of the circular economy.

The development of space tourism works against sustainability trends and international climate agreements such as in Paris in 2015, which sought strong actions to reduce global emissions. The main disadvantage that initial space tourism is forecast as having an increased atmospheric pollution. However, with the assistance of new technological innovations and energy solutions, the levels of such pollution may be limited.

> Many existing space rockets use toxic propellants and enormous amounts of energy in burning kerosene and oxygen. The latest technology is already able to produce rockets that only burn hydrogen and oxygen, rendering them more environmentally friendly. However, energy is also wasted in the production process of the rockets, enhancing the creation of greenhouse gases. The next century is likely to see innovations such as nuclear-powered ships, solar wind sails and space elevators, which may, over times reduce reliance on rockets. (Interview 4)

> If liquid hydrogen and oxygen propellants are used as the future energy source of space rockets, the exhaust product will comprise non-toxic water vapour rather than toxic fumes. Space tourism could become the first large-scale commercial use of hydrogen, breaking the familiarity barrier and advancing its use in aeroplanes and other ground transport systems. (Interview 5)

Entering space requires considerable energy, which ultimately has negative consequences for the climate and environment. At first, space tourism should not have a destructive impact on the earth's environment. New, innovative energy sources and production knowledge are required, as advanced technology is the key to sustainable future developments. A sustainable, lower orbit economy must be created for private companies to operate and utilise these possibilities.

Circular economy

Sustainability issues must be the core of environmental planning, involving everything from the oceans and land to atmosphere and space. The sustainable way of thinking will become increasingly common in space tourism, not simply owing to ethical requirements, but also because every kilogram brought into space costs a lot of money. The technologies developed for space travel and space tourism might also assist in day-to-day living on earth.

> NASA has already developed a number of products for astronauts, such as in food production, which are now used by millions of people. Once space becomes more accessible, contemporary resource wastefulness will no longer be possible, as everything is likely to be created in a recyclable format. (Interview 5)

Education is required for sustainable and resourceful practices, including in space. However, even the dream of the space circular economy may advance sustainable thinking on earth. It highlights concerns regarding the daily living environment and ensuring the protection of the environment, so there will be no need to escape from the planet in order to save humanity.

New circular economy ideas will also emerge after having landed on the moon, given that it offers substantial beneficial metals such as platinum on its surface.

> Everyday items, historically made of aluminium, could eventually be replaced by platinum. With the assistance of the latest technology, it may be possible to transport some materials from space rather than manufacture them on earth. However, if asteroids are brought too close to earth, there will be a risk of collision, destroying part of the planet. The particles transported from asteroids or other planets may also contain unknown bacteria that are dangerous to humans. (Interview 2)

> A global pollution compensation or taxation scheme would ensure greater social and environmental sustainability. (Interview 1)

Energy-sorting quality measurements, and even sustainable certifications, will also be required for future space tourism operators. Such certifications may facilitate greater feelings of safety, as companies will be deemed as taking environmental matters seriously.

Mapping Horizon

Contemporary trends

Current trends in tourism overtly promote sustainability and its transformation into transmodern eco-luxury. Space tourism represents a new era of innovative start-ups, electric cars and the 'show-off styled' experience of sharing via social media platforms. In particular, the findings highlight consumer trends and health space tourism.

The world of experience represents the new luxury, as consumers seek to experience something that is hard to access but not necessarily owned. New trends can be identified in entrepreneurship, such as collaborative consumption-based virtual platforms like Airbnb and Uber.

> At present, it is difficult to forecast how far these can expand without jeopardising feelings of collective security and trust inside a society. If the sharing economy becomes an enduring global megatrend, the private space tourism sector will benefit, if not in virtual existence, then at least in shared physical hubs between private sector parties, the military or space nations. (Interview 1)

> The space environment raises questions about how to eat properly and enjoy music and other sensory cultural activities, whilst remaining in a non-gravity-based environment. (Interview 4)

Providing food culture and other high cultural activities such as music concerts in the same format as on earth will be difficult in space. For example, if Mars were to become inhabited, new solutions to producing tasty culinary experiences or perhaps even making new culinary discoveries that only the space environment could authentically offer will be required.

Health space tourism

In the preferred future, the current trends may become acceptable or remain as visions (Kuusi *et al.*, 2015: 4). The time horizon may vary but, eventually, an activity previously existing only as a vision may become a reality. This may be because of technological developments or other changes in the society.

> Space is an unpleasant environment for the human body, and so pioneer space tourists will not travel to relax but to perform. Astronauts train for many years, and the handful of wealthy tourists who have already visited space spend numerous months in physical training before travelling. Mass space tourists will not have such time. The physical consequences for those without previous physical training remain unknown. (Interview 5)

> Some bed-rest studies simulating blood flow patterns in zero gravity with regard to cardiovascular diseases have demonstrated that it might be beneficial to spend time in zero gravity for an extended period of time. Blood pressure drops to zero and the heart can pump more easily – of benefit for

people with heart diseases. The almost complete absence of gravity pushing on the intervertebral discs in the spine would also be beneficial for people with back problems, as these discs would be allowed to regenerate. (Interview 2)

Even though there is some knowledge of how a human body performs in space conditions, the impacts and long-term consequences for mass tourists or people travelling to Mars yet remains unknown. The space environment could become a new treatment and health tourism option for people with health issues, especially once transportation becomes more appropriate for them. Eventually, if space becomes part of humans living area, the evolution may start changing the human structure to a condition more suitable for space habitation.

Space colonies

Space is a challenging environment for a human body and hence for it to become inhabited. A large number of new technologically advanced innovations will be required. Certainly, the South and North Poles were conquered just over a century ago, yet remain largely inhabited – even though they have the same gravity and greater potential to transport goods such as food.

In order to ensure human survival, such as in the case of a global catastrophe like a comet impact, it would be sensible and advantageous to begin developing space colonies. This will require advanced technological innovations, although some already exist in some capacity. The first permanent settlements will most likely be located on the Moon or Mars. Instead of just admiring the curvature of Earth, future space tourists may also travel to visit friends and relatives living in such settlements. (Interview 4)

The earliest time span estimated for initial settlements on the Moon is 20 years. The development is likely to commence with the erection of greenhouses to ensure food production and to test survival-related technologies. Within the next 100 years, self-sufficient colonies will exist on the moon. Whether colonies will become viable on Mars depends on factors that are currently unknown, such as the impacts of long-term exposure to cosmic radiation while travelling. (Interview 2)

Following an initial period of excitement, space tourism is likely to experience a sharp decline owing to its unpleasant conditions for humans. The industry will advance if space hotels or stations providing gravity are developed, and if permanent colonies on the surface of the moon or Mars support and necessitate regular travel.

Virtual travel and robotisation

Sci-fi franchises such as Star Trek, Battlestar Galactica and Star Wars have provided fascinating and addictive stories about space travel for generations. Their large audiences are clearly interested in the notion of space; accordingly, the first space tourism products will be rooted in people's curiosity.

> Virtual travel will represent the most likely means of 'visiting' space in the future, given that this environment is exceptionally hostile to humans. Although space tourism is physical, it can also be facilitated by nanorobots or avatars. Indeed, nanorobots may advance sci-fi storylines before humans are physically capable of such travel. (Interview 3)

> Robots and artificial intelligence (AI) could be used to help develop future spaceships from the materials of asteroids. (Interview 2)

> 'Second Life' platform studies have already demonstrated that, even though people are afforded the freedom to use limitless imagination in virtual reality, they tend to carefully copy real-life structures. (Interview 1)

People are interested in the idea of space, and the first space tourism products may be based on this curiosity factor. However, the desire to feel at home is likely to remain also in the future and hence influencing both the creation of virtual space tourism experience and designing for future space colonies.

Discussion

This chapter illustrates the futures research traditions for the development of a complex field with various interacting causal processes (Kuusi *et al.*, 2015; Kahn & Wiener, 1967). However, the academic discourse on sustainable space tourism is still taking its first steps; indeed, there are only few existing investigations, against which to consider the current findings.

Duval and Hall (2015) have offered some environmentally-informed points for space tourism as part of a sustainable tourism research agenda. and Toivonen (2017) has introduced a 'Sustainable Future Planning Framework', suggesting that sustainability, planning, weak signals and future scenarios act in synergy with one another to formulate the future aspects of space tourism. The findings of the current chapter can be especially linked under the 'Sustainable Future Planning Framework' and forming new additional categories under that core conceptual structure. Economic impacts and legislation can be placed under 'planning'; alternative energy sources and circular economy under 'sustainability', contemporary trends and health space tourism under 'weak signals' and space colonies and virtual travel and robotisation under 'future scenarios' (Toivonen, 2017: 27).

Within the notion of grounded theory, eight categories generated from the data were divided in Group 1 or Group 2 and placed on the Futures Map, in the section of planning horizon or mapping horizon. Economic impacts, legislation, alternative energy sources and circular economy

(Group 1) placed on the map's 'planning horizon', shared similarity in representing the current ways of living of the developed world. The Group 1 findings especially highlighted the gathering of facts and new knowledge based on the existing traditions; thus, they validated the Futures Maps criteria 3 and 4 (for facts). This kind of gathered knowledge may also prove useful to new commercial space tourism companies, validating the criteria 5 and 6 (for customer needs).

Contemporary trends, health space tourism, space colonies and virtual travel and robotisation (Group 2) shared a similarity in being acceptable trends or possible futures visionary ones; they could thus be placed on the mapping horizon. Group 2, similarly, identified preferred futures developments by pointing out alternative futures for the human race to exist, even though the earth's living conditions may become too challenging. The findings validated criteria 1 and 2 (for possible futures) by pointing out what may be relevant from the point of view of the vision.

Group 1 was directly linked under the subcategory Searching for Knowledge and Group 2 under Future Visioneering. Both subcategories linked to the core variable Making Space Human. Altogether, the findings created a scenario funnel, starting from the 'now' point and defining the range of scenario paths for the future. Making Space Human is already inside the current time horizon, with satellites assisting the living on earth and the international space station providing long-term shelter to the astronauts. Depending on the future technological innovations and legislative allowances, the time horizon for the wildest visioneering could be shorter than yet imagined.

Concluding Arguments

Space tourism has recently turned from science fiction into reality. The original excitement created by Hollywood productions was followed by decades of passivity, until the 2000s saw a number of technological innovations that enabled the space tourism industry to make rapid steps towards the first commercial space flights. Paradoxically, consumer trends in travel simultaneously shifted to more sustainable forms of practice. In order to survive, the space tourism industry needs to respond to this megatrend as well as follow increasingly stringent international climate change agreements. Appropriate regulations to encourage businesses to use a combination of different advanced technologies, so they can avoid having destructive environmental impacts, are required. For the most optimistic or imaginative, space tourism could become a distant 'nature tourism', as seeing the earth from a distance might enhance people's desire to protect the only planet that can offer suitable living conditions for humans.

This chapter has introduced space tourism as a future component of the tourism industry and has presented a future roadmap discussion of some key aspects of sustainable space tourism. Through analysis of interviews and grounded theory, it has been demonstrated that economic impacts, space legislation, alternative energy sources, circular economy, contemporary trends, health space tourism, space colonies, virtual travel and robotisation could constitute some of the sustainable areas for the industry's development.

Current weak signals suggest that travelling to near space is likely to become a trend in the future, especially if space tourism becomes cost-effective for the masses and conscientiously adheres to environmental and safety requirements. The space tourism industry currently appears to be aiming to reach a similar state as early aviation. Ultimately, a customer may simply purchase a ticket at a launching site and shoot up to space to visit their friends living in a moon settlement.

Space tourism could become the first large-scale commercial user of hydrogen, influencing its future use in aeroplanes and other ground transportation systems. The breaking of the familiarity barrier may even enhance the current development of environmental knowledge and protection. The sustainability requirements in space are demanding and must be solved on the surface of the earth before being practised in a space environment. The focus should be on sustainable science, achieved through understanding the psychology of new types of tourism behaviour as well as global environmental agreements.

References

Aalto University (2018) Aalto-1 launch, 21 May 2018. See http://www.aalto.fi/spacecraft/aalto-1-launch (accessed 13 September 2018).

Anderson, E. (2005) *Space Tourist's Handbook*. San Francisco: Chronicle Books.

Ansari XPRIZE (2018) A $10 million competition to usher in a new era of private space travel. See https://ansari.xprize.org (accessed 20 October 2018).

Ashford, D. (2002) *Spaceflight Revolution*. London: Imperial College Press.

Becken, S. (2015) *Tourism and Oil: Preparing for the Challenge*. Bristol: Channel View Publications.

Broderick, J. (2009) Voluntary carbon offsets: A contribution to sustainable tourism? In S. Gössling, C.M. Hall and D. Weaver (eds) *Sustainable Tourism Futures: Perspectives on Systems, Restructuring and Innovations* (pp. 169–199). New York: Routledge.

Blue Origin (2021) Blue Origin safely launches four commercial astronauts to space and back. See https://www.blueorigin.com/news/first-human-flight-updates (accessed 17 August 2021).

Carter, C., Garrod, B. and Low, T. (2015) *The Encyclopaedia of Sustainable Tourism*. New York: CAB International.

Duval, D. and Hall, C.M. (2015) Sustainable space tourism: new destinations, new challenges. In C.M. Hall and S. Gössling (eds) *Routledge Handbook of Tourism and Sustainability* (pp. 450–460). New York: Routledge.

Federation Aeronautique Internationale (2018) Space records. See https://www.fai.org/sport/space (accessed 10 December 2018).

Fawkes, S. (2006) Space tourism and sustainable development. *BIS European Developments in Space Tourism Conference*. 29 November. See https://www.

slideshare.net/guidofawkes/space-tourism-and-sustainable-development-presentation (accessed 18 May 2020).

Glaser, B.G. and Strauss, A. (1967) *The Discovery of Grounded Theory: Strategies of Qualitative Research*. Chicago: Aldine.

Global Information (2018) Global space tourism market size, status and forecast 2025. Market Research Report. See https://giiresearch.com (accessed 12 December 2018).

Hiltunen, E. (2013) *Foresight and Innovation: How Companies are Coping with the Future*. Basingstoke: Palgrave Macmillan.

Insinna, V. (2018) Trump's new space force to reside under Department of the Air Force. DefenceNews. See http://defencenews.com/space/2018/12/20 (accessed 20 December 2018).

Kahn, H. and Wiener. A. (1967) *The Year 2000 A Framework of Speculation on the Next Thirty-Three Years*. New York: Macmillan.

Kuusi, O., Cuhls, K. and Steinmuller, K. (2015) The futures map and its quality criteria. *European Journal of Futures Research* 3 (22), 1–14.

Matos, L. and Afsarmanesh, H. (2004) *In Collaborative Network Organizations – A Research Agenda for Emerging Business models*. Alphenaan den Rijn: Kluwer Academic Publishers.

Mojic J. and Susic, V. (2014) Planning models of sustainable tourism development destination. *Proceedings of International Scientific Conference. The Final and Real Economy. Towards Sustainable Growth*. Faculty of Economics, University of Nis, 17 October.

NASA (1998) General public space travel and tourism. In D. O'Neil, I. Bekey, J. Mankins, T.F. Rogers and E.W. Stallmer (eds) *NASA/STA Joint Study*. NP-1998-03-11-MSFC, March.

NASA (2019) Space tourism and commercialization. See https://www.nasa.gov/johnson/HWHAP/space-tourism-and-commercialization (accessed 3 August 2019).

Ross, S. (2002) *Near-earth Asteroid Mining. Space Industry Report*. US: Virginia Polytechnic Institute and State University.

Ryan, C. (2018) Future trends in tourism research – Looking back to look forward. The future of tourism management perspectives. *Tourism Management Perspectives* 25, 196–199.

SpaceX (2018) Starman in space. See https://www.spacex.com/falconheavy (accessed 20 August 2018).

Tett, G. (2018) America unleashes billionaires to boost the space race. *Financial Times*, 1 June 2018.

The Outer Space Treaty (1967) Treaty on principles governing the activities of states in the exploration and use of outer space. See http://www.state.gov/t/isn/5181.htm (accessed 15 December 2018).

Toivonen, A. (2017) Sustainable planning for space tourism. *Matkailututkimus* 13 (1–2), 21–34.

UBS (2021) Future of space tourism: Lifting off? Or has Covid-19 stunted adoption. See https://www.ubs.com/global/en/investment-bank/in-focus/2021/space-tourism.html (accessed 18 August 2021).

Virgin Galactic (2018) Richard Branson welcomes astronauts' home from Virgin Galactic historic first space flight. See https://www.virgingalactic.com/articles/first-space-flight (accessed 20 December 2018).

Wall, M. (2011) First space tourist: How a US millionaire bought a ticket to orbit. See https://www.space.com (accessed 16 June 2018).

Part 2
Science Fiction

6 A Life Without Limits: Design, Technology and Tourism Futures in *Westworld*

Leon Gurevitch

Chapter Highlights

- *Westworld* is considered as an example of the potential for automation to disrupt one of the few services traditionally assumed to be impervious to automation: human interaction. This invites us to consider the disruption of labour in tourist industry futures and suggests an environment in which user experiences, while apparently freer, are in fact more curated, designed and industrially standardised.
- Notions of the 'tourist gaze' are reconsidered in relation to *Westworld*. Specifically, the show adheres closely to the idea tourism offers a respite to the world of paid work but not without irony: *Westworld* can do so precisely because it utilises automation and artificial intelligence, the ultimate post-work technology.
- Parallels between computer games and future tourist experiences are considered, with *Westworld* as an example of an AI-based service sector operating according to the game logic that runs on algorithmic 'loops' waiting for new branches of possibility to be triggered by human visitors. This, in turn, highlights another feature of an automated tourist future: the imagined primacy of source code and its implementation.
- *Westworld* suggests that future tourist industries using AI will inevitably involve a great deal more orchestration than at present. Depending on the level of automation achievable, *Westworld* suggests that future tourist industries will resemble a hybrid of services presently found in other cultural forms. Just as the banking sector has in recent years become the largest employer of infotech talent, a tourist industry of the future is likely to do the same.

Introduction

In a scene halfway through episode six of the first season of Jonathan Nolan's *Westworld* TV series, one of the artificial human characters (Maeve) wakes up while undergoing repair and demands a walk through the 'back of the scenes' production facility. In a low key but highly impactful scene, as Maeve walks through the facility she is confronted with the truth of her own origin as an artificially created being and the shock that accompanies that. With almost no dialogue at all, the first scripted interaction in some minutes comes as she asks Felix where they are and he replies with one word: 'Design'. Fittingly, the preceding scene has been filled with technologies that underscore Maeve's existence as a designed object. From a 3D printed human presented as a contemporary bio-tech renovation of Da Vinci's famous humanist Vitruvian Man (which I will consider in more detail later) to a sculptor working on the face of a being still to be, the viewer and Maeve are both presented with a radically modified conception of Westworld: not a fantasy Western that recreates the 1860s wild west, but a techno-science fiction in which artificial intelligence is just that. As Maeve and Felix turn to leave a large screen flashes on and an audio track draws Maeve back. 'Welcome to Westworld, Live without Limits' it states as Maeve turns to it and watches herself portrayed in a promotional tourist video for the theme park, a final reminder of her position as an industrial object that exists within the economic confines of a post- (human) labour tourist space in which design (of behaviour, of biology, of experience, of narrative) is the central mechanism whereby agency (or rather, non-agency) and therefore power is negotiated.

In this chapter I will argue that this scene is central to understanding not only *Westworld* as a TV show but also as a popular cultural expression of emergent post-labour futures and its implications for industry. In this case I shall consider *Westworld* as an example of the potential for automation to disrupt the dynamics inherent to the social practices of labour in the service industries. *Westworld* imagines an industry in which AI disrupts one of the few services traditionally assumed to be impervious to automation: human interaction. In so doing it invites us to consider what the consequent disruption in labour will do for the balance of the industry and suggests that its likely implication is an environment in which user experiences, while apparently freer ('Live without Limits' is the sentiment stated by the promotional video and repeated in some form by the other characters throughout) are in fact far more curated, designed and industrially standardised. Throughout *Westworld* there runs a tension between the design of experiences, the disruptive potential of technology to deliver such experiences and the desire of tourists to feel free from the moral, social, economic and political constraints of their daily lives. This chapter considers the implications of this for tourism. More specifically, this chapter considers the relationship between design, post-labour technologies of automation and what sociologist John Urry (1990) called 'the tourist gaze': a set of social conditions under which tourist industries increasingly came to be structured across the 19th and 20th centuries. Here, he argued, consumers sought to engage with the process of the tourist experience first and foremost through the gaze and the anticipated gaze. From the picture of a spectacular tourist destination advertised in a magazine, through to the final gratification received upon reaching a destination and gazing upon it (the Taj Mahal, the Grand Canyon, the Eiffel Tower) the tourist gaze functions as a structuring logic of expectation and satisfaction (or otherwise when the process goes awry for the consumer) according to which the industry operates.

Automation and the Tourist Gaze

In Maeve's epiphanic scene, the dramatic impact builds to a final punch in which her gaze upon a screen featuring a promotional video is structured entirely around the logic of the tourist gaze. The opening welcome, inviting the viewer to 'live without limits' is not intended for her because she is the endless victim of that invitation. Maeve's gaze is dramatic for what it reveals to her about the true logic of a world she has thus far taken for granted. It is no accident that Shakespeare is quoted throughout the series ('all the world's a stage' is not quoted though one could see it as the underlying principle of the show) as numerous characters' psychological trajectories are repeatedly Shakespearian in their discovery of deeper truths both regarding themselves and the wider world at large. When Maeve turns to look at the Westworld promotional reel, her act and her inclusion in the promotional material marks the disruption of the social boundaries of the tourist gaze. As a piece of the apparatus of the Westworld theme park, she is not supposed to be privy to the inner workings of the tourist experience because those workings are dependent upon her not being privy to them: her ignorance of these workings is a key function of the tourist experience. Throughout the series reference is made to the fact that the AI within the park is programmed not to recognise or respond to any allusion to Westworld's artificiality. To do so, it is reasoned by characters central to the construction of its 'narratives', would be to break the spell (in theatrical terms the suspension of disbelief).

This scene, in which a central character is confronted with a fundamental revelation that shifts their perception of reality and of themselves, is not new to Hollywood film or television. In fact, eerily similar scenes have been a staple of the moving image almost from its inception. From the earliest movies made by Edwin Porter (1902) and R.W. Paul (1901) portraying characters perceiving like characters on the screen and coming to the realisation that these representations were not real, through to Don and Kathy from *Singin' in the Rain* gazing upon a promotional advert for the musical they just featured in at the end of the movie, through even to Buzz Lightyear realising his own reality is a construct upon seeing an advertisement for himself on television, popular culture has a long history of epiphany brought on by the realisation that a central characters' reality is not as it seems. In one astonishingly reminiscent scene of Maeve's, Scarlett Johansson's Jordan Two Delta character in the *Island* realises that she is a genetically engineered clone when she is stopped dead in her tracks by a Calvin Klein advertisement she apparently features in. Likewise in the *Westworld* scene, Maeve gazes upon a previous 'narrative' iteration of herself playing in corn fields with the 'daughter' she has dreamed of throughout the season.

In *Westworld* what is additional here is Maeve's disruption of the tourist gaze and her notion of self as a form of artificial life reconfigured through that gaze. In order to fully appreciate the dynamics of what is taking place both here and in the *Westworld* series in general we need to take a closer look at the specifics of Urry's notion of the tourist gaze for what it reveals about the show's construction of potential tourist futures. While insisting on a recognition of historical and sociological variation, Urry nevertheless identifies a number of key features universal to the tourist gaze that he lists as follows:

> 1. Tourism is a leisure activity which presupposes its opposite, namely regulated and organised work… 2. Tourist relationships arise from a movement of people to, and their stay in, various destinations… 3. The journey and stay are to, and in, sites which are outside the normal places of residence and work… There is a clear intention to return 'home' within a relatively short period of time… 4. The places gazed upon are for purposes which are not directly connected with paid work and normally they offer some distinctive contrasts with work… 5. A substantial proportion of the population of modern societies engages in such tourist practices; new socialised forms of provision are developed in order to cope with the mass character of the gaze of tourists… 6. Places are chosen to be gazed upon because there is an anticipation, especially through daydreaming and fantasy, of intense pleasures, either on a different scale or involving different senses from those customarily encountered… 7. The tourist gaze is directed to features of landscape and townscape which separate them off from everyday experience… 8. The gaze is constructed through signs, and tourism involves the collection of signs… 9. An array of tourist professionals develop who attempt to reproduce ever-new objects of the tourist gaze. These objects are located in a complex and changing hierarchy. (Urry, 1990: 2–3)

I quote Urry at length here, not just because he has become a seminal staple of any sociological examination of tourism and its underlying structure, but because Westworld presents a tourist future that both adheres extremely closely to some of these parameters and breaks radically from others. The interplay of Westworld's adherence to, and departure from, the traditional tourist gaze is indicative of the potential for future design technologies to disrupt and remake the tourist experience. For instance, Westworld sticks incredibly closely to an acceptance of the way in which tourism offers a respite to the world of paid work and the constraint it imposes upon contemporary consumers. Yet it is not without irony that Westworld itself can offer 'distinctive contrasts with work' precisely because it is able to utilise the ultimate post work technology: automation and artificial intelligence. Similarly, Westworld breaks from the traditional tourist gaze by relegating the experience of authentic spectacle (something Urry argues all tourists go in search of, frequently in vain given the subjective nature of 'authenticity' and the way in which even the perceived 'authenticity' of the tourist experience is always to some degree a construct) in favour of an unabashedly designed and entirely artificial spectacle. And yet, in doing so Westworld also highlights that tourist experience is constructed. Another way of understanding this is that Westworld is a theme park in the tradition of Disneyland. And yet, while Disneyland does not hide its construction, Westworld trades on a radical erasure of the line separating reality from simulation. Its selling point is that guests can experience the freedom to commit violent and immoral acts in an environment that, while simulated, feels to all intents and purposes real. What Westworld suggests, then, is a future in which tourist experience revolves around a different balance between notions of authenticity and simulation. This different balance, the show suggests, leads to a different focus. While Urry places the outward gaze as the primary motivational driver of the tourist gaze with the result that tourists believe they have enriched their inner self as a secondary by-product of the experience, Westworld is explicit in asserting this new space (facilitated by AI and automation) has an impact first and foremost on the self. William visits Westworld with his brother-in-law on a whim but finds his true self as a psychopath free to treat others as objects and in the process, the show suggests, becomes perfect Captain of Industry material.

Automation, 'Human' Interaction and the Future of Tourism

In the same vein, Westworld suggests that tourism of the future will not involve the same dichotomy of 'escape' from work that Urry describes. If the park can be automated, it is likely that the outside world of work can be automated too. That being the case, the notion of tourism and what it is will likely change radically if it is not defined by its capacity to escape from labour. Indeed, it is likely that it will be marked by a growth of voluntary labour. If an automated future allows for idleness, then tourism would be likely to emphasise the offer of occupation and even some sense of contrasting hardship. In a sense these forms of tourism already exist in the form of adventure tourism involving a managed degree of danger and difficulty; the numbers of people who die on Everest every year are a testament to this. Westworld showcases another version of this, where play and hardship are combined but danger, while seemingly real, is in fact only simulated. Indeed, one of the central characters (William, played in two different time periods by Ed Harris and Jimmi Simpson) spends his life returning to the park, dissatisfied that it offers him everything he wants in the way of brutality, hardship and apparent danger, without ever providing him with genuine danger, the lack of which subsequently mitigates the experience of authenticity (all of which changes at the very end of the last episode of season 1). All of this brings us to the other side of the 'work/leisure' dichotomy that Urry identifies as central to the tourist industry. While the industry is a place of leisure for consumers, it is of course the location of work for those employed in the business of servicing those consumers. The consequence of this, Urry argues, is that social interaction becomes a part of the 'product' being bought and sold:

> ... the quality of the social interaction between the provider of the service, such as the waiter, flight attendant or hotel receptionist, and the consumers, is part of the 'product' being purchased by tourists. If aspects of that social interaction are unsatisfactory (the offhand waiter, the unsmiling flight attendant, or the rude receptionist), then what is purchased is in effect a different service or product. The problem results from the fact that the production of such consumer services cannot be entirely carried out backstage, away from the gaze of the tourists. They cannot help seeing some aspects of the industry which is attempting to serve them. But furthermore, tourists tend to have high expectations of what they should receive since 'going away' is an event endowed with particular significance. People are looking for the extraordinary and hence will be exceptionally critical of services provided that appear to undermine such quality. (Urry, 1990: 40)

Westworld suggests a future in which this dynamic is radically changed. With AI, the show suggests, such social interaction is automated and can therefore be far more tightly managed and controlled. Indeed, a significant proportion of the narrative drama of this series is structured around the question of how this tourist experience would differ if the services on offer were delivered by artificial rather than human means. The suggestion is that, contrary to Urry's claim, such consumer services actually could be entirely carried out backstage while the drama arises from the moments in which the seemingly seamless offering that results breaks down (like live news broadcasts that go awry).

Westworld, then, suggests a future tourist experience in which the service sector of the industry has been automated and outsourced to machines, mitigating the need for human interaction other than between other tourists. A consequence of this is that when operations in the park fail, that failure is chalked up to bad coding, failing automation or malicious human intervention 'behind the scenes'. These notions of 'behind' and 'in front' of the scenes, and control over 'good' and 'malicious' code is revealing in another way. As the first phrase suggests, there is a (varying) degree of theatricality to most tourist industries and in Westworld this is made explicit in the function of the cast of AI characters who are constructed behind the scenes as 'characters' who are given 'narratives' within a network of many other potential and continually interacting narratives. To understand how these supposedly function it is necessary to understand something of the way in which media forms work.

As the oft-cited Lev Manovich has argued (2001), contemporary new media operate according to the interrelated logic of the database and the algorithm; a dynamic best understood by comparing an old media form such as a novel or a play with a new media form such as a computer game. With a novel or a play, a narrative largely is written in a linear manner to be subsequently read or watched in a linear manner. This, as Manovich has argued, is because these forms follow the original logic of the scroll. A game, by contrast, follows the new media logic of the database, which is non-linear. The technology that made this logic possible was the codex (or book) which, by contrast with a scroll, allowed a user to open the codex and interact with the information contained within it in a non-linear manner. The most obvious example of this is the bible with its numerically labelled chapters, verses and books that could allow a well-trained theologian to access the information relevant to any particular situation or moral question immediately and in a non-linear manner. Manovich argues that all modern media essentially follow this pattern; information is contained in a database and can be accessed in a non-linear manner. An example that contrasts with a novel would be a computer game in which players do not have to progress through the game narrative in a linear manner and, indeed, do not even have to progress through any particular narrative at all, presented as they are with many multiple narrative directions they can pursue at any given time. These potential narrative directions are stored in a database and are only set in motion once a player chooses to initiate one. Operating alongside the database is the algorithm, a computational structure that dictates possible outcomes and new potential avenues that open up once a narrative choice has been made by the player. Game logic therefore operates upon the interrelated function of both the database in which multiple potential narrative possibilities are stored, and the algorithm which is/are initiated once a specific narrative direction is chosen.

I provide this level of detail because *Westworld* dramatises the potential significance, possibilities and pitfalls of what happens when databases and algorithms become increasingly central to the functioning of the tourist industry. For the writers of *Westworld* this is an extrapolation out from the present, but in reality we can already see the impact that the logic of the database and the algorithm has had in automating past forms of industry labour out of existence. The contemporary tourist industry has already been forced over the last two decades to respond to Google, one of the first and most widespread database algorithms of recent years to have far-reaching implications for both the automation and distribution of labour. Indeed, the first to recognise this and manoeuvre in the online space ensured their survival in an environment where many travel agents were undercut as their institutional knowledge and specialist expertise was brushed aside by the economies of automated, outsourced and distributed labour that Googles page rank algorithms offered. In this sense Google's page rank outsourced the labour of finding holidays back to the potential tourists themselves who unwittingly accepted extra personal labour in exchange for the apparently democratised networks of information and communication no longer locked up in professionals' hands. Potential holidaymakers have subsequently over the last 20 years started to do the job that was previously the preserve of travel agents. I will set aside the potential benefits or drawbacks of this shift in the interests of brevity, but it is worth noting that *Westworld* sketches out a future in which the database, the algorithm and the automation of labour have a lot further to run. In *Westworld* the 'hosts' run as they are on the logic of the database and the algorithms have replaced human service industry labour.

Algorithmic Automation and the Game Logic of Tourist Experience

Westworld makes explicit the parallels between computer games and the park, with its AI service sector behaving like materialised game characters who operate on 'loops' waiting to be triggered by human visitors in order to operate according to a new behavioural algorithm. From a dramatic point of view, the problems arise from faulty or malicious code rather than as a result of the bad attitudes or poor training of human service industry staff. This highlights another feature of an automated tourist future: the imagined primacy of source code and its implementation. In her article 'On "Sourcery," or Code as Fetish', Wendy Chun has argued that source code and the ability to manipulate it in contemporary culture is fetishised as a modern-day form of sorcery (or 'sourcery' as she aptly spells it). Westworld fetishes source code in precisely this way and it is worth quoting Chun at length as a useful example of what we should avoid imagining when considering tourist futures as well as what possibilities they point to:

> software as source relies on a profound logic of 'sourcery' – a fetishism that obfuscates the vicissitudes of execution and makes our machines demonic. Further, this sourcery is the obverse rather than the opposite of the other dominant trend in new media studies: the valorisation of the

user as agent. These sourceries create a causal relationship among one's actions, one's code, and one's interface. The relationship among code and interface, action and result, however, is always contingent and always to some extent imagined. The reduction of computer to source code, combined with the belief that users run our computers, makes us vulnerable to fantastic tales of the power of computing. To break free of this sourcery, we need to interrogate, rather than venerate or even accept, the grounding or logic of software. (Chun, 2008: 300)

One could not get a better description of what Westworld does than Chun's description here of both the logic of 'sourcery' (deliberately spelt to refer to source code) on the one hand and the subsequent equation of demonic machines on the other. Indeed, in the first episode of the first season, Dolores' 'Father', Peter Abernathy malfunctions on seeing a photograph of Time Square (the robots are programmed not to recognise anything from the outside world). In a mania Abernathy grabs Dolores and states 'Hell is empty and all the demons are here', a line from Shakespeare's play *The Tempest* in which a sorcerer named Prospero spends the duration of the play on an island controlling and manipulating all the characters' actions with his magic. While *Westworld's* very mixed and abundant use of Shakespeare quotes has been criticised, the implication of this particular reference so early on in the series is clear and becomes more so throughout the series; Anthony Hopkins' character, Dr Robert Ford who writes all the code for the AI in the park, is a modern-day Prospero. In Chun's terms he is a quintessential sourcerer of source code. Like Prospero (who Anthony Hopkins once fittingly played in Los Angeles in a John Hirsh production in 1979) Ford works behind the scenes to guide the human and AI creatures of the island (at the beginning of season two we learn that Westworld is literally situated on an island). And while *The Tempest* has been seen by scholars as an autobiographical metaphor for Shakespeare's life, *Westworld* could be seen as an allegorical recast of *The Tempest* with the demons this time reimagined as Chun's machinic entities.

All of this differs from Urry's conception of a tourist gaze that is premised on the voracious appetite for the consumption of perceived authenticity. In *Westworld's* version of an automated tourist future, Design is placed at the centre of an industry that offers a carefully crafted experience which blends the simulation of authenticity with the environmental control of a theme park at the same time. This environmental control extends in both directions with Ford both 'terraforming' the park (literally designing the geography as well as the animals in it) and designing and reforming the psychologies of the hosts and the construction of their 'narratives' so that their behaviours and the experience they can offer the guests is both optimisable and industrially standardisable, at the same time as it can be uniquely tailored to the individual tourist. This, then, is mass tourism offered according to the logic of mass individuation. Small wonder, then, that the signature image of the series is a host body in the process of biologically 3D printed manufacture. The image very deliberately recalls Leonardo Da Vinci's Vitruvian Man: the most iconic representation of the Renaissance shift to both empirical principles for understanding the world and a visual language that captured this understanding. Fundamentally, the Vitruvian Man conveyed the centrality of man (as it was in a patriarchal order) at the heart of the empirical structures of measurement and order underpinning the design and engineering of a new humanistic industrial and philosophical world order. Perhaps unsurprisingly yet somewhat ironically late, Westworld's Vitruvian bot switches from a male body in the first season to a female body in the second season.

Where the emergence of humanism was characterised by empirical science, a philosophical move away from theistic understanding of God's role in the machinations of the world toward an understanding of humanity as the central protagonist uncovering (and then commanding) the laws of nature (and as a result ultimately in charge of their own destiny). Westworld suggests that the emergence of AI marks a new shift. Unsurprisingly, the Renaissance humanist philosophical worldview underpinned Shakespeare's greatest achievement as a playwright: a level of psychological depth and realism in his characters' behaviours and motivations that had never been seen before. In constructing Anthony Hopkins as a Prospero (and therefore Shakespeare's autobiographical figure) that literally writes the code and scripts[1] for both the psychologies and narratives of the hosts, Westworld suggests that future tourist industries using AI will inevitably involve a great deal more orchestration than at present. Depending on the level of automation achievable (and the timeline of that in itself is, as scholars such as Nick Bostrom (2014) point out, an open question), future tourist industries will resemble a hybrid of services presently found in other cultural production. Just as the banking sector has in recent years become the largest employer of infotech talent, a tourist industry of the future could conceivably do the same. Likewise, coders and programmers skilled in the logic of game production could equally likely be employed alongside script writing talent one would currently expect to find in the television industry. Not to mention the artists and engineers designing new objects, landscapes and robotics.

Concluding Arguments

Regardless of the likelihood of Westworld's level of fantasy tech, it is inconceivable that increasing levels of automation and machine learning will not be deployed in the tourist industry because it already is. As this continues, the depth and breadth of the tourist gaze available to consumers will expand significantly. In this sense, *Westworld* doesn't point to where we will end up so much as it points to where we are already heading. Predicting industrial futures (or indeed any future for that matter) based on where we think technology may develop is a notoriously inaccurate and risky undertaking. What is less fraught with the likelihood of inaccuracy is the process of educated extrapolation based on technologies that already exist. *Westworld* is a fascinating case study because it not only conjures a fantasy tourist future and invites the viewer to imagine the possibilities of such fantasy, it also reflects a far more detailed extrapolation from the present than did its filmic predecessor. To take just the technology developments of the past 20 years, it is apparent that tourist industries have already undergone radical change as a result of networked information and the capacity of the automated database and the algorithm to shape the flows of tourists seeking out and purchasing new experiences. Added to this, social media and their distributive power have been equally, if not more, powerful than the tourist industries' advertising efforts in previous decades. The journey *Westworld* invites us to go on is, essentially, a thought experiment to consider what happens to the tourist industry when algorithmic automation continues to undergo as much change over the next two decades as it has over the past.

Note

(1) I use that second word in an awareness that it can be deployed in a computer science sense and a literary sense.

References

Bostrom, N. (2014) *Super Intelligence: Paths, Dangers, Strategies*. Oxford: Oxford University Press.
Chun, W.H.K. (2008) On 'sourcery,' or code as fetish. *Configurations* 16 (3), 299–324.
Manovich, L. (2001) *The Language of New media*. Massachusetts: MIT Press.
Paul, R. (1901) *The Countryman and the Cinematograph*. London: Pauls Animatograph Works.
Porter, E. (1902) *Uncle Josh at the Moving Picture Show*. New York: Edison.
Urry, J. (1990) *The Tourist Gaze*. London: Sage Press.

7 Harry Potter and the Future of Tourism

Ina Reichenberger

Chapter Highlights

- Popular culture plays an important role in shaping societal norms and inducing social change.
- Tourism can harness the power of popular culture and fan activism to address challenges.
- The Harry Potter series explores issues of morality that have contributed to real life activism.
- The narrative of the future of tourism can be changed to the better through the lens of Harry Potter.

Introduction

The Harry Potter phenomenon has captured audiences worldwide. Spanning seven books, eight movies, three theme parks and a further spin-off movie series (*Fantastic Beasts and Where to Find Them*), the magical world J.K. Rowling created remains as popular as ever and appeals to children and adults alike. The dominant themes of the fight between good and evil, facing personal, educational and moral challenges while coming of age, and the complex context of power, authority, courage, love and friendship resonate across countries, continents, cultures and age groups. More than 500 million copies of the books alone have been sold, translated into over 74 languages, uniting people all over the world through a pop cultural phenomenon in an unprecedented manner.[1] In the United States alone at least 54% of the population are familiar with Harry Potter (Statista, 2016), the United Kingdom's Potter-themed Warner Brothers Studio Tour is visited by up to 6000 visitors per day (Sylt, 2017) and the full franchise is estimated to be worth over US$25 billion (Jacobs, 2017). Not surprisingly, elements of the world of Harry Potter have become widely recognisable to the extent that they do not remain within their fictional realm but become meaningful in a variety of real-life contexts. Public reactions to the 2017 US election campaign, for example, compared candidate Donald Trump to Harry Potter's villain Lord

Voldemort, thus influencing contemporary political discourse based on shared perceptions and values that signify what is good and bad in the world of Harry Potter and beyond (Gierzynski, 2014; Nicholson, 2017).

It appears as if Harry Potter readers share similar values, attitudes and thus perceptions shaped through the fictional narrative and now use these to contextualise and make sense of their lived reality, using it as a framework to explore what is acceptable and what is not. While this may seem surprising (it is, after all, only a series of children's books), the relevance of popular culture for societal norms is by no means new. Popular culture goes far beyond simple entertainment but establishes norms, contributes to innovations, and induces social change (Kidd, 2007). This implies that Harry Potter, as the most relevant pop culture phenomenon in the globalised world, has the potential to facilitate much more than the escape into fictional realms, indeed can encourage and contribute to social change. If the magical world of Harry Potter infiltrates world politics and provides us with metaphors to understand political complexities, what can it do for tourism? This chapter will use the Harry Potter phenomenon to highlight the potential impacts of popular culture by addressing readers' perceptions of morality, equality, authority and power and how these can be harnessed to induce social change. Subsequently, it will be shown how challenges in tourism such as inequality, exploitation and power misuse may benefit from the Harry Potter fandom and the power they hold to reshape tourism.

What Harry Potter Tells Us About the World

The universal appeal of Harry Potter stems from the multidimensional storytelling. While on the surface the story is about an orphaned boy who learns that he is a wizard and must subsequently defeat the villain Voldemort who murdered his parents, it goes far beyond this. Rowling addresses social, psychological and cultural issues including social inequalities, love, prejudice, the transition to maturity (Fields, 2007) as well as opposing identities, conflicts, power structures (Franklin, 2006; Vezzali *et al.*, 2015) and mental health (Klonsky & Laptook, 2006). As the story continues, issues become more complex, lines become more blurred and moral judgments are increasingly difficult to make. This section considers the most prominent themes of the novels and highlights their potential and actual impact on readers to provide a solid understanding of how Harry Potter can shape readers' attitudes and perceptions, later to be transferred to a tourism futures context.

'The matrix of popular culture, with its complexly self-referential character, has become perhaps the chief space in which social norms are negotiated and transmitted throughout society' (Kidd, 2007: 76). Children's literature, especially classic fairy tales focusing on a struggle of good versus evil, is no exception, having long been known to impact readers' mind and personal development (Bettelheim, 1976). It has often been argued by educators that Harry Potter especially contributes to readers' social and moral development by exploring power in the context of friendship and courage (Binnendyk & Schonert-Reichl, 2002; Whitney *et al.*, 2005). In the series, characters are continuously faced with decisions that require appropriate choices, taking into account conflicts, multiple agendas and the role of institutions – through this, actors construct their individual moral and ethical systems through negotiations with others (Chappell, 2008). In doing so, Rowling merges two worlds, transferring everyday real-life complexities into the magical to explore readers' interpretations of reality safely in an alternate world (Behr, 2005).

Prejudice and discrimination

One of the themes Rowling explores in the books is racial prejudice – not related to the colour of skin, but to the colour of blood (Lyubansky, 2006). Wizards of 'pure blood' are regarded as superior by characters sympathising with the villain, while those with non-wizard heritage are labelled 'mudbloods' and are subject to discrimination and persecution. 'Pure blood' wizards sympathising with 'mudbloods' are regarded as 'blood traitors', mirroring past and current racial discrimination. 'Muggles', non-magic people, are regarded much the same, with Behr (2005) comparing their treatment to those of colonialists treating indigenous people. 'Intolerance, snobbery and ethnic hatred – all commonplaces of our Muggle world – are reproduced inside the Harry Potter series via wizard-muggle relations' (Behr, 2005: 125). Similarly, the werewolf character Remus Lupin is a societal outcast, stigmatised based on occurrences out of his control and having to resign from his teacher's position after being 'outed'. Parallels can be drawn to the societal status of the LGBTQI+ community or HIV-positive individuals (Naficy, 2006). The role of house elves, 'belonging' to wizarding families and functioning as unpaid housekeepers, is another story line feeding into themes of discrimination and slavery. Hermione Granger's efforts to free them are often met with resistance, both from fellow wizards as well as the house elves themselves, re-contextualising historical 'debates' around the slave trade (Carey, 2003).

Rowling puts these stigmatised characters into positions that highlight their positive personal qualities, clearly displaying discrimination and stigmatisation as actions of the villains. These characters have been successfully used in education and counselling to elicit empathy (Chappell, 2008; Gibson, 2007), and it has been shown that Harry Potter readers across age groups show improved attitudes towards stigmatised groups such as immigrants, refugees and homosexuals (Vezzali *et al.*, 2015). The power of Harry Potter to reduce prejudice in its readers while increasing tolerance and open-mindedness is just one example of how values and attitudes of readers are influenced positively, accepting that what is 'abnormal' is relative and, while frowned upon in one world, is highly regarded or irrelevant in the other (Murakami, 2006).

Power and authority

Power occupies a central role in the Harry Potter series, signified by Lord Voldemort's attempts to regain power and others' attempts to prevent this. While this appears to be a black and white situation, good versus evil, Rowling's storytelling is more complex and continues to highlight the dual capacity for both within the same person throughout the books (Patrick & Patrick, 2006). Through a narrative transformative, character's original perceptions of good or bad change (Behr, 2005). Severus Snape, portrayed as a villain yet working for the good side, uncovered as a member of the Order of the Phoenix dedicated to preventing Voldemort's rise to power. James Potter, Harry's father, beginning as a heroic self-sacrificing character, yet acting as a bully during his youth. The headmaster Dumbledore, wise and inherently good, yet making choices that hurt numerous people around him. Harry Potter discovering increasing parallels and similarities between himself and Voldemort. The search for power within conflicting agendas and situations where no decision is available that benefits all highlight the multidimensionality and blurriness of people both within the books as well as in reality, simultaneously highlighting the challenges of power and authority when the first is sought under the disguise and justification of the latter. Power that corrupts is explored through further narrative transformatives, where originally 'good' people such as Percy Weasley or Minister of Magic Cornelius Fudge struggle with the discrepancy between the respect of authorities and their misuse of power (Behr, 2005).

For readers, these storylines create patterns that highlight the increasing recognition of young characters that society and its underlying mechanisms create structures and frameworks that are not always truthful and beneficial to those living in it (Chappell, 2008). The resistance from Harry Potter and his friends is suggested to encourage readers to critically evaluate and challenge the status quo in their own life that is provided by authorities and those in power, as well as their leadership qualities, to become 'architects of their own agency' (Chappell, 2008: 282). Children can thus develop the courage and capabilities to stand up against the evils they will face in their own life as they mature (Knapp, 2003). After all, it is the characters' choices that determine the story's narrative (Behr, 2005).

Morality

Issues of morality are explored through two themes throughout the Harry Potter series – collective responsibility and the moral values of courage and love, often intertwined and connected to the prior themes of power and authority. 'To act in a moral way, a person must first understand how his or her actions affect the welfare of others, judge whether such actions are right or wrong, intend to act in accord with this judgement and follow through with this intention' (Bear *et al.*, 1997: 14). Kohlberg's six stages of moral development are all evident through the series, at different times through different characters, and thus provide an opportunity for young readers to explore increasingly complex moral challenges and foster their moral growth (Binnendyk & Schonert-Reichl, 2002). Collective responsibility and prosocial behaviour are one example through with moral issues are addressed. This is most visible at Hogwarts, School of Witchcraft and Wizardry, that Harry attends. Students are 'sorted' into four houses and can collect points for their house based on prosocial and obedient behaviour – rule breaking and antisocial behaviour result in a reduction of points. Collective responsibility is thus promoted and proactively reinforced (Binnendyk & Schonert-Reichl, 2002). However, in line with the previous section on critically evaluating power and authority, Rowling highlights that acting morally goes beyond this. Standing up for what one believes is right can and often should override social influence and adherence to rules designed to create prosocial spaces – moral courage is thus required to critically evaluate what shape collective responsibility should take in the long term (Green, 2006). Integrity, courage, friendship and love are portrayed as the core moral values that direct positively portrayed behaviour in the books (Binnendyk & Schonert-Reichl, 2002), overriding self-interest, displaying generosity of character and providing sacrifices for the good of others (Franklin, 2006). One of the core messages of the book is, after all, that love is the strongest kind of magic (Knapp, 2003).

Fan Activism – The Transformative Power of Harry Potter

Popular culture, as mentioned earlier in this chapter, is not only a contributor to social norms but also an inducer of social change – especially in relation to books (Kidd, 2007). Through its focus on discrimination, prejudice, power, authority and moral complexities, the Harry Potter series provides much potential to positively impact its readers and in turn their world. Harry Potter's metaphors and analogies have already found their way in the political discourse (Nicholson, 2017), and especially younger generations have become more active in recent decades – yet less in political and governmental contexts but related to their personal interests (Shresthova & Brough, 2011). Political debates are often framed through a language young people cannot relate to, not addressing issues relevant to them and not encouraged to participate (Buckingham, 2000). Culturally defined solidarity, in turn, is becoming more established and poses a counterpart to more traditional and politically defined solidarity (Shresthova & Brough, 2011). Fandoms, subcultures of fans who share a sense of camaraderie and belonging based on these personal common interests, are built upon a sense of solidarity and shared values, thus providing a sound foundation that 'empowers individuals to make decisive steps towards collective action' (Jenkins, 2015: 206). The Harry Potter fandom, as illustrated above, is united by a shared understanding of what is good and what is bad, with a moral compass that provides a framework to evaluate what is acceptable in achieving one's goals. This collective identity can be mobilised towards collective action in the form of fan activism (Shresthova & Brough, 2011), which has been done through the Harry Potter Alliance (HPA).

The original inspiration for the HPA came from 'Dumbledore's Army', an informal association of students around Harry Potter learning to defend themselves against the dark arts against the orders of the Ministry of Magic – at this time undermined by Voldemort's supporters. The HPA was formed in 2005 and now contains more than 100,000 members in over 70 chapters around the world mobilising fans in campaigns for equality, human rights, literacy and a variety of other philanthropic causes to make a positive difference in the world. Campaigns and engagement are creatively placed within the narrative, terminology and metaphors of the Harry Potter series, thus connecting the fictional content to real-life social justice aims (Jenkins, 2015; Shresthova & Brough, 2011).

Ongoing campaigns of the HPA tackle a variety of issues. 'Accio Books' is an annual worldwide book drive that began in 2009, donating books to communities in need while supporting library infrastructure, advocating within communities for the importance of libraries and reading, and providing literacy and education services. 'Accio Books' is connected to Hermione Granger, one of the main characters, and her desire for knowledge through books that contributed greatly to winning the war against evil. 'Neville fights back' is a campaign based on the character Neville Longbottom, initially timid and shy, but overcoming this for the greater good and eventually standing up for what he knew was right and just despite his fears and even against his friends. 'Neville fights back' is designed to encourage individuals to have conversations around large and seemingly unsurmountable social justice and human rights issues – highlighting that small changes can have large impacts and one person can make a difference. The campaign 'A World #WithoutHermione' acknowledges the female main character without whom Harry Potter would have failed in every book. This campaign is concerned with gender equity in global education contexts, fighting against poverty, gender and racial biases and cultural inhibitors. The HPA also successfully campaigned for official Harry Potter Warner Brothers chocolate merchandise to be fair trade certified, in line with fair and respectful treatment of individuals as portrayed by house elves in the series. Individual HPA chapters also choose their own charitable causes and thus contribute in much greater variety especially on community levels.

Fan-based activism based on popular culture has the potential to induce significant positive changes in the world, and through the immense popularity of the Harry Potter series, the HPA has and continues to achieve just that. Individual dedication to real-life causes that are metaphorically discussed throughout the series is utilised and harnessed through a globally active organisation – all because of a popular culture trend that has remained present and universally relatable for over 20 years. While there is no record of targeted campaigns relating to popular culture that address current and future challenges in tourism, it is a context that lends itself to further exploration. As one of the world's largest industries, facing numerous challenges and causing widespread negative impacts in a variety of contexts, the future of tourism can appear dark at times – yet it is our choices that shape the narrative, so how can it be shaped through the power of popular culture?

Harry Potter and the Future of Tourism

Popular culture and tourism are already intertwined, predominantly through the tourism industry utilising trends in popular culture for product development. Film tourism, literary tourism, theme parks, concerts, events and conventions are just some examples of how trends in pop culture are visible in tourism. These, however, tend to be based on visitors' common interests and exist within restricted spaces only. They often neglect to take into account the potential power that comes with what it is that fandoms identify with on a deeper level and what contributes strongly to the positive and wide-ranging impacts of fan activism. In the case of Harry Potter, highly relevant and dominant themes of the series centre around issues of power, authority, morality, prejudice, discrimination and how the actions of individuals can change the world for the better as well as for the worse. It also consistently highlights that 'doing the right thing' is hardly ever easy, is often met with resistance and restrained by opposing constructs, demands sacrifices, yet every small piece counts and is a crucial component of positive progress and the victory of good over evil. This section will now explore how dominant themes in Harry Potter mirror challenges in tourism, subsequently discussing how the values communicated through the book series allow fans to shape their own future narrative.

Prejudice and discrimination as discussed in Harry Potter are dominant themes in tourism contexts. Racial and gender discrimination (e.g. Campos-Soria *et al.*, 2015; Duffy *et al.*, 2018; Lee & Scott, 2016), accessible tourism (e.g. Michopoulou *et al.*, 2015; Nyanjom *et al.*, 2018), the inclusion of LGBTQI+ travellers (e.g. Kama *et al.*, 2017; Lubowiecki-Vikuk & Borzyszkowski, 2016), staff exploitation (e.g. Armstrong, 2016; Ewart-James & Wilkins, 2015), human trafficking (e.g. Miller, 2017; Neptune, 2016) and sexual harassment/assault (e.g. Kennedy & Flaherty,

2015; Ram, 2015) are just some of the issues that affect both the supply and demand side of tourism. All these real-life issues have counterparts in the Harry Potter universe that stand as examples of immoral, unfair and evil behaviour that will contribute to the victory of the villains if not addressed – racial discrimination of mudbloods, gender discrimination by integrating strong and integral female characters, discrimination of those that are 'different' through stigmatised character Lupin, exploitation and slavery through the treatment of house elves. While no sexual components are directly touched upon by Rowling, the use of the 'cruciatus curse', causing unbearable pain, in connection with power, sadism and entertainment can be used as a parallel to discuss sexual assault and harassment. These are complex and wide-ranging issues, partly embedded in our societal and cultural fabric, partly hidden from visitors' eyes, partly contributed to by tourists, and it may seem impossible to individual tourists for their solitary actions to matter and induce change. Especially those that are not directly affected may be either unaware or ignorant towards the prejudices and discrimination against other tourism stakeholders, in turn contributing to the maintenance of the status quo.

The themes of power and authority, of course, feed into the tourism-related challenges described above and highlight the power that tourists hold. Volunteer tourism may function as an example where tourists have intentions to do good for local communities in need, yet unknowingly contribute to issues such as producing low quality work (Bargeman *et al.*, 2016; Guttentag, 2009), communities developing an over-reliance on volunteers (Guttentag, 2009), and even jeopardising non-government organisation (NGO) work through volunteers' inabilities to adapt to non-Western perspectives (Simpson, 2004). Labelled as 'neo-colonialism' (Palacios, 2010), volunteer tourism has been argued to be built upon unequal relationships relating to power and privilege and thus contributing further to what its participants seek to reduce (McLennan, 2014). By adhering to frameworks and structures created by others in power that may not be beneficial for all stakeholders, tourists may claim ignorance at best, misuse and corruption at worst. The duality of good and bad within the same person and the sometimes intended, sometimes unintended negative consequences for others, are often ignored by tourists while being unaware of the power and potential negative impacts of their behaviour. To this, the knowing and targeted misuse of power can of course be added, exemplified arguably at its worst through child sex tourism (e.g. Davidson, 2004; Tepelus, 2008).

Reading Harry Potter, however, has ingrained messages of personal integrity, moral actions, and collective as well as individual responsibility and their subsequent positive impacts through its dominant morality themes. Those that do not speak up against unjustice, even while not actively supporting it, are drawn as weak, perhaps pitiful characters – Harry Potter and the members of the Order of the Phoenix however are not, and their fans hope not to be either. Not only do they reflect upon the impact of their own actions to bridge the potential gap between good intentions and bad impacts, they also take action against others immoral behaviour. To address issues like those mentioned above, tourists are required to make moral judgements and act accordingly based on what they believe is right or wrong. If proactive moral agency is not utilised, it is replaced by moral disengagement, where people exhibit unethical behaviour without guilt (Detert et al., 2008). Prohibitors of moral agency may be justifying the immoral, the use of sanitised language to reduce the negative afflictions, the displacement and diffusion of personal responsibility, the disregard of consequences and/or dehumanisation of those impacted by consequences (Bandura, 2002).

Popular culture in the form of Harry Potter provides metaphors that can contribute to the reduction of tourists' moral disengagement, allowing them to critically evaluate their own actions in the context of shared values and perceptions. Fan activism through the HPA has focused on recognising current societal challenges, drawing parallels to fictional situations relating to prejudice, discrimination, power and authority, thus harnessing the power of Harry Potter readers for the greater good through encouraging collective responsibility and prosocial behaviour by allowing them to follow into the footsteps of fictional characters. Much the same can be done in a tourism context, where the Harry Potter fandom can make choices based on moral agency that change the current narrative. On the one hand tourists are required to critically evaluate the impact of their actions and alleged moral behaviour so as not to fall into the trap of accidental corruption and misuse of power, on the other hand they must stand up against what is wrong, even if they are not directly affected and may think their potential impact small.

Issues of prejudice and discrimination, with the foundation of Harry Potter readers showing improved attitudes towards stigmatised groups, higher tolerance and open-mindedness (Vezzali et al., 2015), can be discussed in the context of mudbloods, Remus Lupin, house elves and others, where Rowling clearly highlighted the immorality of issues while including space to discuss and discard potential justifications. Issues of power and authority, intended (e.g. the sexual exploitation of children in travel and tourism, sexual harassment) or unintended (e.g. volunteer tourism), can be tackled through the perspectives of either the villains or Percy Weasley and Cornelius Fudge, characters who hope to do good, yet are unable to recognise their wrongdoing, trusting in an authority that fails to protect. Language as a powerful tool is required to clearly communicate the impacts these issues have on disadvantaged groups, while highlighting the need for personal and collective responsibility by drawing connections to the fictional characters who challenged and successfully changed the status quo through the courage to stand up (Chappell, 2008; Knapp, 2003) by acknowledging consequences and personalising those affected by them.

Concluding Arguments

While at first glance it seems unlikely that a series of children's books may indeed change the future of tourism, popular culture's impact on social norms and social change is by no means new (Kidd, 2007). Neither is fan activism, providing individuals united by common interests, passions and values with an opportunity to induce change (Shresthova & Brough, 2011). Harry Potter is considered one of the most significant pop cultural phenomena of all time and provides fans across all age groups and cultures with a relatable tool to translate its values of love, courage, friendship and tolerance to real life. This power has already been harnessed in a variety of charitable and social activism contexts and holds great potential to be expanded to help reduce the negative impacts of tourism by drawing attention to a variety of issues that may not always be visible, not within tourists' direct responsibility, or seemingly unsurmountable.

Prohibitors of moral agency were utilised throughout Rowling's series to reduce moral disengagement and thus allow for translation into different contexts. Opening fans' eyes to issues in tourism and the potential impact of their travel behaviour and showing them how to act in a way that mirrors Harry Potter, Neville Longbottom, Hermione Granger or other beloved characters, may allow us to continue to change the narrative. As Dumbledore said when discussing Harry's inner turmoil about whether he is a good person or not – it is our choices that show what we truly are.

Tourism's contribution to social change through transformational tourism and its potential impact on social justice, equality and human rights based on love and hope have been addressed previously (Reisinger, 2013, 2015). Although most research is concerned with the transformation of the tourists' 'self', some attention has also been paid to the transformation of the 'other' through social change. Previously explored in contexts such as ethnic and rural communities (Diekmann & Cloquet, 2015; Schweinsberg *et al.*, 2015) and poverty (Freire-Medeiros & Cohen, 2015), stakeholders are able to encourage, support and facilitate transformative tourism through the Harry Potter phenomenon. Fans themselves are able to communicate challenges in tourism to a large audience, drawing comparisons to themes or characters within the books to provide visitors with the necessary context and guidelines to act morally. Destinations themselves are able to do the same – Thailand, for example, is hoping to draw visitors' attention to human trafficking as a problem related to its sex tourism industry (Li, 2018). While this may seem abstract and unrelated to many tourists who visit for other reasons, drawing upon certain themes in the Harry Potter series could allow destinations or organisations to instil a sense of awareness and responsibility among those who may otherwise not be responsive to the campaign. One crucial factor in moral disengagement is the diffusion of personal responsibility, something that has successfully been overcome by the HPA's campaign 'Neville fights back' and shows great potential to utilise fans' identification with the series for transforming tourism for the better.

Note

(1) http://www.wikisummaries.org/wiki/Harry_Potter

References

Armstrong, R. (2016) Modern slavery: Risks for the UK hospitality industry. *Progress in Responsible Tourism* 5 (1), 67–78.
Bandura, A. (2002) Selective moral disengagement in the exercise of moral agency. *Journal of Moral Education* 31 (2), 101–119.
Bargeman, B., Richards, G. and Govers, E. (2016) Volunteer tourism impacts in Ghana: A practice approach. *Current Issues in Tourism* 21 (13), 1486–1501.
Bear, G.G., Richards, H.C. and Gibbs, J.C. (1997) Sociomoral reasoning and behaviour. In G.G. Bear, K.M. Minke and A. Thomas (eds) *Children's Needs II: Development, Problems and Alternatives* (pp. 13–25). Bethesda: National Association of School Psychologists.
Behr, K.E. (2005) 'Same-as-difference': Narrative transformations and intersecting cultures in Harry Potter. *Journal of Narrative Theory* 35 (1), 112–132.
Bettelheim, B. (1976) *The Uses of Enchantment: The Meaning and Importance of Fairy Tales*. Harmondsworth: Penguin.
Binnendyk, L. and Schonert-Reichl, K.A. (2002) Harry Potter and moral development in pre-adolescent children. *Journal of Moral Education* 31 (2), 195–201.
Buckingham, D. (2000) *The Making of Citizens: Young People, News and Politics*. London: Routledge.
Campos-Soria, J.A., García-Pozo, A. and Sánchez-Ollero, J.L. (2015) Gender wage inequality and labour mobility in the hospitality sector. *International Journal of Hospitality Management* 49, 73–82.
Carey, B. (2003) Hermione and the house-elves: The literary and historical contexts of J.K. Rowling's antislavery campaign. In G.L. Anatol (ed.) *Reading Harry Potter. Critical Essays* (pp. 103–115). Westport: Praeger.
Chappell, D. (2008) Sneaking out after dark: Resistance, agency, and the postmodern child in JK Rowling's Harry Potter series. *Children's Literature in Education* 39 (4), 281–293.
Davidson, J.O.C. (2004) 'Child sex tourism': An anomalous form of movement? *Journal of Contemporary European Studies* 12 (1), 31–46.
Detert, J.R., Trevino, L.K. and Sweitzer, V.L. (2008) Moral disengagement in ethical decision making: A study of antecedents and outcomes. *Journal of Applied Psychology* 93 (2), 374–391.
Duffy, L.N., Pinckney, H.P., Benjamin, S. and Mowatt, R. (2018) A critical discourse analysis of racial violence in South Carolina, U.S.A: Implications for traveling while Black. *Current Issues in Tourism* 22 (19), 1–17.
Ewart-James, J. and Wilkins, N. (2015) The staff wanted Initiative: Preventing exploitation, forced labour and trafficking in the UK hospitality industry. In L. Waite, G. Craig, H. Lewis and K. Skrivankova (eds) *Vulnerability, Exploitation and Migrants. Migration, Diasporas and Citizenship* (pp. 256–268). London: Palgrave-Macmillan.
Fields, J. (2007) Harry Potter, Benjamin Bloom and the sociological imagination. *International Journal of Teaching and Learning in Higher Education* 19 (2), 167–177.

Franklin, N. (2006) The social dynamics of power and cooperation. In N. Mulholland (ed.) *The Psychology of Harry Potter* (pp. 169–174). Dallas: Benbella Books.

Gibson, D.M. (2007) Empathizing with Harry Potter: The use of popular literature in counselor education. *Journal of Humanistic Counseling, Education and Development* 46 (3), 197–210.

Gierzynski, A. (2014) How 'Harry Potter' shaped the political culture of a generation. *The Washington Post*. See https://www.washingtonpost.com/posteverything/wp/2014/08/19/how-harry-potter-shaped-the-political-culture-of-a-generation/?utm_term=.aa0014713ff9 (accessed 18 December 2018).

Green, M.C. (2006) Resisting social influence. Lessons from Harry Potter. In N. Mulholland (ed.) *The Psychology of Harry Potter* (pp. 299–310). Dallas: Benbella Books.

Guttentag, D.A. (2009) The possible negative impacts of volunteer tourism. *International Journal of Tourism Research* 11 (6), 537–551.

Jacobs, E. (2017) How J.K. Rowling built a $25bn business. See https://www.ft.com/content/a24a70a6-55a9-11e7-9fed-c19e2700005f (accessed 15 December 2018).

Jenkins, H. (2015) Cultural acupuncture: Fan activism and the Harry Potter alliance. In L. Geraghty (ed.) *Popular Media Cultures* (pp. 206–229). London: Palgrave Macmillan.

Kama, A., Ram, Y. and Mizrachi, I. (2017) A genuine gay-friendly city – The touristic benefits of LGBT inclusion. *Critical Tourism Studies Proceedings* 2017 (80), 1–2.

Kennedy, K.M. and Flaherty, G.T. (2015) The risk of sexual assault and rape during international travel: Implications for the practice of travel medicine. *Journal of Travel Medicine* 22 (4), 282–284.

Kidd, D. (2007) Harry Potter and the functions of popular culture. *The Journal of Popular Culture* 40 (1), 69–89.

Klonsky, E.D. and Laptook, R. (2006)'Dobby had to iron his hands, Sir!' – self-inflicted cuts, burns, and bruises in Harry Potter. In N. Mulholland (ed.) *The Psychology of Harry Potter* (pp. 189–203). Dallas: Benbella Books.

Knapp, N.F. (2003) In defense of Harry Potter: An apologia. *School Libraries Worldwide* 9 (1), 78–91.

Lee, K.J. and Scott, D. (2016) Racial discrimination and African Americans' travel behavior. *Journal of Travel Research* 56 (3), 381–392.

Li, B.L. (2018) Thailand taps tourists to fight human trafficking and keep the country smiling. See https://www.reuters.com/article/us-thailand-trafficking-tourism/thailand-taps-tourists-to-fight-human-trafficking-and-keep-the-country-smiling-idUSKBN1HO1D2 (accessed 16 December 2018).

Lubowiecki-Vikuk, A.P. and Borzyszkowski, J. (2016) Tourist activity of LGBT in European post-communist states: The case of Poland. *Economics & Sociology* 9 (1), 192–208.

Lyubansky, M. (2006) Harry Potter and the word that shall not be named. In N. Mulholland (ed.) *The Psychology of Harry Potter* (pp. 233–248). Dallas: Benbella Books.

McLennan, S. (2014) Medical voluntourism in Honduras: 'Helping' the poor? *Progress in Development Studies* 14 (2), 163–179.

Michopoulou, E., Darcy, S., Ambrose, I. and Buhalis, D. (2015) Accessible tourism futures: The world we dream to live in and the opportunities we hope to have. *Journal of Tourism Futures* 1 (3), 179–188.

Miller, L. (2017) Child sex tourism and its relationship to global human trafficking. *University of St. Thomas Journal of Law and Public Policy* 12 (1), 1–5.

Murakami, J.L. (2006) Mental illness in the world of wizardry. In N. Mulholland (ed.) *The Psychology of Harry Potter* (pp. 175–188). Dallas: Benbella Books.

Naficy, S.T. (2006) The werewolf in the wardrobe. In N. Mulholland (ed.) *The Psychology of Harry Potter*. Dallas: Benbella Books.

Neptune, R.E. (2016) Sexual exploitation of children and adolescents, human trafficking and mega sporting events: A case study from Brazil. *Transformation: An International Journal of Holistic Mission Studies* 33 (3), 218–224.

Nicholson, R. (2017) He who must not be named: How Harry Potter helps make sense of Trump's world. *The Guardian.* See https://www.theguardian.com/books/2017/mar/13/he-who-must-not-be-named-how-harry-potter-helps-make-sense-of-trumps-world (accessed 15 December 2018).

Nyanjom, J., Boxall, K. and Slaven, J. (2018) Towards inclusive tourism? Stakeholder collaboration in the development of accessible tourism. *Tourism Geographies* 20 (4), 675–697.

Palacios, C.M. (2010) Volunteer tourism, development and education in a postcolonial world: Conceiving global connections beyond aid. *Journal of Sustainable Tourism* 18 (7), 861–878.

Patrick, C.J. and Patrick, S.K. (2006) Exploring the dark side. Harry Potter and the psychology of evil. In N. Mulholland (ed.) *The Psychology of Harry Potter* (pp. 221–232). Dallas: Benbella Books.

Ram, Y. (2015) Hostility or hospitality? A review on violence, bullying and sexual harassment in the tourism and hospitality industry. *Current Issues in Tourism* 21 (7), 760–774.

Reisinger, Y. (2013) *Transformational Tourism: Tourist Perspectives.* Wallingford: CABI.

Reisinger, Y. (2015) *Transformational Tourism: Host Perspectives.* Wallingford: CABI.

Shresthova, S. and Brough, M.M. (2011) Fandom meets activism: Rethinking civic and political participation. *Transformative Works and Cultures* 10. See https://doi.org/10.3983/twc.2012.0303 (accessed 18 May 2020).

Simpson, K. (2004) 'Doing development': The gap year, volunteer-tourists and a popular practice of development. *Journal of International Development* 16 (5), 681–692.

Statista (2016) Have you read any of the Harry Potter books or seen any of the Harry Potter movies? See https://www.statista.com/statistics/549724/familiarity-with-harry-potter-books-movies-usa/ (accessed 17 December 2018).

Sylt, C. (2017) Harry Potter tour conjures up $435 million of revenue for Time Warner. Available from: https://www.forbes.com/sites/csylt/2017/12/06/harry-potter-tour-conjures-up-435-million-of-revenue-for-time-warner/#5577ad0269c1 (accessed 14 December 2018).

Tepelus, C.M. (2008) Social responsibility and innovation on trafficking and child sex tourism: Morphing of practice into sustainable tourism policies? *Tourism and Hospitality Research* 8 (2), 98–115.

Vezzali, L., Stathi, S., Giovannini, D., Capozza, D. and Trifiletti, E. (2015) The greatest magic of Harry Potter: Reducing prejudice. *Journal of Applied Social Psychology* 45 (2), 105–121.

Whitney, M.P., Vozzola, E.C. and Hofmann, J. (2005) Children's moral reading of Harry Potter. Are children and adults reading the same books? *Journal of Research in Character Education* 3 (1), 1–24.

8 Wildlife Tourism in 2150: Uplifted Animals, Virtual and Augmented Reality and Everything In-between

Giovanna Bertella

Chapter Highlights

- Scientific and science-fictional technological advancements trigger deep reflections.
- Technology can help solve long-debated challenges in wildlife tourism.
- The use of technology in tourism has important practical and ethical implications.
- Could the most important drivers for change in wildlife tourism be within us?

Introduction

This study presents and discusses a possible scenario for 2150. The scenario is based on the challenges of wildlife tourism in terms of animal protection and educational potential and on expected future technological advancements. The aim of the study is to offer some points of reflection for today's tourism scholars and practitioners.

Thinking about the future of tourism is a useful exercise that can help us become more aware of the connection between past, present and future (Yeoman & Postma, 2014). In this context, science fiction can be a valuable source of inspiration (Yeoman, 2012; Yeoman & Mars, 2012). Unlike fantasy, the imaginary aspect of science fiction is rationally rooted in plausible developments of current technologies. This study examines possible developments in biotechnology and virtual and augmented reality.

Science fiction can stimulate deep reflection, something that some scholars believe is lacking in wildlife tourism, especially in regard to ethics

(Fennell & Nowaczek, 2010; Yeoman & Postma, 2014). Through narrative, science fiction stories can engage us more than other types of communication. Science fiction concerns the impacts of science on our world and can therefore encourage us to reflect deeply on our responsibilities (Vint, 2008). Thus, this study aims to provoke reflections, not only about future developments in wildlife tourism, but also about our underlying view of wildlife and our relationship with animals and nature. Science fiction can enable us to look forward and also to look deeply into wildlife tourism.

This chapter begins by presenting the challenges of wildlife tourism addressed in this study and the drivers of change in a futuristic scenario. It continues by describing this future scenario in more detail. It then discusses certain elements related to future wildlife tourism and potential tools for education and behavioural change. Finally, some conclusions are drawn, including the author's opinion about a possible desirable future path.

Wildlife Tourism Today

Experiences that centre on the interaction of tourists with wild animals, such as wildlife watching tours, hunting tours and zoo visits, present several challenges (Newsome *et al.*, 2005; Reynolds & Braithwaite, 2001). These include the possible short- and long-term negative impacts on the focal wildlife and their habitats as well as the implicit message that wild animals are commodities and that nature is a playground for humans (Burns, 2015; Castley, 2016; Green & Giese, 2004). In the case of captive wildlife, these challenges also often include the limited effects of programs supporting environmental education and the recovery and breeding of endangered species (Carr & Cohen, 2011; Tribe & Booth, 2003). Tourism and leisure scholars have also discussed various positions on animal ethics and highlighted the extreme complexity of the issue (Carr & Young, 2018; Fennell, 2012; Lovelock, 2007).

Thinking about possible futures of wildlife tourism, it is also important to consider the issue of species losses. Ceballos *et al.* (2015: 15) observe that 'our global society has started to destroy species of other organisms at an accelerating rate, initiating a mass extinction episode unparalleled for 65 million years'. We humans are causing enormous environmental damages, compromising not only our own lives but also the lives of numerous animal species. This means that the future of wildlife tourism is compromised as well by the potential disappearance of some species and the extremely fragility of the remaining ones.

Bertella (2018a) argues that it is time to re-think wildlife tourism. She suggests that a new kind of wildlife tourism could emerge from critical consideration of the underlying anthropocentric assumptions of wildlife tourism and by replacing these assumptions with reflections based on the values of entangled empathy and care. A suggestion like this implicitly refers to a cultural change, something that, realistically, will take a long time – if it happens at all. Can we afford to wait?

Scientific and Science Fictional Drivers of Change

Factual and fictional science can be important sources of inspiration about what tourism might look like in the future. Taking an imaginative approach to wildlife tourism, Mackenzie Wright (2018) reflects on the possibility of cloning animals for safari zoos and sport hunting. Mackenzie Wright (2016) explores the latter type of tourism in a quite provocative way, imagining a future in which wealthy tourists hunt humans.

The current study takes a similarly imaginative approach that begins with technological advancements in biotechnology and in virtual and augmented reality. The next sections address scientific advancements and science fictional predictions for these two types of technology.

Biotechnology and animal uplifting

Biotechnology can be used to genetically change organisms; these changes can occur within and across species. Some studies suggest that animals' cognitive capacity can be enhanced through neurological modifications. Since the 1990s, these possibilities and projects exploring them have been debated. This debate pays particular attention to the ethical implications of such work, its underlying assumptions of human superiority, and possible regulations around biotechnology (Dvorsky, 2008; Ferrari, 2015; Parry, 2010).

In science fiction, the term 'uplift' refers to the use of biotechnology (and eventually other methods) to change animals into more sentient beings (Vint, 2008). *The Island of Dr Moreau* by H.G. Wells, written in 1896, is a milestone in science fiction on animal uplifting. Along with its movie adaptations, *The Island of Dr Moreau* has often been referenced in studies on topics such as animal ethics and scientific understanding (Jörg, 2003; Vint, 2007). The story concerns the shipwrecked Prendick, who lands on an island where a scientist, Dr Moreau, and his assistant experiment with the creation of human-like animals. In addition to satisfying the scientist's curiosity, the aim of these experiments is to create servants. These servants are visualised in the 1996 movie adaptation of the novel; Dr Moreau, played by Marlon Brando, has a team of helpers, varying from ape-like waiters to tiny ridiculed companions.

Animal uplifting is a central component of H.G. Wells' novel. Presumably, the author chose this topic because of his concern about the possible effects and the ethical aspects of the emerging (at that time) practice of animal vivisection. One striking aspect of this novel is the sensitivity and deep reflection of its view of animals. In particular, the novel blurs the line between animals and humans, along with the related assumptions of cultural dominance (Vint, 2008).

Another literary work of science fiction about animal uplifting is David Brin's *Uplift* series. In these books, uplifting experiments are the standard method by which any species increases in sentience. In Brin's universe, humans and dolphins have collaborative relations. The latter are fundamentally different to the ones described by H.G. Wells who appears to be very concerned about the tendency of humans to view themselves as god-like compared to nature and animals (Vint, 2007).

Some of the most popular science fiction stories of animal uplifting are the films, books, television series and comics inspired by Pierre Boulle's 1963 novel *Planet of the Apes*. Here, humans and apes are equally intelligent and interact in collaborative and conflictual ways. Rupert Wyatt's 2011 film *Rise of the Planet of the Apes* addresses the issue of co-existence and raises interesting questions about animal rights and personhood (Hamilton, 2016).

Virtual and augmented reality

Virtual reality (VR) can be defined as an electronic simulation of an environment that users can enter and explore which creates real-time experiences using one or more of the user's five senses (Guttentag, 2010). Closely related to VR is augmented reality (AR). Through AR technology, synthetic images can be superimposed over real images so that the environment surrounding the user is a combination of reality and artificiality (Wu *et al.*, 2013). VR and AR technology can create alternative, enriched realities.

Schott (2015) examines the potential of VR and AR to educate people about complex phenomena such as sustainability. VR and AR allow digital immersion and active learning, and 'virtual fieldtrips' can be important arenas where students – and tourists – engage with their minds and their actions. Based on established pedagogies of active and experiential learning (for example Dewey, 1963), Schott argues that such technology-based experiences can enhance understanding and lead to lifelong learning.

Some scholars have also investigated this technology as a potentially important component of the tourism experience. Here, the main focus has been on VR and the close interplay between tourists' bodies and technological devices as a way to influence and shape tourists' attitudes and behaviours (Guttentag, 2010; Tussyadiah *et al.*, 2017, 2018). VR also has potential at the destination level and could be used in strategies to attract and retain tourists (Marasco *et al.*, 2018).

So far, few studies have focused on AR. Jung *et al.* (2015) explore the use of AR in theme parks, and Kourouthanassis *et al.* (2015) investigate the emotional and behavioural effects of a mobile AR travel guide. He *et al.* (2018) review recent scholarly contributions and argue that AR can enhance tourists' experiences, improving their attitudes and behavioural intentions.

Science fiction contributions inspired by VR and/or AR are quite numerous as the idea of parallel and/or alternative worlds accessible through technology is one of the main themes of this genre. Many science fiction characters use technology to travel through time, sometimes in utopian universes, sometimes in alternative universes where past, present and future overlap (Tegmark, 2003). Some examples are the 1895 novel *The Time Machine* by H.G. Wells and the film trilogy (1985–1990) *Back to the Future* by Robert Zemeckis (Fhlainn, 2016).

More recently, some episodes of the TV series *Black Mirror* (2011–2017) by Charlie Brooker describe the possibility of changing a person's perception of reality through implants; the AR device becomes a part of the person. In 'Arkangel', a device is implanted in a child's brain to monitor her actions and pixelate all images that could cause fear or stress. In 'Men Against Fire', the soldiers' neural implants provide them with data via AR, enhancing sensory stimuli and turning them into effective killers of those who disagree with the dominant political system.

Wildlife Tourism in 2150

In 2150, many wildlife tourism practitioners and scholars will agree that the sector's challenges of wildlife protection and environmental education have been overcome and that these successes are due to the ability to uplift wild animals.

Initially, animal uplifting projects were performed secretly in laboratories located in isolated, peripheral areas. At the end of the 2090s, the first encouraging results of these experiments were made public. This marked the beginning of a heated debate over the ethics of such experiments as well as practical issues such as human safety. The anti-uplifting movement took various forms: academic discussions, non-violent street demonstrations, undercover investigations of the laboratories, and vandalism.

Wildlife tourism was one of the first sectors to take advantage of this development. Soon, all wildlife tourism encounters involved uplifted animals who behaved in line with the tourists' expectations and desires. These encounters occurred in large areas accessible only by tourism operators or in enclosed areas such as zoos. Tourists participating in wildlife tours with guaranteed views and sometimes guaranteed interaction with the animals reported high levels of satisfaction with the experience. Zoos and aquaria regained their popularity as institutions of entertainment and education. Social media overflowed with pictures of happy tourists close to wild animals, sometimes feeding, petting and playing with them. Wildlife hunting tourism was also affected by the new technology: the animal employees were trained to appear shy and sometimes dangerous but to always succumb to the hunters.

Some anti-uplifting protests targeted the new wildlife tourism sector. In 2100, the sector responded to critics by devoting part of its profits to protecting natural wild animals. These animals, especially endangered ones, could continue to live naturally, protected from human presence in large preserves and sanctuaries financed by the new wildlife tourism sector.

It is now 2150. The anti-uplifting movement has lost its strength. The vast majority of people have become accustomed to the idea of using biotechnology to turn animals into employees. Uplifted animal employees enable the engaging, educational and memorable wildlife tourism experiences. Recent studies show that the new form of wildlife tourism is a significant contributor to shifts toward greener, more animal-friendly attitudes and behaviours among tourists.

What the general public doesn't know is that, a decade after the first experiments, all uplifting projects were discontinued. After initial enthusiasm, the international scientific community agreed that the ethical problems of the experiments and of the possible developments could not be dismissed. The experiments were then replaced by AR research projects. Although the tourism industry refers to the uplifting process to explain the behaviour of wild animals at tourist attractions, no uplifted servant animal has ever been used, in tourism or in any other sector.

Wildlife tourism is flourishing, and it is built on a lie.

Learning from the Future

The 2150 scenario described above raises several questions that could encourage reflections about today's wildlife tourism and possible future developments. These questions and related considerations address: (a) the use of wild animals in tourism; (b) the adoption of the precautionary principle and the collaboration between academia and industry; and (c) the staging of wildlife experiences.

Wild animals in tourism

In 2150, tourists accept that uplifted animals are created and used as employees – more specifically, as actors. A first point of reflection is to consider how far this scenario is from today's wildlife tourism.

Some animals in today's wildlife tourism are tame enough to be approached by and, sometimes, to interact with humans, including trainers, guides and tourists. This concerns wild animals in captivity, whose training encourages them to behave in specific ways, sometimes performing shows for the tourists. This practice is quite widespread, but it is also not immune from criticism by scholars, practitioners and animal activists. An example is the use of captive cetaceans such as dolphins and pinnipeds such as sea lions and walruses in aquaria (Bertella, 2018b; Desmond, 1999).

It is not impossible to imagine that, just as modern tourists accept and appreciate tamed, trained wild animals, tomorrow's tourists might accept and appreciate uplifted animals working as actors. On the other hand, animal uplifting differs from modern training methods as it involves enhancing animals' cognitive capacity. More precisely, it would make their cognition closer to ours, facilitating inter-species communication. How would tourists view such animals? What would determine their attitudes and behaviour towards uplifted animals? Would it be a question of how much humanity were 'injected' into the wild animals? Or a question of the animals' cognitive capabilities? Where would a tourist draw the line between human and animal?

It is not the purpose of this chapter to delve into these questions, although they have important ethical implications. Still, it might be appropriate to reflect on the power mechanism underlying wildlife tourism, especially its lethal forms. For example, rather than provoking deep reflections in the tourists, would the possibility of uplifted animals ultimately be reduced to a question of power? As mentioned in the third section, Mackenzie Wright (2016) imagines a future in which wealthy tourists hunt humans. Would the tourists of 2150 enjoy slaughtering human-like animals? Would they enjoy such a use of power? While Mackenzie Wright's (2016) scenario highlights discrimination across social classes, the scenario described in this chapter considers discrimination based on species. In *Rise of the Planet of the Apes*, the chimpanzee Caesar leads a rebellion against abusive human oppressors. A future in which wildlife tourism uses uplifted animals might become the battlefield for the rights of a new marginalised class.

Finally, the use of uplifted animals in tourism, particularly in close encounters and interactions, would imply the possibility of communication. Would all uplifted animals follow the scripts that their human employers had designed for them? Would free communication occur? If so, what would happen? Would we be willing to listen? If so, what are the chances that we might not like what the animals could tell us?

Precaution and collaboration between academia and industry

In the 2150 scenario described here, uplifting projects meet some criticism and, eventually, are discontinued by the international scientific community. This decision stems from the issue's complexity, which is considerable, especially from an ethical point of view. The precautionary principle discussed by Fennell and Ebert (2004) could be useful for future-focused developments in tourism that relate to sustainability challenges such as environmental impacts. Precaution could also be the guiding principle for ethical issues in tourism. Bertella (2018a) applies this idea to the case of close encounters with cetaceans. She argues that, if we accept the possibility that cetaceans might have extremely sophisticated cognitive capacities as well as emotional lives to the extent that it is plausible to talk about cetacean culture, then we should reconsider the way we conduct and promote close encounters with them.

Several questions arise here. How do today's scientists view the precautionary principle? To what extent is this principle perceived as supporting sustainable, responsible development? To what extent is it perceived as a conservative barrier to development? The scientist from *The Island of Dr Moreau* would probably reply that scientific curiosity cannot be held back by caution. What would today's scientists in the field of tourism say? And are scientists' perceptions of precaution in line with those of wildlife tourism practitioners?

The latter question leads to the roles of academia and industry in shaping wildlife tourism. Studies on academic engagement suggest that numerous individual, organisational and context-related factors influence the extent to which academics seek contact and collaborate with industry actors (Perkmann *et al.*, 2013). In the field of wildlife tourism, the question is: To what extent and in what ways are modern scholars from the social and natural sciences engaged in the development of this sector?

It is the author's experience that some practitioners have some clear expectations that scholars should play a more active role. After a whale-watching seminar where the author gave a lecture, one practitioner commented explicitly on this topic, saying, 'It's about time that academics enter the field!' This is in line with the expected role of universities in the co-creation of sustainable solutions for industries and communities (Rubens *et al.*, 2017).

Once these expectations are recognised, the next question is to what extent today's academics are willing and prepared to enter the practical field. Do we have an open, collaborative attitude towards practitioners? Moreover, should individual academics freely determine their own level of engagement, or should it be encouraged through tailored PhD programs and considered as important as typical academic achievements such as publications and titles?

Staged wildlife experiences

The third topic raised by the 2150 scenario concerns the staging of wildlife experiences enhanced by technology, including animal uplifting and VR and AR.

Although animal uplifting did not actually occur, the tourists of 2150 believed and gladly accepted that their experiences were staged by human directors and performed by animal actors. In this 2150 scenario, animal encounters are designed to satisfy any desire a tourist might have, and wildlife tourism is reported to have important educational impacts. Such education might reinforce a utilitarian and anthropocentric worldview, which could be criticized (Bertella, 2018a). This is an issue in wildlife tourism experiences in 2150 and also today, especially when captive wild animals are used as performers. What is the underlying, not always explicit, message of such experiences?

Also, in the reality of the 2150 scenario, wildlife tourism experiences are staged using not animal uplifting but AR. Like the neural implants in *Black Mirror*, technological devices could change tourists' perceptions and enhance memorable, meaningful experiences. The considerations discussed around uplifting regarding staging and education are still valid in this case. In addition, since the adoption of AR was kept secret in the scenario, it is worthwhile to consider the appropriateness of misleading tourists in order to promote positive change. To begin, it could be argued that lying for the greater good, although ethically debatable, is acceptable in an emergency. For the long-debated risks and challenges of wildlife tourism, technology could be an effective solution – maybe the only one.

On the other hand, the secret use of technology could be viewed as a shortcut solution to a problem that would be better solved by education. This raises another question: Would something be lost in the attempt to reach the desired goals faster? As mentioned in the introduction, how wildlife tourism could be re-conceptualised and practiced in the future is, ultimately, a matter of cultural change, something that is usually considered achievable through education. In the 2150 scenario, technology does not support education; its use could be called mind manipulation. Could the secret use of AR be counterproductive? Would it enhance or discourage critical thinking? Can such a use of a sophisticated technology be an alternate solution if cultural change is too difficult to achieve? If so, who would be in charge of setting the desired goals?

Concluding Arguments

This chapter has described a futuristic scenario that began with a recognition of the need to re-think wildlife tourism. The future of wildlife tourism might include the use of technology to enhance tourist experiences; this could have a considerable impact on our society and on wildlife. These impacts could be positive, such as wildlife protection and environmental responsibility. The use of technology might also lead to undesirable situations, such as the creation of a marginalised class of human-like animals. The use of technology in wildlife tourism might employ processes that are problematic from an ethical point of view, including animal manipulation and misleading practices.

This chapter has raised many questions, showing that imagining futuristic scenarios based on scientific and science fictional technological advancements can trigger important reflections about the future and also the present. Moreover, when the issues under consideration involve animals, these reflections tend to engage us deeply. Behind the relationship between humans and animals – which is the core of wildlife tourism – lies the unresolved question about the blurred line between 'us' (human animals) and 'them' (non-human animals).

It is the author's opinion that a desirable path for the development of wildlife tourism is a path surrounded by encounters that adopt technological solutions such as VR and AR in various ways. Some encounters might completely rely on this technology and help solve problems related to animal welfare, environment protection and overtourism. On the other hand, other encounters could adopt technology to design human–animal interactions from a less anthropocentric perspective. A critical use of VR and AR might help shed some light on the aforementioned question of 'us' and 'them'. Tourists might not only meet wild animals in technology-supported settings, but also feel how the animals perceive these encounters and, more in general, the world.

It can be noted that this kind of critical use of technology would imply a cultural change and would, in essence, be based on a reconceptualisation of animals. In his book *The Outermost House*, naturalist and writer Henry Beston uses his knowledge and sensibility to reflect on the animal world, a world so close to us and, at the same time, mysterious and fascinating as the worlds created in science fiction. According to him, this reconceptualisation might consist of the realisation that '(w)e need another and a wiser and perhaps a more mystical concept of animals. Remote from universal nature, and living by complicated artifice, man in civilisation surveys the creature through the glass of his knowledge and sees thereby a feather magnified and the whole image in distortion. We patronise them for their incompleteness, for their tragic fate of having taken form so far below ourselves. And therein we err, and greatly err. For the animal shall not be measured by man. In a world older and more complete than ours they moved finished and complete, gifted with extensions of the senses we have lost or never attained, living by voices we shall never hear. They are not brethren, they are not underlings; they are other nations, caught with ourselves in the net of life and time, fellow prisoners of the splendour and travail of the earth'. (Beston, 1988: 24–25)

References

Beston, H. (1988) *The Outermost House: A Year of Life on the Great Beach of Cape Cod*. New York: St. Martin's Press.
Bertella, G. (2018a) Sustainability in wildlife tourism: Challenging the assumptions and imagining alternatives. *Tourism Review* 74 (2), 246–255. See https://doi.org/10.1108/TR-11-2017-0166 (accessed 18 May 2020).
Bertella, G. (2018b) An eco-feminist perspective on the co-existence of different views of seals in leisure activities. *Annals of Leisure Research* 21 (3), 284–301.
Boulle, P. (1963) *Planet of the Apes*. New York: Vanguard Press.
Brin, D. (2018) The uplift novels. See http://www.davidbrin.com/books.html (accessed 12 December 2018).

Burns, G.L. (2015) Animals as tourism objects: Ethically refocusing relationships between tourists and wildlife. In K. Markwell (ed.) *Animals and Tourism: Understanding Diverse Relationships* (pp. 44–59). Bristol: Channel View Publications.

Carr, N. and Cohen, S. (2011) The public face of zoos: Images of entertainment, education and conservation. *Anthrozoös* 24 (2), 175–189.

Carr, N. and Young, J. (2018) *Wild Animals and Leisure. Rights and Wellbeing*. New York: Routledge.

Castley, J.G. (2016) Wildlife tourism. In J. Jafari and H. Xiao (eds) *Encyclopaedia of Tourism* (pp. 1020–1021). Cham: Springer International Publishing.

Ceballos, G., Ehrlich, P.R., Barnosky, A.D., García, A., Pringle, R.B. and Palmer, T.M. (2015) Accelerated modern human-induced species losses: Entering the sixth mass extinction. *Sciences Advances* 1 (5). See https://advances.sciencemag.org/content/1/5/e1400253 (accessed 18 May 2020).

Desmond, J. (1999) *Performing Nature: Shamu at Sea World*. Chicago: University Chicago Press.

Dewey, J. (1963) *Experience and Education*. New York: Simon & Schuster.

Dvorsky, G. (2008) All together now: Developmental and ethical considerations for biologically uplifting nonhuman animals. *Journal of Evolution and Technology* 18 (1), 129–142.

Fennell, D. (2012) *Tourism and Animal Ethics*. London: Routledge.

Fennell, D. and Ebert, K. (2004) Tourism and the precautionary principle. *Journal of Sustainable Tourism* 12 (6), 461–479.

Fennell, D. and Nowaczek, A. (2010) Moral and empirical dimensions of human–animal interactions in ecotourism: Deepening an otherwise shallow pool of debate. *Journal of Ecotourism* 9 (3), 239–255.

Ferrari, A. (2015) Animal enhancement: Technovisionary paternalism and the colonisation of nature. In S. Bateman, J. Gayon, S. Allouche, J. Goffette and M. Marzano (eds) *Inquiring into Animal Enhancement: Model or Countermodel of Human Enhancement? Health, Technology and Society* (pp. 13–33). London: Palgrave Pivot.

Fhlainn, S.N. (2016) 'There's something very familiar about all this': Time machines, cultural tangents, and mastering time in H.G. Wells's The Time Machine and the Back to the Future trilogy. *Adaptation* 9 (2), 164–184.

Green, R. and Giese, M. (2004) Negative effects of wildlife tourism on wildlife. In K. Higginbottom (ed.) *Wildlife Tourism. Impacts, Management and Planning* (pp. 81–97). The Gold Coast: Common Ground Publishing Pty Ltd and Cooperative Research Centre for Sustainable Tourism.

Guttentag, D.A. (2010) Virtual reality: Applications and implications for tourism. *Tourism Management* 31 (5), 637–651.

Hamilton, S.N. (2016) 'Human no like smart ape': Figuring the ape as legal person in Rise of the Planet of the Apes. *Law and Humanities* 10 (2), 300–321.

He, Z., Wu, L. and Li, X.R. (2018) When art meets tech the role of augmented reality in enhancing museum experiences and purchase intentions. *Tourism Management* 68, 127–139.

Jung, T., Chung, N. and Leuea, M.C. (2015) The determinants of recommendations to use augmented reality technologies: The case of a Korean theme park. *Tourism Management* 49, 75–86.

Jörg, D. (2003) The good, the bad and the ugly – Dr. Moreau goes to Hollywood. *Public Understanding of Science* 12, 297–305.

Kourouthanassis, P., Boletsis, C., Bardakia, C. and Chasanidou, D. (2015) Tourist responses to mobile augmented reality travel guides: The role of emotions on adoption behaviour. *Pervasive and Mobile Computing* 18, 71–87.

Lovelock, B. (2007) *Tourism and the Consumption of Wildlife. Hunting, Shooting and Sport Fishing*. London: Routledge.

Mackenzie Wright, D.W. (2016) Hunting humans: A future for tourism in 2200. *Futures* 78–79, 34–46.

Mackenzie Wright, D.W. (2018) Cloning animals for tourism in the year 2070. *Futures* 95, 58–75.

Marasco, A., Buonincontri, P., van Niekerk, M., Orlowski, M. and Okumus, F. (2018) Exploring the role of next-generation virtual technologies in destination marketing. *Journal of Destination Marketing and Management* 9, 138–148.

Newsome, D., Dowling, R.K. and Moore, S.A. (2005) *Wildlife Tourism*. Clevedon: Channel View Publications.

Parry, S. (2010) Interspecies entities and the politics of nature. *The Sociological Review* 58 (1), 113–129.

Perkmann, M., Tartari, V., McKelvey, M., Autio, E., Broström, A., D'Este, P., Fini, R., Geuna, A., Grimaldi, R., Hughes, A., Krabel, S., Kitson, M., Llerena, P., Lissoni, F., Salter, A. and Sobrero, M. (2013) Academic engagement and commercialisation: A review of the literature on university–industry relations. *Research Policy* 42 (2), 423–442.

Reynolds, P.C. and Braithwaite, D. (2001) Towards a conceptual framework for wildlife tourism. *Tourism Management* 22 (1), 31–42.

Rubens, A., Spigarelli, F., Cavicchi, A. and Rinaldi, C. (2017) Universities' third mission and the entrepreneurial university and the challenges they bring to higher education institutions. *Journal of Enterprising Communities: People and Places in the Global Economy* 11 (3), 354–372.

Schott, S. (2015) Digital immersion for sustainable tourism education: A roadmap to virtual fieldtrips. In G. Moscardo and P. Benckendorff (eds) *Education for Sustainability in Tourism: A Handbook of Processes, Resources, and Strategies* (pp. 213–228). Berlin: Springer-Verlag Berlin Heidelberg.

Tegmark, M. (2003) Parallel universes: Not just a staple of science fiction, other universes are a direct implication of cosmological observations. *Scientific American* 288, 40–51.

Tussyadiah, I.P., Jung, T.H. and tom Dieck, M.C. (2017) Embodiment of wearable augmented reality technology in tourism experiences. *Journal of Travel Research* 57 (5), 597–611.

Tussyadiah, I.P., Wang, D., Jung, T.H. and tom Dieck, M.C. (2018) Virtual reality, presence, and attitude change: Empirical evidence from tourism. *Tourism Management* 66, 140–154.

Tribe, A. and Booth, R. (2003) Assessing the role of zoos in wildlife conservation. *Human Dimensions of Wildlife* 8 (1), 65–74.

Yeoman, I. (2012) *2050 – Tomorrow's Tourism*. Bristol: Channel View Publications.

Yeoman, I. and Mars, M. (2012) Robots, men and sex tourism. *Futures* 44, 365–371.

Yeoman, I. and Postma, A. (2014) Developing an ontological framework for tourism futures. *Tourism Recreation Research* 39 (3), 299–304.

Vint, S. (2007) Animals and animality from the island of Moreau to the uplift universe. The yearbook of English studies. *Science Fiction* 37 (2), 85–102.

Vint, S. (2008) 'The animals in that country': Science fiction and animal studies. *Science Fiction Studies* 35 (2), 177–188.

Wells, H.G. (1996) *The Island of Dr. Moreau*. New York: Modern Library.

Wells, H.G. (2018) *The Time Machine*. London: Legend Press.

Wu, H.-S., Lee, S., W.-Y., Chang, H.-Y. and Liang, J.-C. (2013) Current status, opportunities and challenges of augmented reality in education. *Computers and Education* 62, 41–49.

9 Tears in the Rain: Tourism in the World of *Blade Runner* and *Total Recall*

Peter Bolan

Chapter Highlights

- The parallels between *Blade Runner*, *Total Recall* and how tourism is evolving.
- Digital connectivity, tourist tastes and virtual experiences.
- Benefits of AI (artificial intelligence) versus threats to tourism and society.
- Lessons for tomorrow's tourism from science fiction influenced technology.

Introduction

The worlds portrayed in the films *Blade Runner* and *Total Recall*, based on the literary work of renowned science fiction author Philip K. Dick allow us to experience realms populated by replicants (artificial humans) and memory implants of experiences and places that were never actually visited. Such works highlight the inherent dangers of such advances in technology as well as the benefits. In our own present world, we are already seeing airlines and hotels experimenting with robots for check-in, online chatbots to deal with customer queries, keyless check-in through smartphones in hotels, the increasing use of VR (virtual reality) and AR (augmented reality), developments in AI (artificial intelligence) and increasing connectivity among myriad devices (dubbed the Internet of Things – 'IOT'). In *Total Recall* (based on the novel 'We can remember it for you Wholesale') the character Douglas Quaid partakes in the process of having a vacation to Mars implanted directly into his memory. How far are we from such a phenomenon where we cross the line from technology helping to market and promote destinations to actually replacing them with something completely virtual and yet completely lifelike. In *Blade Runner* the creation of artificial humans called replicants has gone too far and the dangers they present are considered so great they have to hunted down and destroyed. Already we are seeing the threat that robots, AI and related digital advances are becoming to people's jobs and livelihoods. How far will this go, even in an industry like tourism and hospitality that has always been a people industry, we are already seeing such technology being developed and embraced. This chapter explores how tourism is evolving via technology through the lens of the science fiction films and literary works *Blade Runner* and *Total Recall* and what lessons we must learn from these works to ensure tomorrow's tourism industry thrives and benefits rather than succumbs to the dangers we could be exposed to.

Tourism and Technology in the Early 21st Century

When it comes to holiday and destination choice the core influencing factors such as climate, visitor attractions, good beaches, activities etc. are well known. That traditional picture has already changed however. Increasingly, it's about connectivity through digital. The tourist wants to stay connected through their smartphones, tablets and a myriad of other digital devices. That has become the new modern essential for the holidaymaker. Tourists are using their smartphone to check their itineraries, for researching local attractions, looking up local events, accessing maps, checking out hotel and restaurant reviews (often in-trip these days and no longer up front). Increasingly we are seeing the use of mobile apps to also provide interpretive information at visitor attractions and event venues – and through advances in AR and VR offering a more immersive experience for the user. The tourist also wants to share their experiences, their images and increasingly video through social media. A recent UK-based survey (Deloitte, 2017) found that as many as one in three tourists place access to Wi-Fi as the most desired 'comfort' while on an international trip. In the same survey a staggering 54% stated they would ask a restaurant if it provided Wi-Fi before deciding to eat there. Technology and digital connectivity have become the latest 'must-have' overriding other considerations for many tourists.

This is giving rise to 'smart' facilities: smart hotels, smart restaurants, smart visitor attractions, smart event venues etc. In fact, we need to start moving much more seriously towards this on a larger scale with smart resorts and smart cities. This is now required to meet the wants and needs of today's tourist. On the tourism provider side of the equation, it gives businesses a strong vantage onto visitor behaviour that is in many respects unparalleled in terms of customer touch-points and on-site evidence of traveller sentiment. Indeed, the consumer, even in their own home, is becoming increasingly comfortable conversing with AI programs and personas such as Alexa, Siri and Cortana and allowing such artificial personas to make choices and suggestions on what they do, what they purchase and where they travel to. Trends in digital transformation in such ways are only the beginning and will strongly influence tourist behaviour and in time the very nature of tourism itself.

What we have all seen portrayed in science fiction has in many respects now become reality in our world and the pace of development and change occurring means this will continue to be the case, influencing many industries, and tourism in particular. The literary works of the science fiction author Philip K. Dick (and the movies derived from these such as *Blade Runner* and *Total Recall*) are centrally concerned with the question of what is real. In other words, a perspective on the question 'what is fake?' and, if you can make fake seem authentic enough, does it matter? In other words, where does authenticity lie within this. What is authentic? What does it mean to the consumer and to the tourist?

Authenticity in Tourism: What is Real?

It was arguably MacCannell's influential 1973 paper on staged authenticity and social space in tourist settings that gave impetus to the topic of authenticity in tourism and set the scene for subsequent studies. He proposed that '... touristic consciousness is motivated by its desire for authentic experience' (1973: 101). MacCannell felt that modern society (at least in the 1970s when his paper was written) was in-authentic and to an extent alienating and therefore served to drive tourists in search of the authentic – a fruitless quest as MacCannell argues because he believes that all that exists out there is staged authenticity, that the tourist can never really find a truly authentic place or experience.

Steiner and Reisinger (2006) suggest that as part of a postmodern society we constantly search for stimulation through events and images. Butler (1990), almost two decades ago, stated that we may be entering an era where people's views and knowledge of the world is based on something inherently false that they have gleaned through various media forms such as movies and fictional literature (such as the science fiction works *Blade Runner* and *Total Recall*). Researchers such as Jansson (2007: 5) believe that all forms of media, especially digital media '... influence perceptions of place, distance, sociality, authenticity, and other pre-understandings that frame tourism'.

According to MacCannell (1973: 590), tourists are motivated by the need for experiences more profound than those associated with the '... shallowness of their everyday lives'. Such is the central tenant of the film *Total Recall* (based on the novel 'We can remember it for you Wholesale'), where people have 'fake' vacation experiences implanted directly into their memory. As such they then remember their holiday as if they had actually lived and experienced it. Such blurring of realities through advancing technology raises all kinds of questions on where we may be heading.

Blade Runner: Replicants, AI, Holograms and Flying Cars

In the film *Blade Runner*, the main protagonist Roy Batty (a replicant played by Rutger Hauer) delivers the famous line 'I've seen things you people can't possibly imagine'. When the film was released in 1982 it presented a vision of the future that indeed showcased things most viewers could 'only' imagine. From replicants (artificial humans) to AI holograms and the use of flying cars to beat congestion and overcrowding on the ground (for those that could afford them). As we near the third decade of the 21st century, some of these aspects are now actually with us and others may not be far away.

According to Iwashita (2006: 59) most tourist-generating nations are what can be termed '… post-industrial, postmodern, globalised societies in which media dominate people's everyday life by providing a vast amount of information, representations and images of the world on a global scale'. We see this everywhere in *Blade Runner*. Huge neon-lit moving adverts (hundreds of feet high) on the side of buildings, some of the adverts changing in response to people as they pass by (tailored to their individual preferences and tastes). The static billboards we would have been used to in the early 1980s have given way to this kind of advertising in our own world. Companies such as Amazon and Facebook track people's online behaviour and preferences relentlessly, utilising such data to tailor or 'personalise' the adverts, news and products that the consumer sees.

What has become known as 'smart tourism' draws together many aspects of what is under examination here. Smart tourism is all about the increasing reliance of tourism destinations, their industries and their tourists on emerging forms of digital technology that allow for massive amounts of data to be transformed into value propositions (Gretzel *et al.*, 2015). Recent studies in smart tourism recognise a number of key aspects of how technology can be employed to enrich traveller experiences, for example through experience co-creation (Neuhofer, 2016), gamification (Xu *et al.*, 2013), context-aware services (Lamsfus *et al.* 2015) and AR (Disztinger *et al.*, 2017). We see much of this 'smartness' manifested in the realm of digital connectivity. This IOT allows all of our devices, gadgets and appliances to be connected and share information in today's world, from our smartphone to our car, from our tablet to our television, from our laptop computer to our home heating system. Everything can be 'joined' and use that data so we can control things from a distance without even being there, or increasingly that such data can control us, making suggestions for what we do, what we purchase and so on.

When we converse digitally (especially with a business) often that is no longer with a human being. Chatbots have now become commonplace in tourism. They are essentially algorithms powered by the latest AI technology, which can mimic a customer care member of staff to answer specific customer queries and problems. They even operate through the most popular social media and mobile communication platforms such as Facebook Messenger, WhatsApp and Skype as well as dedicated online business sites. This can save businesses money (normally spent on human staff) while at the same time enabling quick responses to the customer on a 24/7 basis. Something today's tourist expects through digital communication. Such digital-human interfacing common in the world of *Blade Runner* has now entered our world too.

Chatbots not only replace traditional customer support, but can also drive sales as well. Using access to huge amounts of data the AI bot can suggest a last-minute upgrade offer or cross-sell other related travel products such as insurance, hotel rooms etc. Since the chatbot already knows the client's preferences and accepted price ranges, it can actually create better offers than human agents who usually don't have access to such a wealth of data. Chatbots can also provide a personalised one-to-one experience with the customer every time, without showing any signs of irritation or growing tired of the client's indecisiveness. Very much on the surface a positive win for both tourists and tourism provider, but with the downside of job-losses in terms of how such aspects have been handled traditionally by human agents. Could this be the start of replicants taking over human tasks and jobs just as in *Blade Runner*?

Advances in holographic technology may provide yet another avenue in creating more immersive experiences for tourists through technology. In *Blade Runner* holographic technology abounds and is common place in everyday life. Again, we now see this echoed in our own world as the use of holograms continues to grow and develop. In 2018, some 500 residents of Brussels were turned into virtual 3D statues using holographic imaging technology and put on public exhibition for a major tourism campaign. Following the 2016 terrorist attacks it was felt that portraying the friendly, open vibe of the city's population in such a way would help welcome visitors back to the city (Gianatasio, 2018). China has also recently showcased developments, utilising the technology in Shanghai to create a landscape featuring a holographic moon overhead a myriad of holographic people and holographic creatures of all kinds to an audience of than 700,000 visitors to the city (Chen, 2018). On the entertainment side we have seen holograms of Elvis, Michael Jackson and a host of other performers no longer with us, seemingly brought back to life to perform on stage as holograms in highly realistic form. What science fiction works like *Blade Runner* once made fantastical and out of reach have now come to pass. As holographic technology develops further, we will see in the next decade or two a level of realism that will fundamentally change the nature of how we interact with the world around us, with profound implications for tourism.

Uber held its second annual flying taxi conference in Los Angeles in May 2018. Ironically in the very city where *Blade Runner* is set. Prototypes for the these flying cars already exist and a further myriad of concept designs have flooded the market from companies as diverse as Tesla to Rolls Royce, Boeing to Bell (formerly Bell Helicopters). Businesses such as Uber (which have taken the taxi world by storm in their harnessing of digital technology) are helping to fuel the race to create proper fully functioning and operational flying cars for passenger transportation, to avoid the huge issues of congestion and overcrowding of conventional vehicles on surface roads (just as in the Los Angeles of *Blade Runner*). Through such recent conferences they are seeking partners that can meet the necessary technological specifications – electric-powered, minimal noise, with vertical take-off and landing capabilities, as well as a company that can scale production to build tens of thousands of vehicles to meet the demand of Uber's on-demand digital service. What was once pure science fiction has now almost become reality for tourists and business commuters and will fundamentally change the nature of transportation in some of the world's busiest cities.

Eating habits also now reflect what was predicted in movies such as *Blade Runner*. In the film, what is essentially a huge, heavily urbanised, sprawling, bleak vision of the future, street food is the choice for many, and is seen everywhere. Indeed, when we first meet Harrison Ford's character, Rick Deckard, he is eating at an outdoor Japanese noodle bar (which is part of whole array of similarly themed street food outlets in an overcrowded part of the city). In the almost 40 years since the film's release, such street food has now become the choice for many travellers who want to experience authenticity of food and drink in the destinations they visit. So, we see the penchant for street food featured in science fiction has now indeed come to pass in our own reality, especially with tourists.

Total Recall: Vacation Memory Implants, Robots and Self-driving Cars

In *Total Recall*, it has become common place for people to visit a company like Rekall Inc. that plants false memories into people's brains, in order to experience the thrill of various holidays and vacation experiences that they never actually had. A common theme of Philip K. Dick's literary work (on which the film is based) being what is reality and what is fake? Where does authenticity actually lie? If the fake seems authentic enough, does it become real? In essence here we are in the realms of what can be termed hyper-reality.

Hyper-reality can be seen as a condition in which 'reality' has been replaced by simulacra. The notion of hyperrealism can be seen as a symptom of postmodern culture. Two of the most prominent hyperrealists have been Jean Baudrillard and Umberto Eco. In 'Simulacra and Simulation' (1995) following on from his previous work (1973), Baudrillard argues that our postmodern culture is a world of signs that have made a fundamental break from reality. As such Baudrillard postulates that various forms of media abandon 'the real' for the hyper-real by presenting an increasingly real simulation of a comprehensive and comprehendible world (just as in the memory implants of *Total Recall*).

Authors such as Iwashita (2006: 59) strongly support the work of Eco (1990), Butler (1990), Baudrillard (1995) and Tzanelli (2004) on such concepts stating 'It can be argued that it is those media representations and images that people actually consume rather than realities...through which they understand the world'. In other words, people's notion of what is real is increasingly formed through media and digital media in particular. Today's global media culture '... enables people to travel mentally and emotionally without moving in physical geography' (Jansson, 2002: 430). What Jansson is saying is that people can travel to other places without the need to leave their home area and they can do so primarily through media forms such as film, television and the internet (through increasing levels of connectivity across multiple devices and platforms).

Another feature of society, travel and work in *Total Recall* is the prevalence of robots. In our own reality, the Henn-na Hotel in the city of Sasebo, Japan has already begun using robots. Only two human staff are employed alongside some 10 life-like robots who handle everything from greeting, check-in, carrying bags, cleaning rooms to handling guest queries (as well as speaking several languages). Other hotel chains such as Starwood Hotels have been using a system through guest smartphones to provide keyless check-in via a dedicated app that will also allow features such as booking a meal, ordering drinks and answering queries. Examples like the Henn-na Hotel are simply taking that to the next level. The guest interaction is still digital (incorporating elements of AI) but the interface appears more realistic and physical because in occurs with robots face to face and not just through the smartphone.

On the airline side, Air New Zealand have trialled robot check-in at Sydney airport. The robot called Chip can check passengers in to their flight and point them in the right direction. This is building further on Air New Zealand's use of a chatbot called Bravo Oscar Tango that can answer over 900 questions per day. Building further still, the airline (in conjunction with Soul Machines) have also created 'Sophie', a so-called 'digital human'. Sophie uses AI to play an ambassador role for New Zealand, answering questions travellers have about the country as a destination (Fernando, 2017). Through incorporating a wide range of emotional responses, expressions and behaviours, combined with access to real-time information and data, digital concierges or virtual customer agents like Sophie are becoming more widely accepted (building on the widespread use and acceptance of AI devices at home such as Alexa and Siri). Again, what was previously science fiction (as portrayed in the world of *Total Recall*) has now become a reality.

The world depicted in *Total Recall* is also one where self-driving cars and taxis are common place in the urban environment. They are now with us today. In our world, companies such as Tesla, BMW, Volvo and Ford continue to test actual working self-driving cars. In fact, it is not just traditional car manufacturers that are striving to perfect this technology.

Tech companies such as Apple have now entered the race as well. Such autonomous vehicles as depicted in the worlds like *Total Recall* tap into society's insatiable thirst for technology to remove human effort and indeed thinking, allowing humans to concentrate on other things while travelling. While providing a useful function in some regards in this way, there are dangers to removing all human control from such scenarios. Indeed, we have seen accidents occur already while testing such vehicles. As such, there will be a huge element of trust to be overcome before the average person will accept this. Furthermore, the wider dangers do not relate just to physical accidents occurring with machine-controlled vehicles, but rather one more step in the continual giving up of control to machines, for our thinking, our decision-making, our travel, our experiences, our way of life.

Tears in the Rain or a Bright Future for Tourism?

As this chapter has shown, advances in technology, people's addiction to connectivity through their smartphones, the prevalence of digital across society and people's everyday lives, the growing acceptance of allowing machines to collect data on us and actually make choices for us is fuelling a world that no longer seems all the different from much of what we see in the once science fiction depictions of *Blade Runner* and *Total Recall*. When we add to that developments in areas such as AI, robot technology, holograms and self-driving cars then we truly are travelling through such worlds in our own reality.

For all their advanced technology, the worlds depicted in *Blade Runner* and *Total Recall* are ones of huge dystopian urban sprawls, filled with overcrowding, congestion and climates out of control. Concerns about climate change are also with us now in our own world and are well known. We have also seen an increase in what has become known as 'overtourism'. In its recent context, overtourism can be seen as the phenomenon of a popular destination or sight becoming overrun with tourists in an unsustainable way. The first attempt to define the term overtourism was in August 2016 when Rafat Ali, the CEO and founder of Skift, wrote a foreword to an article about the impact of tourism in Iceland. Ali (2016) described overtourism as something which '…represents a potential hazard to popular destinations worldwide, as the dynamic forces that power tourism often inflict unavoidable negative consequences if not managed well'. Perhaps the technology from *Blade Runner* and *Total Recall* can help provide the solutions to such huge challenges of congestion, overcrowding and anti-social behaviour that overtourism can bring, providing tourism with a brighter future.

Flying cars could help avoid congestion on the ground. AI could be used to help predict tourist flows and anticipate and alleviate periods of peak congestion before they happen. Memory implants of 'fake' vacations (if realistic enough) could replace actual travel for some people at least some of the time. Whether or not any of this is a good thing is more open to debate. Are we changing things for the better or are we opening ourselves up to changes that may result in inherent dangers and loss of our decision-making and indeed identity? We are moving evermore into the realm of smart tourism (as discussed by authors such as Gretzel *et al.*, 2015 and Neuhofer, 2016) but if we consider the dangers then just how 'smart' is it to do so? If we rush too far, too fast, without heeding some of the consequences then it may indeed (as Roy Batty would have put it) be tears in the rain.

Concluding Arguments

Blade Runner and *Total Recall*, based on the literary work of renowned science fiction author Philip K. Dick, present us with fantastical vistas and lives intertwined with technology in myriad ways. Our own world has caught up with such depictions in many respects. We accept such technology, indeed many crave it and seem unable to exist in a fulfilling way without it (think of the addiction to social media and other smartphone applications). As such many people now spend a sizeable proportion of their time in a digital realm, not many steps away from what we have seen in these science fiction worlds. Such developments have brought many advantages, including personalisation, convenience, connectedness and so on, but at what cost?

Is it worth the loss of control, the choices made for us, the increasing threat to jobs, the fact that machines and AI programs know more about us and our behaviour than perhaps we do ourselves. Such developments are helping tourism to develop, to grow, to create new kinds of experiences but at what point do the technological developments cross the line from helping create and enhance tourism to actually posing a threat to tourism as we still currently understand it. Will vacation memory implants supersede the need to actually travel somewhere in physical reality. Will developments in holographic technology combined with AR and VR enable people to experience lifelike destinations while still in the comfort of their own living room. That would certainly solve some of our overtourism challenges. It would alleviate security fears and problems such as lost luggage and delayed flights. But will it be 'real'? Will it be 'authentic'? We come back full circle to the fundamental question of what is reality and how technology, as seen and influenced though science fiction worlds, alters the perception of what the tourist finds fulfilling.

References

Ali, R. (2016) *Exploring the Coming Perils of Overtourism*. New York: Skift.
Baudrillard, J. (1973) *Toward a Critique of the Political Economy of the Sign*. St. Louis: Telos.

Baudrillard, J. (1995) *Simulacra and Simulation*. Ann Arbor, MI, University of Michigan Press.

Butler, R. (1990) The influence of the media in shaping international tourist patterns. *Tourism Recreation Research* 5 (2), 46–53.

Chen, L.Y. (2018) Holograms in China captivate Tourists' eyes, and wallets. See https://www.bloomberg.com/news/articles/2018-02-12/holographic-models-virtual-jellyfish-are-the-new-draw-in-china (accessed 18 May 2020).

Deloitte (2017) *Holiday Survey: Retail in Transition*. London: Deloitte.

Disztinger, P., Schlögl, S. and Groth, A. (2017) Technology acceptance of virtual reality for travel planning. In R. Schegg and B. Stangl (eds) *Information and Communication Technologies in Tourism 2017* (pp. 255–268). Cham: Springer Inter-national Publishing.

Eco, U. (1990) *Travels in Hyperreality*. Fort Washington: PA Harvest Books.

Gianatasio, D. (2018) Holographic statues of 500 locals help boost both tourism and morale in Brussels. See https://www.adweek.com/creativity/holographic-statues-of-500-locals-help-boost-both-tourism-and-morale-in-brussels/ (accessed 18 May 2020).

Fernando, I. (2017) The human face of artificial intelligence. See https://www.ibm.com/blogs/ibm-anz/digital-humans/ (accessed 18 May 2020).

Gretzel, U., Sigala, M., Xiang, Z. and Koo, C. (2015) Smart tourism: foundations and developments. *Electronic Markets* 25 (3), 179–188.

Iwashita, C. (2006) Media representation of the UK as a destination for Japanese tourists: Popular culture and tourism. *Tourist Studies* 6 (1), 59–77.

Jansson, A. (2002) Spatial phantasmagoria: The mediatisation of tourism experience. *European Journal of Communication* 17 (4), 429–443.

Jansson, A. (2007) A sense of tourism: New media and the dialectic of encapsulation/decapsulation. *Tourist Studies* 7 (1), 5–24.

Lamsfus, C., Wang, D., Alzua-Sorzabal, A. and Xiang, Z. (2015) Going mobile: Defining context for on-the-go travelers. *Journal of Travel Research* 54 (6), 691–701.

MacCannell, D. (1973) Staged authenticity: Arrangements of social space in tourist settings. *American Journal of Sociology* 79 (1), 589–603.

Neuhofer, B. (2016) Innovation through co-creation: Towards an understanding of technology-facilitated co-creation processes in tourism. In R. Egger, I. Gula and D. Walcher (eds) *Open Tourism – Open Innovation, Crowdsourcing and Co-creation Challenging the Tourism Industry* (pp. 17–33). Berlin: Springer-Verlag.

Steiner, C.J. and Reisinger, Y. (2006) Understanding existential authenticity. *Annals of Tourism Research* 33 (2), 299–318.

Tzanelli, R. (2004) Constructing the 'cinematic tourist': The 'sign industry' of The Lord of the Rings. *Tourist Studies* 4 (1), 21–42.

Xu, F., Weber, J. and Buhalis, D. (2013) Gamification in tourism. In Z. Xiang and I. Tussyadiah (eds) *Information and Communication Technologies in Tourism 2014* (pp. 525–537). Cham: Springer International Publishing.

10 Destination of the Dead: The Future for Tourism

Mairead McEntee, Ruairi McEntee,
Una McMahon-Beattie and Ian Yeoman

Chapter Highlights

- The popularity of the zombie genre shows no signs of dying...
- Both tourism and zombism entail a permanent altering of the person.
- Negative impacts of overtourism are juxtaposed to the impact of a zombie plague.
- Can management of a fictional zombie plague offer solutions to overtourism?

Introduction

One of the few similarities between the book, *World War Z: An Oral History of the Zombie War* (Brooks, 2006) and the 2013 film was the suggestion that Israel would quarantine itself, choosing to close its borders to protect the country from the zombie plague. 'At Taba, we were taken off the bus and told to walk, single file, past cages that held very large and fierce-looking dogs. We went one at a time' (Brooks, 2006: 40). This checkpoint was to ensure only the non-infected could enter the quarantine zone. Similarly, in Spring 2018, over the bank holiday between Saturday 28 April and Tuesday 1 May, Venetian authorities proposed turnstiles at two key entry points to the city: the Calatrava Bridge at Piazzale Roma and Lista di Spagna. As the city filled with visitors, these checkpoints would be used to deny entry to tourists, with only Venetians permitted to pass through (Brunton, 2018). This use of checkpoints, in both the fictional example and in real life, was to deny entry to the infected/tourists and exemplifies how an 'invasion' of tourists may be potentially treated in the same way as an invasion of zombies.

Since the early 2000s there has been a remarkable increase in the popularity and occurrence (Bishop, 2009) of the zombie genre. Films such as the *Resident Evil* series, the remake of *Dawn of the Dead* and *Shaun of the Dead*, combined with new books, comics (*The Walking Dead* comic was first published in 2003), games, TV programmes and even fitness apps such as *Zombies, Run!*, has resulted in zombism going mainstream. Grossman (2009) in *Time* magazine identified zombies as the new vampires (interestingly, an argument could be made that tourists who travel to hedonistic destinations for example, Ibiza, where they party all night and sleep all day mimic the nocturnal activities of vampires in films, such as *The Lost Boys*...) and as Drezner identified, 'By any observable metric, the living dead have become the hottest paranormal pop culture phenomenon of this century' (2014: 825).

At the same time, local communities around the world have begun to fight back against a new threat, that of overtourism. Goodwin (2016: 1) noted that, 'Overtourism describes destinations where hosts or guests, locals or visitors, feel that there are too many visitors and that the quality of life in the area or the quality of the experience has deteriorated unacceptably'. Overtourism may be considered as the inevitable evolution of non-sustainable mass tourism, given the continued unprecedented growth (UNWTO, 2017) of a remarkable global phenomenon (Weaver, 2011). As global tourism continues to grow and new tourist-generating regions emerge or mature, those honey pot destinations, such as Venice and Barcelona are suffering from overtourism.

Bell *et al.* (2013: 7) proposed 'the use of alternative future scenarios ... as a means of opening up dialogues with stakeholders about how things might be'. In Turner and Ash's seminal book, *The Golden Hordes* (1975), tourism in Europe was compared to an invasion of barbarians. Given the extraordinary growth in tourism since 1975, this chapter considers how tourism can be compared to a new horde, not golden but rather a shambling, shuffling and decaying zombie horde. Within the zombie genre, zombies are used to underline how society collapses due to an external threat (Drezner, 2014) or change, as George A. Romero himself discussed with McConnell (2008) 'They [zombies] don't represent anything in my mind except a global change of some kind. And the stories are about how people respond or fail to respond to this.' In our future scenario this external threat or change is from overtourism. Just as Bishop (2009: 201) argues that zombies 'are a grotesque metaphor for humanity itself', overtourism may be a warning parable for destinations regarding the development of tourism. This metaphor of tourists as zombies is, therefore, intended to stimulate debate about the potential overwhelming negative impacts of overtourism if not managed correctly by stakeholders.

The 'Otherness' of Being

Before exploring zombism and overtourism, it should be recognised that there is a fundamental similarity between being a tourist and becoming a zombie. Conceptually while the outward persona remains unchanged (until at least, in the case of the zombie, decay sets in) there is an 'otherness' that results from being a tourist or indeed a zombie. By its very nature tourism has an impact upon the participant. 'Tourism involves for the participants a separation from normal "instrumental" life and the business of making a living and offers entry into another kind of moral state in which mental, expressive, and cultural needs come to the fore' (Graburn, 1983: 11). Nash (1996) in citing Graburn (1989) further developed this thinking, to make the assertion that tourists enter a non-ordinary sacred high before re-entering their ordinary, profane, workday existence. That is, they enter this state of 'otherness'. McAlister (2012: 476) similarly argues zombies represent an otherness, 'a monstrous version of the human', they are human but something other than human.

While, tourists are still human the very concept of being a tourist requires a person to be something other than their normal selves when on holiday resulting in a persona change which has a lasting impact. Just as zombism changes the person, tourists are also altered irrevocably by their holiday experiences.

Impacts of Tourism and Zombism

It is not just the participant who is changed by tourism. Any tourist activity has an effect upon the economy, social and cultural fabric (Urry, 2009) and environment of the host destination (Mason, 2016).

The Environmental Zombie

When overtourism occurs then the negative impacts of tourism outweigh the positive, resulting in, from the local community point of view at least, a dystopian scenario. While the use of dystopian or apocalyptic may seem far-fetched, consider that familiar plot device of a small band of humans battling against the overwhelming, unrelenting hordes of zombies. Compare this to the stands being taken by residents of Barcelona or Venice against the hordes of tourists, that unstoppable collective (Bishop, 2009) of visitors to these destinations. When overtourism descriptions are analysed the similarity between tourism and zombism may not seem so far-fetched. Furlough (1998) describes tourists as 'herd-like', while Urry (2009) describes St. Tropez as being swept away by a black tide of human filth and suggests leaving it to the tourists (or invaders as he called them) much like the Redeker solution to the zombie out-break identified in *World War Z*. Even the words of the local Venetian press who dubbed the 2018 long spring weekend that would bring the 'invasion' of tourists a 'nightmare', could be effortlessly included in any description of a zombie horde. It is easy to imagine tourists shuffling through Venice during this weekend 'as a slow-moving and cumbersome horde', words used by Schum (2015) to describe the undead. Similarly, Bishop (2009: 19) identified that zombies were unrelenting, 'the more you kill, the more keep popping up', similar, one would imagine to how Venetians must feel when, yet another cruise ship prepares to disembark its passengers in to the city.

Conversely, the negative environmental impact of tourism has long been recognised, Cooper and Hall (2013) for example, cited both Bramwell (2004) and Stamboulis and Skayannis (2003) in identifying the impacts on the environment (natural and man-made) from mass tourism, which overwhelms local environments resulting in destruction and stress on the destination. This is perhaps typified by the situation at Maya Bay on Ko Phi Phi Leh island in Thailand. This destination made famous by its appearance in the 2000 film, *The Beach* is now closed indefinitely to tourists to enable an environmental recovery (Ellis-Petersen, 2018). Francis (2017: 1) in discussing overtourism identified that 'When narrow roads become jammed with tourist vehicles, that is overtourism. When wildlife is scared away, when tourists cannot view landmarks because of the crowds, when fragile environments become degraded – these are all signs of overtourism'. These words could be easily describing scenes from any number of zombie-based films or books. Take for example the Atlanta road scene from the *Walking Dead* comic and TV series with the road out of city jammed with vehicles, filled not with tourists but with survivors fleeing the city. The main location used in *Dawn of the Dead* (both the original and the remake) has a zombie horde surrounding a mall, making access impossible; or the final fight scenes from Zombieland, within these scenes the survivors cannot view the amusement park due to the zombie crowd. Fragile degraded environments abound in the zombie genre, consider the London detailed in *28 Days Later* (For the sake of accuracy, the authors recognise that in *28 Days Later*, the cause of the plague is the rage virus rather than a zombie virus.) or the environment illustrated in the computer game *The Last of Us* and as per *World War Z* the destruction of many species and environments due to various government's responses to the zombie plague.

The Social and Cultural Zombie

Just as various stories in the zombie genre revolve around the total breakdown of society, as noted by Bishop (2009: 18), zombie movies 'presented a grim view of a modern apocalypse in which society's infrastructure breaks down'. Tourism too can have significant negative effects on the culture and society of a tourist destination (Mason, 2016).

In *Dawn of the Dead* (both the original and the remake) people hide in a shopping mall to avoid the crowds of zombies and to reinforce the main theme of 'rampant consumerism' (McAlister, 2012: 473). Regi *et al.* (2016) citing various authors found that as an activity, tourism is often defined through the framework of consumption. Reflecting again, the overlap between tourism theory and zombie activity. In several scenes, in both of the *Dawn of the Dead* films, zombies are shown massing against the doors, banging and hammering to get in. It is a familiar scene that is played out in many zombie-based movies, books and programmes. In *Shaun of the Dead*, the main protagonists '... go to the Winchester, have a nice cold pint, and wait for all of this to blow over,' (*Shaun of the Dead*, 2004). However, their proposed solution to the zombie apocalypse does not work out for them, as the Winchester is soon attacked and invaded by a zombie horde.

These scenes in both *Dawn of the Dead* and *Shaun of the Dead* and in various other examples, may be compared to tourists invading areas or facilities in a city in which the residents are simply trying to live their daily lives. For example, the local residents who shop at Barcelona's famous La Boquería market may consider themselves attacked and invaded as they try to use their local facilities. Mishan (1969), argues with regard to tourists 'what a few may enjoy in freedom the crowd necessarily destroys for itself' (as cited by Urry, 2009: 40). Or as Peter in the original 1978 version of *Dawn of the Dead* points out 'When there's no more room in hell, the dead will walk the earth'.

Society is facing other issues arising from overtourism. A combination of low-cost flights and the proliferation of platforms such as Airbnb has resulted in cities 'infested with tourists' rather than lived in by families. The University of Siena's Laboratory of Socio-Geographical Research (Ladest, 2018) calculates that, in 2017, 19% of the entire housing stock, within the medieval walled centre of Florence, is listed on Airbnb (a 60% increase over the same period in 2015). Athens experienced a similar increase, in 2017, with 5127 listings reserved at least once on Airbnb in Athens — a 56.5% increase from 2015 and some neighbourhoods, such as Koukaki subsequently experiencing either a spike in rental prices or loss of long-term rental units. If this situation continues it may leave the local population fighting to retain some semblance of social infrastructure and cultural values against unrelenting hordes of tourists. A situation similar, it may be argued, to that experienced by the communities identified by Dillard in his discussion of *Night of the Living Dead*. Dillard in this analysis identified 'a group of people... [who] struggle to maintain their rationality and their values' in a world that has gone berserk Bishop (2009: 199 citing Dillard).

Often, in zombie stories, the first time a zombie is encountered the protagonist tries to talk to the zombie, only to be greeted with an incoherent groan. In *The Walking Dead* comic series, the first time Rick (the main protagonist) encounters the zombies (It is noted that in The Walking Dead series, zombies are known as walkers.) (Kirkman, 2003); he opens a door to be greeted by the walking dead, one moves towards him and grabs his t-shirt, they fall through a door and down a flight of stairs. All the while Rick is begging for the zombie not to touch him and asks, 'Do you not understand me?' Of course, the zombie cannot reply to Rick. Indeed, most tourists cannot reply in the language of their host nation, so they resort to pointing and nods in agreement or shakes of the head to say no. Armed with a guide book and some simple phrases a well-meaning tourist can easily stumble into an embarrassing situation. Meade (2014) gives the simple, yet apt, example, of a tourist wanting to tell someone that they had bought 'mtumba' (used clothes) at a market in Kenya, but instead claimed they had 'bought matumbo' (intestines). Interestingly, the TV programme, *Z Nation* in its 2018 series introduced zombies that are now capable of communication – suggesting that zombies, unlike some tourists, have the ability to evolve and learn.

Throughout the zombie genre, it is recognised that zombies have one basic desire, that is to eat human brains and flesh and that this controls their every move. Regardless of the dangers surrounding the zombie, this desire motivates (if, indeed, a zombie can be motivated) and is often depicted by zombies encountering a 'fatal' accident, for example the zombie falling into the meat grinder on the TV show of *The Walking Dead*. Tourists motivated by social media now begin to resemble these unthinking, unfeeling, unfearing zombies. Instead of putting the camera down for a few minutes and enjoying the experience and ambience of a historical sight, the tourist shuffles along searching for the next social media like to ensure internal gratification (Kang & Schuett, 2013). Social networking sites alongside smart phones have allowed individuals to upload their photographs quickly and easily (Belk, 2013), helping to give rise to this phenomenon. Posing in front of a landmark, a selfie is taken and posted online as a statement saying 'I've been here' (Lyu, 2016: 185). Along with this statement, there is also an element of attention seeking (Lyu, 2016) behind posting selfies on social networking sites. This in turn can lead to the tourist seeking a more extreme place to take a selfie as to garner a bigger response on social media, more hits and likes. Which may in turn lead the tourist into dangerous situations, where they are focusing on their camera phone on the end of a selfie stick instead of where they are going. A quick internet search for 'selfie tourist fall' brings up dozens of links to news articles featuring hapless holidaymakers who have fallen from various tourist attractions while trying to take selfies from as close to the edge as possible. From falling down a staircase in the Taj Mahal (Anon., 2015) to being gored by wild bison in Yellowstone Park (Millar, 2015), the tourist selfie taker plunging down holes has become such a worldwide phenomenon that has led to some authorities to erect signs warning of the danger of taking selfies and using selfie sticks. The official tourism site for Iceland has even produced a YouTube guide for safe selfies, as part of its Iceland Academy classes (Anon., 2016).

Or perhaps, this search for the perfect selfie, regardless of the danger is something more natural? Toxoplasma gondii is one of a number of parasites which can turn 'victims into zombies' (Knight, 2013: i) and may infect both humans and warm-blooded animals. Its definitive host is the domestic cat, with rodents being the intermediate host. The parasite has the ability to change the rodent's behaviour, more specifically it makes the rodent unafraid of cats. This is a strategy that the parasite uses to increase its chance of being consumed by a cat and therefore completing its life cycle. Experiments carried out by Berdoy *et al.* (2000), showed that not only does the Toxoplasma gondii parasite make the rodent unafraid of cats; it actually attracts the rodent to the cat. The rodent becomes zombie-like, being driven along by a foreign invader which has hijacked its nervous system for its own benefit. Humans who are infected with Toxoplasma gondii have been shown to have a significantly increased risk of traffic accidents than non-infected. This increase in risk of traffic accidents is because toxoplasmosis leads to prolongation of reaction times and brings about changes in personality profiles as a result of 'the parasite's manipulative activity' (Flegr, 2013: 128). This raises the question about the selfies tourists discussed above, are they also controlled by parasites? Or indeed the change in personality experienced by all tourists. As noted previously, people on holiday display different behavioural characteristics from when in their usual environment. Therefore, is there an unseen enemy, within tourists forcing them to change their normal behaviour, to take unnecessary risks, while it tries to complete its life cycle? Perhaps, this parasite lies dormant for most of the year until, the necessary habitat change for the tourist awakens it. As Moore (2002) concluded we may not even know when our behaviour is being manipulated by parasites...

The starkest example of the impact upon society and culture by tourism, may relate to the impact of sex tourism and specifically the sexual exploitation of children. Tepelus (2008) identified that while adult sex tourism may be considered immoral but not illegal, the sexual exploitation of children in travel and tourism, that is 'Tourism, for the purpose of sexual relation with a minor is a crime' (UNWTO, 2004 as cited by Tepelus, 2008: 103). Dirks (2018) describe zombies as 'walking dead' creatures, that are destructive, malevolent and prey on human flesh. Surely this description could also be used for those tourists who visit destinations for the sole purpose of engaging in the sexual exploitation of children.

The Economic Zombie

Mason (2016) argues that over-dependence on tourism is a result of government's emphasis on tourism as a tool for development. Economic over-dependence therefore leaves regions at a significant risk of overtourism. Following, the 2008 economic collapse, Iceland used tourism as a key tool for revitalizing the economy. However, Iceland is now considered to be at serious risk of overtourism (Tourtellot, 2017). Intriguingly, according to *World War Z*, Iceland did not fare well during the zombie crisis. As Brooks (2006) envisaged the scenario, infected refugees overran the country, cramming their way onto the island. This description of a zombie invested Iceland can be compared to current discussions regarding a tourism invested Iceland. There is limited infrastructure in Iceland capable of dealing with a million tourists each year (Chapman, 2017). As a nation of about 350,000 citizens, Icelanders are not used to thousands of foreigners visiting, or cramming into, small communities each week (Sheivachman, 2017) with facilities at key attractions struggling to cope.

In discussing the impact of tourism, Beckerman, (1974) argued that the mega rich tourist is 'safe' from the negative impacts of mass tourism due to their exclusive access to resorts, islands and facilities. The 2015 zombie genre-based film, *The Rezort*, takes this notion of exclusivity and applies it to the concept of a safari park, where the rich are free to hunt the undead. However, and as made clear in other sources, such as the *World War Z* book and the 2005 Romero film *Land of the Dead*, nobody, not even the wealthy, and nowhere is safe from the threat of zombism or indeed overtourism. As demonstrated in numerous zombie stories, geographical space is a finite resource and the protagonists can never outrun the zombie plague. Similarly, destinations that were once considered exotic and 'out-of-reach' are now mundane. Florio *et al.* (2017: 1) identified how Travel & Leisure described Thailand in 1986 as a 'wildly exotic new frontier'. In 2019, a record 41.1million tourists are expected to visit Thailand (Yuvejwattana, 2019).

What Happens Next?

Is it beyond the realms of science fiction that those regions that become reliant upon tourism are facing a dystopian future because of overtourism? Once extremely popular destinations that have, subsequently, fallen out of tourist favour, such as the Catskills, or Atlantic City, may not be the dystopian environment beloved in zombie pop culture, but desolate images suggest that a total collapse is more likely than a revival. The Salton Sea, albeit due in part to environmental issues, is perhaps the best example and one could easily imagine any post zombie world resembling this locality. Other examples included the abandoned stadia and superstructure used for the 2004 Athens Olympics (as highlighted in Figures 10.1 and 10.2) which could conceivably be used for the next series of *The Walking Dead*.

As discussed at the beginning of the chapter, the solution imposed by the Venetian authorities replicated that identified, by Israel, in *World War Z*. Therefore, if overtourism mimics the effects of a zombie horde descending upon a region perhaps the solution is to be found in fiction. The *World War Z* book further identified The Redeker plan as the solution to the zombie threat. This plan utilised as many natural barriers as possible to create a safe zone, to which the government and a selected percentage of population would retreat, the rest of the country would then be abandoned to the zombies. Controversially in the book, some pockets of population were to be left to distract the zombie hordes. Is this the solution for countries with specific honeypot destinations such as Iceland? Leave the

Destination of the Dead: The Future for Tourism 127

Figure 10.1 Paint and old posters peel from the walls of a section behind the basketball arena, built for the 2004 Athens Olympics. (Photograph authors' own)

Figure 10.2 The swimming facilities, while still in use, are slowly falling into disrepair. (Photograph authors' own)

Blue Lagoon to the tourists and retreat Icelandic society to those less popular destinations and there the sociocultural norms of the country can remain untainted by tourist/zombie infection. Should the Spanish government forsake Barcelona to the hordes and concentrate on preserving other tourist destinations in the country. Or maybe just pockets of the city, e.g. abandon La Boquería market to the tourists and create an alternative for the local populace.

Drezner (2014) discussed the need for more stories in which humans demonstrate an innovative solution to the zombie issue. Innovative solutions are also required for overtourism. Destinations cannot wait on the local population 'rebelling' against tourists, the relevant government should be tackling the issue before it reaches this stage. As Drezner (2014: 833) further postulates 'drawing from popular culture allows for greater creativity in the response to new challenges or new situations'.

Concluding Arguments

Just as the characters in zombie stories cannot escape the relentless hordes of undead, destinations cannot escape the inexorable march of the tourist and for many destinations this may inevitably result in overtourism.

Tourist destination organisations may not advocate shooting visitors in the head (the accepted lore is that the only guaranteed way to stop a zombie is to destroy the brain). Other solutions, however, as identified in the zombie genre such as distraction and evasion may be appropriate. As noted previously, is the solution to overtourism to 'sacrifice' honeypot destinations in the hope that tourists ignore the rest of the region/country? Or perhaps some sort of subterfuge would be more appropriate. Destination Management Organisations (DMOs) could choose not to sell their tourism product to tourists, marketing campaigns, by these DMOs, would suggest that tourists would not want to visit that destination becoming effectively Destination Deterrent Organisations. This obviously would mean forsaking any potential benefits from tourism or offering very limited tourism. However, just as the protagonists in *Resident Evil: Apocalypse* discovered, as well as myriad other examples, once the zombie horde is released its very difficult to control, so the same could be said of tourism.

Ultimately, many zombie stories end up with humankind living in a quasi-pre-industrial society, the most recent season of *The Walking Dead* tv series for example. Therefore, perhaps the way to address overtourism is to treat tourism as it would have been considered in the past. That is, tourism should not be considered as a right for the consumer but as a privilege afforded to only a few.

Finally, as Romero himself points out, zombie stories are a way of describing how the world responds to a global issue – the continued unprecedented growth in international tourism and how governments and populations react is therefore, perhaps, the ultimate zombie story.

References

Anon. (2015) *BBC News*. See http://www.bbc.com/news/world-asia-india-34287655 (accessed 5 April 2018).
Anon. (2016) *Inspired by Iceland*. See https://www.inspiredbyiceland.com/icelandacademy/a-guide-to-safe-selfies-in-iceland (accessed 10 April 2018).
Beckerman, W. (1974) *In Defence of Economic Growth*. London: Jonathan Cape.
Belk, R. (2013) Extended self in a digital world. *Journal of Consumer Research* 40, 477–500.
Bell, F., Fletcher, G., Greenhill, A., Griffiths, M. and McLean, R. (2013) Science fiction prototypes: Visionary technology narratives between futures. *Futures* 50, 5–14.
Berdoy, M., Webster, J. and Macdonald, D. (2000) Fatal attraction in rats infected with Toxoplasma gondii. *Proceedings of the Royal Society of London. Series B: Biological Sciences.*
Bishop, K. (2009) Dead man still walking: Explaining the zombie renaissance. *Journal of Popular Film & Television* 37 (1), 16–25.
Brooks, M. (2006) *World War Z*. New York: Crown Publishing Group.
Brunton, J. (2018) Venice poised to segregate tourists as city braces itself for May Day 'invasion'. *The Guardian*, 1 May. See https://www.theguardian.com/travel/2018/may/01/venice-to-segregate-tourists-in-may-day-overcrowding (accessed 2 May 2018).
Chapman, M. (2017) Sustainable tourism in Iceland. *Guide to Iceland*. See https://guidetoiceland.is/history-culture/sustainable-tourism-in-iceland (accessed 2 May 20180.
Cooper, C. and Hall, C.M. (2013) *Contemporary Tourism: An International Approach* (3rd edn). Oxford: Goodfellow Publishers.
Dawn of the Dead (1978) [Film]. Directed by George A. Romero. United Film Distribution Company.
Dirks, T. (2018) Greatest Zombie films. *FilmSite*. See https://www.filmsite.org/zombiefilms.html (accessed 19 June 2018).
Drezner, D. (2014) Metaphor of the living dead: Or, the effect of the zombie apocalypse on public policy discourse. *Social Research* 81 (4), 825–849.
Ellis-Petersen, H. (2018) Thailand Bay made famous by The Beach closed indefinitely. *The Guardian*, 3 October. See https://www.theguardian.com/world/2018/oct/03/thailand-bay-made-famous-by-the-beach-closed-indefinitely (accessed 5 October 2018).
Flegr, J. (2013) Influence of latent Toxoplasma infection on human personality, physiology and morphology: Pros and cons of the Toxoplasma–human model in studying the manipulation hypothesis. *The Journal of Experimental Biology* 216, 127–133.
Florio, E., Goldstein, N., Suozzi, M.A., Farley, A. and Lindberg, P.J. (2017) Top travel destinations: 1971–2011. *Travel + Leisure*, 2 February, See https://www.travelandleisure.com/articles/top-travel-destinations-1971-2011 (accessed 26 February 2018).
Francis, J. (2017) Overtourism – what is it and how can we avoid it? *Responsible Travel*. See https://www.responsibletravel.com/copy/what-is-overtourism (accessed 5 February 2018).
Furlough, E. (2008) Making mass vacations: Tourism and consumer culture in France, 1930s to 1970s. *Comparative Studies in Society and History* 40 (2), 247–268.
Goodwin, H. (2016) Managing tourism in Barcelona. Responsible Tourism Partnership working paper 1. Responsible Tourism Partnership. See http://haroldgoodwin.info/RTWP/01%20Managing%20Tourism%20in%20Barcelona.pdf (accessed 14 April 2018).
Graburn, N.H. (1983) The anthropology of tourism. *Annals of Tourism Research* 10, 9–33.
Grossman, L. (2009) Zombies are the new vampires. *Time Magazine*, 9 April, See http://content.time.com/time/magazine/article/0,9171,1890384,00.html (accessed 22 April 2018).

Kang, M. and Schuett, M.A. (2013) Determinants of sharing travel experiences in social media. *Journal of Travel and Tourism Marketing* 30 (1–2), 93–107.

Kirkman, R.T.M. (2003) *The Walking Dead*. Portland: Image Comics

Knight, K. (2013) How pernicious parasites turn victims into zombies. *Journal of Experimental Biology* 216, i–iv.

Ladest (2018) *State of AirBnB 2018* Siena: Università di Siena. See http://ladestlab.it/maps/82/state-of-airbnb-2018 (accessed 18 December 2018).

Lyu, S.O. (2016) Travel selfies on social media as objectified self-preservation. *Tourism Management* 54, 185–195.

Mason, P. (2016) *Tourism Impacts, Planning and Management* (3rd edn). Oxon: Routledge.

McAlister, E. (2012) Slaves, cannibals, and infected hyper-whites: The race and religion of zombies. *Division II Faculty Publications 115*. See https://wesscholar.wesleyan.edu/cgi/viewcontent.cgi?article=1114&context=div2facpubs (accessed 26 February 2018).

McConnell, M. (2008) Interview: George A. Romero on diary of the dead. See https://www.cinemablend.com/new/Interview-George-Romero-Diary-Dead-7818.html (accessed 2 May 2018).

Meade, S. (2014) *Rosetta Stone*. See http://www.rosettastone.co.uk/blog/6-most-funny-and-embarrassing-language-mistakes/ (accessed 21 May 2018).

Millar, M.E. (2015) *The Washington Post*. See https://www.washingtonpost.com/news/morning-mix/wp/2015/07/23/bison-selfies-are-a-bad-idea-tourist-gored-in-yellowstone-as-another-photo-goes-awry/?utm_term=.11f0c0352d6f (accessed 5 April 2018).

Moore, J. (2002) *Parasites and the Behavior of Animals*. Oxford: Oxford University Press.

Nash, D. (1996) *Anthropology of Tourism*. Oxford: Elsevier.

Régi, T., Rátz, T. and Michalkó G. (2016) Anti-shopping tourism: A new concept in the field of consumption. *Journal of Tourism and Cultural Change* 14 (1), 62–79.

Schum, A.C. (2015) Zombies: The undying genre. *The Artifice*, See https://the-artifice.com/zombies-undying-genre/ (accessed 10 April 2018).

Shaun of the Dead (2004) [Film]. Directed by Edgar Wright. London: Universal Pictures

Sheivachman, A. (2016) Iceland and the trials of 21st Century tourism. *Skift*. See https://skift.com/iceland-tourism/ (accessed 12 June 2018).

Tepelus, C.M. (2008) Social responsibility and innovation on trafficking and child sex tourism: Morphing of practice into sustainable tourism policies? *Tourism and Hospitality Research* 8, 98–115.

Tourtellot, J. (2017) Overtourism: Too much of a good thing. *National Geographic*. See https://www.nationalgeographic.com/travel/features/overtourism-how-to-make-global-tourism-sustainable/ (accessed 12 June 2018).

Turner, L. and Ash, J. (1975) *The Golden Hordes: International Tourism and the Pleasure Periphery*. London: Constable & Co.

UNWTO (2017) *UNWTO Tourism Highlights* (2017 edition). Madrid: UNWTO

Urry, J. (2009) *The Tourist Gaze* (2nd edn). London: Sage

Weaver, D. (2011) Can sustainable tourism survive climate change? *Journal of Sustainable Tourism* 19 (1), 5–15.

Yuvejwattana, S. (2019) Thailand expects a record 41.1million foreign tourists. *Bloomberg*, 28 January. See https://www.bloomberg.com/news/articles/2019-01-28/thailand-expects-a-record-41-1-million-foreign-tourists-in-2019 (accessed 2 February 2019).

Part 3
Disruption

11 Holidays with Inspector Maigret: Mixed Reality Adventures as Value Drivers in Future Tourism

Stephan Bingemer

Chapter Highlights

- Based on Simenon's *Maigret and the Dead Girl* we describe a complex scenario for a mixed reality (MR) adventure that integrates into a city travel experience.
- A precise picture of how it could feel to play a MR adventure is developed thus identifying crucial aspects for an appropriate design of an MR adventure.
- This chapter stands out by introducing the reader into a new thought world of what could be possible in future and bringing technology and creative potential together.
- The integration of MR adventures into tourism products and the potential for future product development is discussed.

Introduction

In 2016, Pokémon Go gave us a first impression of how mixed reality (MR) might affect us in future (Clark & Clark, 2016). Pokémon Go players appeared in specific locations (i.e. Pokéstops and Arenas) that they were guided to via their smartphone's Pokémon Go app. The app overlaid reality with elements of the virtual game (Eßler, 2016).

We extend this idea by analysing how a future would look like that allows us to melt our reality with scenes of our favourite books, thus making us participating observers of the plot. In this chapter, we will create a futuristic vision of how MR adventures could take place and how this might affect the travel and tourism industry. We define MR adventures as a real-life adventure that overlays scenes of a fictional story (e.g. a novel) to the customer's reality (e.g. using virtual reality (VR) glasses).

We suggest that such deep-dive-adventures of our favourite novels might become an integral part of a city tour similar to the present hop-on-hop-off-tours, visits of historic monuments and museum visits. We use a 'belletristic deep-dive' with a concrete scenery to 'feel into the topic' and to analyse how a touristic MR product could look. By doing so, we extract key decision elements to build such a story. For this deep-dive, we selected George Simenon's masterpiece 'Inspector Maigret' series as we believe that its mixture of novel and detective story makes it an ideal candidate for the construction of an MR adventure. Compared to other thrillers, Simenon's Maigret is less about 'finding the murderer' and more about 'understanding the social context of the people involved'. This means that in contrast to classical PC adventures, the focus is less on finding the murderer but much more on 'sensing the social context in the original surrounding described in the novel'. Simenon's Maigret possesses characters that appear in every episode of the story, making it interesting to lovers of the series to see how those characters develop over time and how they interact. From a practical standpoint, many of the Maigret novels are set in streets of Paris that still can be found today (partially with new names) and with building that already existed at the time in which the novel is set.

We assume that to get into the play, a MR adventure has been booked in advance at an online or offline travel agency, together with transfer and hotel. After booking the adventure, a confirmation is sent with the precise coordinates and time to start the adventure. The necessary equipment (in our case we assume the equipment to be light MR VR glasses and earphones) is delivered to your hotel and waiting to be used as soon as the story starts.

Theoretical Background

The purpose of this book is to shed light on various perspectives of the future. In this chapter, one detailed scenario is developed that allows us to gain a deeper insight into how one potential future might look. Of course, no one can guarantee if this future will really come true. We see three main streams of literature that provide theoretical support, namely scenario building, the literature field of gamification and storytelling and MR interactive learning.

We follow Schwartz (1991) in his understanding of scenarios as being about perceiving long-term futures in the present. Scenarios can be applied in order to learn from a variety of alternatives scenarios or they can be focused on raising questions, triggering valuable debates and challenging conventional wisdoms (Greeuw *et al.*, 2000; Guimarães Pereira *et al.*, 2001; Lobo *et al.*, 2005). In this chapter we focus on the latter approach. This chapter is not written with the purpose of providing many scenarios that result in very different futures, but we cover one specific scenario in depth. Our attention is to provide the reader with a perspective that enables discussion, allows us to understand need for change and that helps to further develop our current viewpoint on how MR adventures are designed.

This chapter not only covers a scenario, but we discuss an adventure or more general – a game. Thus, gamification and storytelling literature is a useful starting point for guiding the construction of the scenario we describe. Giakalaras (2016) states that a game must have four essential elements: players, actions, payoffs and information (abbreviated as PAPI). Our scenario is created with the idea that a potential tourist is the player. As we assume that the MR adventure follows the storyline of a novel, action and information are embodied in the storyline of the novel. The payoff is mainly the satisfaction to be part of the story and to enjoy a city from a completely different perspective.

In the context of MR usage in museums, Jung and Tom Dieck (2017) have found that MR exhibits are adding value to the overall user experience. This is even more the case if a personalisation to the user's interest is happening (Marshall *et al.*, 2016; Papagiannakis *et al.*, 2018). Vassiliadi *et al.* (2018) have suggested a 'Literature-based MR presence' consistent with our approach in a learning context. They argue that literature excerpts linked to real heritage sites can evoke deeper sensations.

Creating Novel Customer Experiences in Travel and Tourism

Customer experience is gaining special attention in the domain of travel and tourism (Gopalan & Narayan, 2010). One reason might be that the holiday experience (including pre- and post-experiences) is the key value that customers take away from their travel investment. Therefore, it is utmost important to the travel and tourism industry to develop innovative ways to create events and experience that have long-lasting impact on the customers and that foster customer loyalty. The MR adventure format that we describe in this chapter has the potential to create a completely new access to customers' travel experience. Using MR, the adventure is a mixture of reality and fiction. It only works being at the right time and place, thus making the user curious about the adventure to come. In MR, a fictional story is overlaid to the travellers' reality. This allows real-world locations to be 'part of the game', thus helping to create true immersion. Restaurants, hotels, train stations, bistros, postal offices, public places and many more can design spaces that allow watching a specific part of the story. Restaurants can create specific rooms where a part of the story can be followed alone or with other players (e.g. a bar where the police officers – i.e. fictional characters and MR adventurers – meet). City tourist offices can support such MR adventures by creating spaces in the city that support the story of the MR adventure. While commuting from one scene location to another scene location, users will get in touch with the city in a different, much deeper way. In contrast to the classical touristic monument visit, the story will bring the user to places that he or she would not have visited without the story plot. Thus, the overall experience of a city visit might change having played an MR adventure. If MR adventures are to be played for multiple days, the quality of the adventure is a critical component. Only if the immersion is excellent, will customers take away a great experience and not leave the story plot. In this chapter, we support the development of such MR adventures by pointing out critical decision items throughout the play.

Quantum Leap MR Adventure Preview Based on *Maigret and the Dead Girl*

To display our vision of MR adventures, we carefully selected the context of Simenon's novel *Maigret and the Dead Girl* (Simenon, 2017) to serve as the plot for the described adventure.

The scene starts in Paris at night, nearby the famous Moulin Rouge. We start to 'feel into the story':

> *Imagine you being in the city of Paris (France). The adventure starts in the ninth district, Place Adolphe Max, formerly known as Place Vintimille. Through your VR glasses, you see her lying on the street – a young girl, murdered. An old-fashioned police car from the 1930s is arriving. As the door opens, you suddenly recognise him – the star of the many novels you read: Inspector Maigret. Today, you will be part of his story. You chose the 45th novel of Simenon as a scenario: Maigret and the dead girl. Your vacation was planned to match the winter season. You needed to wake up at 2 o'clock as the scene at Place Vintimille happens around that time and you did not want to miss the first scenes. Maigret is calling you. You cross the street and now you are in the play.*

To start the adventure, the customer needs to match the coordinates and time (and perhaps even the season) sent by the MR provider to start the adventure. This is done to catch a similar ambience as the one described in the novel. Moreover, if the novel describes a scene at night, there is a better immersion if the customer adventure starts at a similar time. We believe that such MR adventures have the potential to become a top-up of a city trip as well as they might attract heavy users. One central aspect is the duration of such an MR adventure. One-day adventures seem to be easier to control and might very well serve as a teaser to attract more users. Moreover, it seems more likely that tourists that travel a city for 3–4 days will pick short runtimes. However, all depends on the story that is underlying the adventure. Many of Simenon's Maigret novels have a plot that covers more than one day of time, thus true immersion calls for a multiple-day adventure. Nevertheless, there are shorter Maigret stories, too, that might be adopted to match a one-day timeframe. Reflecting the potential user groups, we recommend a runtime of the adventure of 2–3 days within one city. We feel that plots that include larger trips (e.g. there are Maigret novels that include domestic and European travels) are more interesting to heavy users and early adopters than to the average tourist.

In order to move to the different scenes, mobility needs to be designed into the adventure. This can be realised by combining the offer with a ticket for public transport, with a bike, car or motorbike rental or by providing a driver service, e.g. with a car that matches the time of the story (i.e. in the case of our Maigret novel, a black Citroën from the 1920s would be a perfect fit).

> *Maigret approaches you and shortly wraps up the scenery adding his personal perceptions and opinions. Whilst in reality you are standing on Place Adolphe Max alone, only some tourist dropping by on their way to the next bar, in your glasses you see the forensic pathologists doing their business on the dead girl surrounded by police officers that hide the action from the public.*

One challenge that becomes obvious here, is that we need to ensure that the adventurist can watch the plot in the VR glasses without being exposed to risks from real life (e.g. by traffic that he might not be aware of as he is focused on watching the plot in the VR glasses). One approach could be to show areas that indicate secure watching positions in the VR glasses. Another question is whether scenes can be rewound, accelerated or stopped. On the one hand, this would ensure comfortable playing conditions and it would match our experience world from PC adventures and recorded movies. On the other hand, having the possibility to stop the story makes it less special and decreases immersion. Another decision that the provider of the adventure needs to carefully decide is if the adventurist is allowed to go everywhere in the scene and what details are made available to him (e.g. shall the adventurist be able to see the murdered girl, her clothes full of blood, or shall the surrounding police hinder the adventurist from getting too close?). Derived from this question we therefore feel that such adventures must undergo a certain degree of age control.

> *Following the plot of the novel, Inspector Lognon, the luckless inspector of the district arrives. As in all Maigret novels, he wants to solve the case on his own and quicker than his boss. Therefore, after being send to interview the night porters, he is following his own investigations without providing any update to Maigret. The novel plot follows Maigret to the forensic institute where he receives additional information on the dead girl. Maigret asks for the girl's age and is answered that the girl is 19–22 years old. Then he waits with Inspector Janvier for an hour for the forensic analysis and discusses with him and you the case. Maigret asks Janvier to fetch the clothes of the dead girl. In the dress, he finds a label that says 'Salon Irene' in the Rue de Douai near Place Vintimille.*

Technically, it is interesting how the change of scenes is realised in the MR adventure. It all starts with the question how the adventurist notices that he or she needs to be at another place. Of course, one could argue that the adventurist shall listen to the characters and he would be integrated into that plot. Nevertheless, what happens if there is unexpected noise from real-life, e.g. an ambulance siren disturbs the scenery? The adventure needs to be prepared for the unexpected, unpredictable real-life scenarios. It seems useful to have a 'playback' and a 'postpone' or 'freeze' option in order to recover situations where the scenery is disturbed or an unexpected event appears (e.g. a manifestation in the street does not allow for the adventure to be followed at the defined time). Again, we need to keep in mind that any element that allows the user to playback, postpone or freeze a scene is lowering the immersion. In order to cover unexpected events, a time buffer needs to be planned into the plot. The same applies for having a drink or food. Either these natural events can be integrated into the plot, which would foster immersion, or eating and drinking are done in an 'adventure break' to recover from longer walks or a stressful scene. In case of a break, the adventure needs to integrate a freeze option that allows getting back into the play after the break is finished. If the adventurist is aware, where he or she has to go next, the question is what transportation mode is used and how the adventurist is guided to the next place. The combination of different transport modes (e.g. bus, taxi, furniculaire, boat, rental car) could foster sales for tickets tailored to the MR adventure. In case a driver is driving you to the next location, he needs to get clear GPS coordinates in order to know where to go next and how long it will be until the adventurer needs to be there. In a more sophisticated integration, the MR adventure could order taxi services for the player – even without the knowledge of the adventurer of his or her final destination. As traffic is unpredictable, it makes sense to integrate a functionality that stops the plot if getting from scene A to scene B takes longer than it would usually do (especially in cities with a lot of traffic jams like Paris). This functionality would allow changing place without being in a hurry all the time and thus ensuring safe travel between locations.

> *Doctor Paul finalises his forensic analysis and discusses his findings with Maigret, Janvier and you. In an intense exchange of information, Doctor Paul debriefs especially Maigret. Then Maigret and Janvier say they go to bed. The scene restarts next day in Maigret's apartment where he has a phone call with Inspector Lognon before he rushes to Salon Irene at rue de Douais.*

We recommend the adventurist's character to be designed rather as a side character. The adventurist is most likely a reader of the novel that the plot is based upon. Therefore, we expect that customers want to enjoy the scenery and the plot. If the adventurist needs to do a lot of interaction (like in the early PC adventures), there is a risk that the customer will perceive this as stressful. This shows a central value creation element of such a new form of adventure: the adventurist experiences a MR and has the opportunity to become immersed in a piece of literature. Therefore, the plot of the original story (in our case *Maigret and the Dead Girl*) needs to be sufficiently recognizable even if the plot is reduced to match timeframes and to keep the adventure easy to follow. This means that the MR adventure relates to its underlying novel as a cut movie version to an overlong movie version. As Maigret and Janvier go to bed, the question is if this is a good breakpoint to go to bed and to ensure that the next morning the plot is starting at a new meeting point. The fact that the next day starts at Maigret's apartment points us to the fact that we need to anticipate places that we cannot enter. From the novel, we know that Maigret is living in Boulevard Richard Lenoir and with a little investigative work; we find number 132 to be the house with Maigret's apartment. To make a MR adventure realistic, a way to treat the problem is to rent an apartment in the house and make it Maigret's apartment. Another way to circumvent the issue of not being able to access the location could be to install a fictive place where Maigret's apartment can be watched, ideally close by the location described.

> *Maigret meets Inspector Lognon with Inspector Lucas in Salon Irene. Maigret is interviewing Irene and is receiving the information that the dead girl has rented her dress. Irene hands over the dress that has been left in the Salon Irene in return for the rented dress. Then Maigret and Lucas return to their office at Quai des Orfèvres. An anonymous call referring to the picture of the dead girl in the newspaper informs Maigret that the dead girl lived in Rue de Clichy, 122. Lognon again is splitting apart and is trying to solve the case on his own. Later in the story, it seems as if Lognon has disappeared.*

From this point, the storyline continues to lead us to several locations and the adventurer learns more and more about the dead girl. The picture of the dead girl gets clearer and clearer. We learn her name, Louise, and that she is born in Nice as daughter of a dancer and a con artist. We learn that the police in New York investigated her father for his criminal activities and that he might have collected a significant amount of money from those activities together with his accomplice in New York. In Rue de Clichy, we learn that Louise has been at the wedding of her friend that hosted her in her apartment in Rue de Clichy, 122. This explains the rented dress. At the wedding, she learned that someone searched for her. The person has left a message in an envelope that was placed in the Pickwick bar on her behalf:

> *Maigret enters the Pickwick bar with you. The waiter, Falconi, a known gangster, confirms that the accomplice of her father deposited an envelope on behalf of Louise after her father died in prison in the U.S. When asked for the girl, he describes her different to the picture that Maigret has collected throughout the plot. Falconi tries to point Maigret to an American who has gone to Brussels but Maigret realises the trap, as he is not convinced that Falconi's description of the dead girl matches his social studies of her. Maigret combines that the American who is going*

to Brussels is just a lie and takes him to the office in Quai des Orfèvres. It looks as if Inspector Lognon has followed the wrong trace and this is why he disappeared. There he confirms that a partner of him wanted to betray Louise and steel her ID card in order to retrieve the father's money in New York on her behalf. However, when he wanted to steel the ID, his bracelet interlocked with the handbag of the girl. In order to free himself he knocked her down with his slayers.

Here, a last element comes into play: How is the final scene to be designed? Of course, this is the scene rounding up the adventure. One way to round up the story is a last conversation with Maigret, a last review of the case and his opinion on what has happened. This could happen in a bar, a restaurant or at a location chosen by the adventurer. This occasion might be a good moment for cross selling other adventures, e.g. Maigret could mention other cases he has already solved and ask you if you want to go through that the next time. After both have spoken and exchanged opinion, Maigret and you are leaving the location in two different directions. After the end, an evaluation of the adventure appears and you receive information on where the MR glasses will be retrieved. As a last element, the adventure could recommend a good site to have lunch or dinner in order to reflect the case.

Discussion of the Impact on Travel and Tourism

As we can see from this chapter, MR has the potential to disrupt the way we do city travels and how we can combine fictional stories (e.g. from novels we read) with touristic stays. Travel and tourism might profit from such MR adventures in multiple ways. First, the existence of a highly immersive technological experience that combines city tours with an attractive plot has the power to increase the amount of tourist arrivals in the city where the adventure takes place. Second, in a MR adventure, tourists are guided to new locations within the city, e.g. in other districts. This can help to grow new touristic businesses, restaurants, bars, shops in districts that beforehand were not in the focus of tourists, at all. Third, the adventure itself calls for an integration of local tourist spots to be integrated as part of the story. In the case of Pokémon Go, clever bartenders directed tourists into their bars by ensuring that virtual Pokémon were available in their place and thus gamers came into the bar in their search for Pokémon. This concept could be put into a bigger context. Restaurants, bars, hotels, shops could be integrated into the story as far as it is supported by the story plot. This would lead to another way to attract people into selected locations.

Despite being disruptive, the context of this chapter is not that far away from a possible future. Pokémon Go has shown us that MR can help to direct tourist into specific places and to leverage the power of a virtual adventure. We think that if story, technical realisation and ease of use come together, such an adventure has the power to become an attraction. However, we believe that for true immersion a high degree of excellence needs to be reached. We therefore believe that well-designed MR adventures require a very professional planning and implementation framework. The provider of such a MR adventure needs to network with touristic places within the adventure city, thus making city marketing companies and destination marketing companies (DMCs) excellent partners for such an activity. Apart from designing the MR adventure per se, the adventure needs to be marketed and sold. This could become an interesting field for city tour packages. Including an MR adventure in a touristic package would allow for a completely new experience that excels the experience from a 'monument visit only'. Such a product could include the novel that has served as the basis for the plot as a print version sent to the customer or delivered as an audio book to prepare the adventure. For touristic use, it is interesting to know whether such adventures could be experienced in a group of travellers. Moreover, it would be interesting to know which group size could serve as a maximum limit. From our Maigret example, we would see no objections for groups of up to four persons. We think up to eight persons is the maximum for such a group. Bigger groups could be broken down into smaller groups.

One main element that determines the success is the content of the plot. Maigret is an ideal content as it describes in precision the locations that are visited in Paris, it provides detailed description of each spot and many of the novels are playing in a limited area of not more than 50 km. Moreover, Maigret is an art piece in describing human characters, thus making it especially interesting to listen to the story, even if the adventurer is not actively integrated into the plot. The adventurer gets the opportunity to sense the story by watching what Maigret is doing and how he is doing it. Of course, other subjects would also work. In order to develop touristic value, we think that the plot itself needs to be sufficiently known and interesting. Other novels that could serve as story plots could for example be the novels of Dan Brown or Henning Mankell, as their plots are closely connected to specific locations, too. Different to many of the Maigrets that play in Paris, many novels of Dan Brown touch multiple cities thus fostering the need for longer adventure duration and more complex transfer (e.g. by plane). Moreover, Dan Browns plots are plotted as a chase – making it difficult to integrate a third party into the plot. In contrast, Mankell's Wallander is playing in and around the Swedish city of Ystad, making it a lot easier to locate an MR adventure. In the case of Mankell, many locations of the plot are exchangeable (e.g. a scene in the field as in Mankell, 2011), making it easier to find appropriate spots for a MR adventure. As Maigret, Wallander is not a pure criminalistics book but it makes readers feel into the plot. Many more contents could serve as a plot for an MR adventure: A historic novel, a city tour, a play of Goethe, a movie – there are few limits to the content with one exception: They must give sufficient detail on the location and should allow for the scenario to be played, maximum seven days. In consequence, contents that cover very long periods (if they cannot be condensed into few decisive days) are not recommended as the basis for such an MR adventure.

Finally, we need to point out that transparency is an important element. It must be made very clear in the buying process of such a ticket what precisely is included. If a restaurant needs to be visited for the plot, is it to be paid by the traveller? Is transportation included in the MR adventure? Is the hotel room included or am I booking only the adventure itself as a single service? Is the rent for the equipment included? Only if the game conditions are clear and no unexpected costs will occur, are customers likely to trust this special form of adventure.

Concluding Arguments

This chapter presents a vision for MR adventures of the future. Setting up such adventures means a true disruption. In order to realise such an adventure many different skills are needed including storytelling, MR technology, theatre play, knowledge on PC adventures, sound, cinema and many more. By the high potential of immersion, MR means a quantum leap in the way tourists can dive into culture and history of a tourist city.

One managerial implication is that as soon as the VR technology has become sufficiently mature, touristic players, such as DMCs, city tourist offices, city marketing, tour operators and technology providers shall search for partners to develop such opportunities. One signal for this maturity would be the development of MR into every-day eyewear (i.e. without the burden to carry massive VR glasses). Another managerial implication is to define ways to search, book and deliver such an experience to the customer. Moreover, a pricing strategy for this new kind of offer has to be decided. From a tourism perspective, tour operator packages including MR adventures need to be created. Providers of MR adventures need to reflect how their offer can be included into standard TO packages as a value element.

This chapter also has some implications for further research. Of course, this outlook is a very early projection that is still some steps away from a future reality to come. The closer this future reality approaches, the more scientific research question will need to be covered: what is the market for such a product? How do MR adventures change the perception and experience of a city as compared to classical monument visits? How is an MR adventure changing the customer experience? Moreover, to anticipate such a disruption, it would be interesting to see how tourists would perceive such an upcoming technology and getting more insights on their willingness-to-pay. It could be interesting to see how this attitude towards MR changes with augmenting technological matureness of the underlying technologies. In order to be able to answer such questions, time series analyses that start before such a technology eventually drops in could bring important insights into the antecedents of the diffusion of ground-breaking MR experiences.

Limitations

As all scientific work, this chapter is subject to several limitations. First, it is obvious that no one is able to predict the future to come. In an attempt to get close to a future reality, we created a vision on how a future adventure might look, based on what we know and what we have available today. Of course, this does not mean that this potential future will ever happen. Nevertheless, having a look at the changes that have appeared in the past and the technological development we can summarise, there is a fair chance that the described scenario could happen. Second, as we touch into the future, we are unable to apply many of the classical research methods. In consequence, the presented work has descriptive and visionary character. It is meant to help readers to detect turning points that converge into the direction of our vision. Third, by using a storytelling approach, we wanted to make readers immerse into the story. This is by no means a representative technique. Representative research techniques would come into play once those technologies are developed and real users can be interviewed on their perceptions.

References

Åstrøm, J.K. (2017) Theme factors that drive the tourist customer experience. *International Journal of Culture, Tourism and Hospitality Research* 11 (2), 125–141.

Clark, A.M. and Clark, M.G.T. (2016) Pokémon Go and research: Qualitative, mixed methods research, and the super complexity of interventions. *International Journal of Qualitative Methods* 15 (1), 1–3.

Eßler, P. (2016) *Wie restaurants und cafés an Pokémon Go mitverdienen*, 16 July. See https://www.sueddeutsche.de/wirtschaft/pokemon-go-wie-restaurants-und-cafes-an-pokemon-go-mitverdienen-1.3080864 (accessed 15 September 2018).

Giakalaras, M. (2016) Gamification and storytelling. Available at: file:///C:/Users/e78557/OneDrive%20-%20Ulster%20University/Downloads/Gamificationandstorytellingv2.pdf (accessed 18 August 2021)

Gopalan, R. and Narayan, B. (2010) Improving customer experience in tourism: A framework for stakeholder collaboration. *Socio-Economic Planning Sciences* 44 (2), 100–112.

Greeuw, S., van Asselt, M., Grosskurth, J., Storms, C., Rijkens-Klomp, N., Rothman, D. and Rotmans, J. (2000) Cloudy crystal balls. An assessment of recent European and global scenario studies and models. *Environmental Issues Series 17*. Copenhagen: European Environmental Agency.

Guimarães Pereira, A., von Schomberg, R. and Funtowicz, S. (2001) Foresight knowledge assessment. *International Journal of Foresight and Innovation Policy* 3 (10), 53–75.

Jung, T.H. and Tom Dieck, M.C. (2017) Augmented reality, virtual reality and 3D printing for the co-creation of value for the visitor experience at cultural heritage places. *Journal of Place Management and Development* 10 (2), 140–151.

Lobo, G., Costa, S., Nogueira, R., Antunes, P. and Brito, A. (2005) A scenario building methodology to support the definition of sustainable development strategies: The case of the Azores region. *Proceedings of the 11th Annual International Sustainable Development Research Conference,* 6–8 June, Helsinki, Finland.

Mankell, H. (2011) *Faceless Killers.* Vintage Crime/Black Lizard, New York: USA.

Marshall, M.T., Dulake, N., Ciolfi, L., Duranti, D., Kockelkorn, H. and Petrelli, D. (2016) Using tangible smart replicas as controls for an interactive museum exhibition. In *Proceedings of the 10th International Conference on Tangible, Embedded, and Embodied Interaction,* February, pp. 159–167.

Papagiannakis, G., Geronikolakis, E., Pateraki, M., Bendicho, V.M., Tsioumas, M., Sylaiou, S., Liarokapis, F., Grammatikopoulou, A., Dimitropoulos, K., Nikos, G., Partarakis, N., Margetis, G., Drossis, G., Vassiliadi, M., Chalmers, A., Stephanidis, C. and Thalmann, N. (2018) Mixed reality gamified presence and storytelling for virtual museums. See https://www.researchgate.net/publication/322941247_Mixed_Reality_Gamified_Presence_and_Storytelling_for_Virtual_Museums (accessed 2 January 2019).

Schwartz, P. (1991) *The Art of the Long View.* New York: Currency Doubleday.

Simenon, G. (2017) *Maigret and the Dead Girl.* Translated by Curtis, H. Westminister, London: Penguin Classics.

Vassiliadi, M., Sylaiou, S. and Papagiannakis, G. (2018) Literary myths in mixed reality. *Frontiers in Digital Humanities* 5 (21), 1–8.

12 Digital Destinations and Avatar Tourists: A Futuristic Look at Virtual Reality Tourism and Its Real-World Impacts

Daniel Guttentag

Chapter Highlights

- Virtual reality (VR) eventually will offer experiences nearly indistinguishable from real life.
- VR is a uniquely effective medium for creating powerful, memorable experiences.
- VR tourism experiences will compete with real ones and shift tourism purchase patterns.
- The emergence of VR tourism will have numerous implications for the tourism sector.

> A year here and he still dreamed of cyberspace ... he'd still see the matrix in his sleep, bright lattices of logic unfolding across that colorless void.
>
> William Gibson, *Neuromancer* (1984)

Introduction: Dawn of a New Tourism

Neuromancer (Gibson, 1984), the classic science fiction novel quoted above, is credited with providing one of the first visions of the sophisticated virtual worlds that are becoming increasingly achievable through virtual reality (VR). Thinking back to its publication in 1984, one perhaps must be reminded that this was a time well before the first web browsers had popularised the internet, a time right around Apple's release of its first Macintosh desktop computer and a time before the term 'virtual reality' had even been introduced. Today, a mere 35 years or so later, VR has evolved into a multi-billion-dollar industry, and with nothing more than a cheap cardboard headset and a smartphone, anyone can quickly immerse him/herself into resplendent and limitless VR worlds. This progress not only highlights the rapid pace of modern technological development, but also underscores the value in occasionally peering beyond the immediate horizon with a futuristic look at the more distant future.

This chapter offers such a futuristic imagining of VR tourism over the coming centuries. It draws on inspiration from science fiction stories, while still being rooted in existing VR technology and cutting-edge research. The chapter begins with an overview of present-day VR technology, which is followed by an examination of the futuristic directions this technology is heading, positing that VR experiences eventually will become nearly indistinguishable from real life. Next, the chapter highlights the distinct ability VR experiences have to elicit powerful emotional responses. With such powerful experiences in mind, the chapter then discusses the phenomenon of VR tourism, with a focus on why it is reasonable to anticipate VR tourism experiences eventually will compete with real tourism experiences and alter tourism purchase patterns more generally. Finally, the chapter examines various positive and negative implications of this phenomenon, as related to issues such as destination economies and data privacy.

Genesis 2.0: Today's VR Technology

VR can be defined as an artificial, three-dimensional (3D) environment that renders in real time in response to a user's movements, and immerses the user such that feelings of 'presence' (i.e. actually being there) are produced (Rubin, 2018). VR technologies have existed for several decades, but over the past several years the industry has begun to truly enter the mainstream thanks to the release of numerous affordable consumer-oriented VR products – in particular, the Oculus Rift, the HTC Vive and Sony PlayStation VR. These systems all involve a VR headset that is 'tethered' by cable to a computer or video game console. Additionally, the consumer market has been inundated with countless very basic VR headsets, generally ranging in price from $10 to $40, into which a user can slide a smartphone that provides the audiovisual VR content. While smartphone-generated experiences cannot match the quality of the more advanced and expensive tethered devices, the future of VR is undoubtedly untethered. In fact, Oculus recently introduced untethered Go and Quest headsets, and HTC introduced a wireless adapter for its Vive system. Discussion of VR technology tends to focus on the visual component, but the more advanced systems also provide 3D ('binaural') audio, and include handheld controllers for manual input and tactile (generally vibratory) output ('haptic feedback').

VR today represents a multi-billion-dollar industry and the technology is growing in popularity. Although VR certainly is not yet a ubiquitous household item, and much of its current popularity centres on the video gaming community, VR is continually penetrating further into mainstream culture as its applications diversify. For example, VR serves as a training tool for jobs ranging from surgeon to football quarterback, a learning tool for classroom education and a therapeutic tool for the treatment of psychiatric disorders like phobias or post-traumatic stress (PTSD) (Bailenson, 2018). VR also serves as a powerful storytelling platform that has been adopted by news companies including The New York Times and CNN, and in 2017 a VR film (Carne y Arena) even was awarded a Special Achievement Academy Award. Moreover, VR has been used as a marketing tool by major companies including Coca-Cola, McDonald's and Hyundai. Finally, there are several VR social worlds (i.e. 'metaverses') that enable interaction between avatars. These include High Fidelity and Rec Room, the former of which is being developed by the founder of Second Life, an early metaverse that was designed for desktop use and peaked in popularity around 2007.

The tourism sector has been fairly quick to realise potential uses for VR. Most often, VR has been adopted as a marketing tool for destinations and tourism suppliers, and tourism research similarly has focused chiefly on VR's tourism marketing potential (Yung & Khoo-Lattimore, 2017). Nonetheless, other tourism applications have included using VR for entertainment, heritage preservation, training and education (Guttentag, 2010; Guttentag *et al.*, 2018). Examples of VR use within tourism include VR videos promoting travel to British Columbia (Zimmerman, 2015), rollercoaster riders in *Six Flags America* donning VR headsets and speeding through a virtual Superman story (Popper, 2016), detailed 3D scans preserving endangered historical sites as part of the Open Heritage Project (Statt, 2018), Best Western Hotels & Resorts staff receiving training via VR simulations (Mest, 2016) and visitors to Florida's Dali Museum exploring a 3D version of one of the artist's paintings (Harpaz, 2016).

Minds in the matrix: Tomorrow's VR technology

The VR tourism applications listed above demonstrate what can be done with modern VR technology, but this potential expands dramatically when considering the future of VR. This evolution admittedly is impossible to know for certain, yet much can be gleaned from current VR development work on all five senses. To begin, one can assume audiovisual rendering quality will continue advancing rapidly, and future VR devices will become untethered, less cumbersome, and likely merged with augmented reality (that overlays digital imagery on a real-world view). Haptic feedback devices soon will surpass today's rudimentary handheld controllers. For example, developers already are working on gloves and jackets that can transmit more complicated tactile sensations, such as squeezing a rubber duck or receiving a hug (Lai, 2016; Tangermann, 2018). Moreover, researchers have developed a VR accessory that can mimic temperature and wind, via a temperature module attached to the user's neck and a fan in front (Revell, 2017). Likewise, another VR accessory can emit smells to a VR user (Horwitz, 2018). Finally, even virtual digital food is being experimented with via electrodes and thermal stimulation of the tongue to mimic certain flavours, paired with stimulation of jaw muscles to mimic texture (Turk, 2016).

More futuristic VR developments may include stimulating the nervous system directly, potentially via nerve-stimulating implants, which would allow nearly limitless opportunities for sensory stimulation (Caughill, 2017; Mokey, 2016). Such implants may sound outlandishly invasive, except many people already have accepted microchip implants that merely handle such mundane tasks as buying drinks (Brooks, 2017). The next frontier for such implants would be the brain, and neural implants already are being developed by the United States military and companies including Elon Musk's Neuralink, plus are being experimented with in the treatment of mood disorders and physical disabilities (Caughill, 2017; Reardon, 2017). For example, Pentagon researchers taught a quadriplegic with neural implants to fly a plane in a flight simulator using only her mind (Atherton, 2015). Furthermore, non-invasive brain–computer interfaces also are being developed; for example, researchers proved subjects in a functional magnetic resonance imaging machine could control a VR avatar using just their thoughts (Cohen *et al.*, 2014), and the VR videogame *Awakening* is played solely with one's mind (Leary, 2017).

When one considers the stunning VR capabilities already in development, in combination with the rapid pace of technological advancement (i.e. Moore's Law) in VR and other related areas (e.g. artificial intelligence [AI]), it seems inevitable that over the next several hundred years VR will come to look a lot different than it does today. Firstly, VR eventually should be capable of creating experiences that are nearly (or even completely) indistinguishable from real life. Secondly, VR gradually should become embedded into the fabric of everyday life, much like computers and cell phones today. These notions elicit thoughts of science fiction movies like *The Matrix*, *Inception* and *Avatar* that have explored the blurring between real-life and artificial experiences. Nevertheless, the notion of artificial realities can be found in stories dating all the way back to Plato's Allegory of the Cave, which highlighted how humans' sensory perceptions could create limited and false impressions of reality. The consequences of a future with highly realistic VR experiences would be profound. As Madary and Metzinger (2016) stated, 'VR technology will eventually change not only our general image of humanity but also our understanding of deeply entrenched notions, such as "conscious experience," "selfhood," "authenticity," or "realness."' Yet perhaps the Morpheus character in *The*

Matrix encapsulated this notion even more clearly when he asked Neo, 'What is "real"? How do you define "real"? If you're talking about what you can feel, what you can smell, what you can taste and see, then "real" is simply electrical signals interpreted by your brain.'

Fake Worlds, Real Emotions: The Power of VR Experiences

Modern VR experiences are not nearly realistic enough to be unmistakable from reality, yet they still can produce a sense of 'presence' that a user is actually there, along with feelings of 'embodiment' in a virtual avatar. Such feelings result from the immersiveness of VR, as it fully envelops the user and eliminates external signals of the outside world, along with the interactive nature of VR, as the experience adjusts in concert with the user's movements. These characteristics make VR fundamentally distinct from other forms of media, like movies or books – in VR, 'Things are not just happening, they are happening to us' (Hudson, 2016). Movies, books and other traditional media certainly can impact audiences too, but the unique characteristics of VR give it a particular capacity to elicit emotional responses.

These responses perhaps most obviously can be observed in myriad YouTube videos of terrified people crossing a virtual plank high in the virtual air. Recounting some of the more memorable VR reactions he has observed, Bailenson (2018) described a federal judge who 'dove horizontally into a real table in order to catch an imaginary ledge after he "fell" off the virtual platform' and a rapper who 'crawled across the plank on his hands and knees' (2018: 3). Nevertheless, VR can provoke a whole constellation of emotions, ranging from awe to spirituality to intimacy to empathy (Chirico *et al.*, 2018; Murdoch & Davies, 2017; Rubin, 2018).

Indeed, the powerful emotional impact of VR experiences is why VR has been dubbed 'the ultimate empathy machine' (Milk, 2015), and researchers have found it can encourage empathy towards issues including racial bias, colour blindness and the acidification of coral reefs (Bailenson, 2018). It also is why, as was described previously, VR has proven useful in the treatment of various psychological disorders like PTSD. Furthermore, it is why VR is uniquely useful as an educational technology, as research suggests that VR forms memories in a manner akin to experiential learning, in which the individual forms memories of doing something rather than just learning about it (Rubin, 2018). Finally, it is why VR video games tend not to feature the same degree of gory first-person violence often found in traditional videogames, as VR's realism renders such violence too disturbing and unpleasant (Bailenson, 2018).

Headsets not Suitcases: VR as Tourism

The memorable realism that VR experiences provide lends itself well to VR tourism experiences. Hobson and Williams (1995) recognised this alignment very early on, noting that both VR and tourism involve seeking experiences that transcend everyday reality by providing escape to a temporary alternate reality. The authors subsequently considered the prospect of virtual tourism experiences replacing real experiences, which is a theme other tourism scholars have explored as well (e.g. Cheong, 1995; Dewailly, 1999; Guttentag, 2010). While other VR tourism applications, such as VR as a tourism marketing tool, easily can be investigated, examining VR as a tourism substitute inherently involves thinking about future technologies and therefore requires a more futuristic lens. Indeed, extant research on this topic unsurprisingly has found VR experiences are not perceived as acceptable substitutes for real travel (Mura *et al.*, 2017; Sussmann & Vanhegan, 2000). These studies have helped highlight various noteworthy limitations of VR tourism – it is inauthentic, there is no social interaction, VR experiences must be brief, food cannot be eaten, and souvenirs cannot be collected. However, by examining each of these limitations independently, it becomes apparent that, as VR technology and attitudes towards it evolve, these limitations will fade.

Beginning with perhaps the most obvious limitation of VR tourism, it is true that visiting a destination/attraction in VR does not represent an 'authentically real' visit. Nonetheless, as authors like Hobson and Williams (1995) and Guttentag (2010) have recognised, tourists regularly accept inauthentic simulacra in place of the 'real thing,' whether that be an indoor ski slope in Dubai, a replica Lascaux Cave (containing fragile prehistoric art) in France, or a 'traditional' Balinese dance that has been commodified for tourists. Much like a modern tourist would visit one of these attractions and feel they had been skiing, had seen Lascaux Cave or had observed a traditional Balinese dance, despite knowing the artificiality of each one, in the future people may perceive VR experiences as acceptably real despite their artificiality. Once the sensations and emotions of visiting somewhere in VR become essentially indistinguishable from visiting it in real life, the distinction between the two should become less dichotomous and more a matter of variations. As was mentioned previously, it is not unreasonable to foresee future interpretations of 'authenticity' and 'realness' shifting in response to VR advancement (Madary & Metzinger, 2016).

The social component of travel, whether involving one's travel party, other travellers, or locals, is central to tourism experiences, yet most current VR is experienced individually. Nonetheless, there is little question that social interaction will be central to future VR technology. Indeed, VR likely represents the future of modern online social networks (O'Brolcháin *et al.*, 2016), as evidenced by Facebook's recent $2 billion acquisition of Oculus. Facebook Spaces actually already allows for VR social interaction and, as described previously, numerous VR metaverses are currently being developed. VR also already has been used for a multi-performer live theatre show (Gallagher, 2018) and a rave (Ohanesian, 2016). Such interactive experiences will evolve considerably as the intricate subtleties of human communication are reproduced more accurately in VR, and the possibility for human touch can be more faithfully transmitted. As Bailenson (2018) noted, '[VR] is going to become a must-have technology when you can simply talk and interact with other people in a virtual space in a way that feels utterly, unspectacularly normal' (2018: 175).

Even as a ubiquitous technology, VR experiences still likely will tend to be counted in hours rather than the days that comprise a real vacation. Nevertheless, as VR equipment becomes decreasingly cumbersome (e.g. imagine VR 'headsets' more like contact lenses), the possibility of spending comparatively longer durations in VR seems likely. Indeed, given the extended periods some people already spend in front of computers doing work or playing videos games, shifting to extended VR sessions seems inevitable, and numerous individuals already have spent over 24 consecutive hours in VR (e.g. Pangburn, 2016). Also, it is worth noting that VR experiences need not be sedentary, and to accommodate space limitations virtual worlds can be designed with visual tricks that make a person believe he/she is exploring a complex path system when he/she is really just essentially walking in circles (Vanderbilt, 2016).

With regards to food, as was described previously, digital culinary experiences already are under development. As another example of this potential, Ranasinghe *et al.* (2017) transmitted 'virtual lemonade' over the internet by capturing the colour and acidity of real lemonade and transmitting it to a water-filled tumbler with electrodes around the rim that mimicked the sourness of the lemonade, and lights that mimicked the colour. Furthermore, the company Project Nourished is developing a generic food substance designed to be eaten in VR, and by manipulating attributes like smell and chewiness the company aims to replicate the sensation of eating real foods (as seen in the VR experience) (Swerdloff, 2015).

Finally, with regards to souvenirs, 3D printers will allow individuals to obtain (i.e. recreate) goods found in VR. The prospect of more general commerce taking place in virtual worlds was proven by Second Life, in which real economic transactions (buying land, virtual goods, etc.) were common. In fact, in 2009 the Second Life economy was reportedly worth over $500 million, and in that year over $50 million in virtual money was cashed out for real money (Takahashi, 2010). A modern parallel to this is the blockchain-based Decentraland Genesis City, where plots of land/pixels currently are being purchased for tens of thousands of dollars (Russon, 2018). Moreover, this general phenomenon can be observed in modern consumers' expansive appetite for in-app purchases within video games; for example, both FarmVille and Fortnite have generated over $1 billion each in such purchases (Ha, 2013; Moore, 2018).

In sum, looking at current trends in consumer behaviour, tourism behaviour and VR technology shows that each of the supposed limitations of VR tourism will become increasingly less hindering as VR technology progresses. On the other hand, the potential benefits of VR tourism, as described by various tourism scholars (e.g. Cheong, 1995; Dewailly, 1999; Guttentag, 2010), are clear. Firstly, VR tourism would allow people to effortlessly access on-demand experiences without suffering the myriad undesirable aspects of tourism, including financial and times costs, crowds and lines, inclement weather, health risks and physical discomforts. Secondly, VR tourism would permit people to access experiences that are otherwise inaccessible, due to factors like cost, distance, visa restrictions, danger or physical limitations. Thirdly, VR allows users to experience not just virtual replicas of the real world, but also limitless worlds that surpass what a tourist could experience in real life. For example, instead of just seeing the ruins of the Roman Colosseum, in VR one could experience the Roman Colosseum brought back to life and full of battling gladiators, or instead of just swimming with dolphins, in VR one could experience being a dolphin and seeing the ocean through its eyes. Fourthly, VR tourism would not produce the negative environmental and social impacts caused by tourism, such as overtourism and carbon emissions.

While such benefits underscore the potential VR has to disrupt the tourism sector, it is worth clarifying that the notion that VR tourism will compete with real-world experiences and influence travel decisions more broadly should not be misconstrued as a claim that all travel for all people will be replaced by VR. As Guttentag *et al.* (2018) proposed, different people will accept different VR tourism substitutes, and the likelihood of an individual accepting a VR tourism substitute will be determined by three key factors – personal characteristics of the user, characteristics of the real-world experience being replicated, and the quality of the VR experience. Personal characteristics would relate to the individual's attitudes towards technology (high or low technology affinity), the motivations behind his/her interest in an experience (relaxation, thrill-seeking, social bonding, etc.) and the barriers he/she faces towards engaging in the experience in real life (financial means, physical limitations, etc.). Characteristics of the real-world experience would relate to the drawbacks associated with real-world engagement (danger, site fragility, etc.), and the complexity of the sensory and emotional stimulation produced by the real-world experience (e.g. surfing versus observing the Statue of Liberty). Finally, the quality of the VR experience would relate to how richly it can be produced in VR, in terms of high-quality imagery and other sensory outputs.

Also, when thinking about the prospect of VR tourism, as was explained by Guttentag *et al.* (2018), scholars' common preoccupation with the authenticity of VR reproductions is somewhat misguided. Aligning with the concept of disruptive innovation (Christensen, 1997), Guttentag *et al.* (2018) argued, 'VR experiences do not need to be perfectly authentic simulacra in order to be used as substitutes, they simply need to be "good enough" ... for the individual to spend his or her scarce travel time and money elsewhere.' Moreover, consumer decisions regarding VR tourism often will not involve consciously choosing between real-life visitation or VR visitation, but will instead involve VR more subtly shifting tourism purchase patterns. For example, a tourist in Rome may not perceive a VR Roman Colosseum as 100% authentically equal to the actual Roman Colosseum, yet, because of that individual's past experiences or future opportunities to visit the Colosseum in VR, they may be more drawn to spend their limited time and money on other potential experiences in Rome. As an actual example of this basic substitution phenomenon, some studies have found television broadcasts negatively impact live attendance at sporting events (e.g. Allan & Roy, 2008; Buraimo *et al.*, 2010). Needless to say, these viewers do not perceive watching a game on television as identical to watching it at the stadium, but it is apparently good enough to render the stadium less desirable (presumably due to cost, inconvenience, etc.).

Taken together, the observations made in this section demonstrate that the notion of VR tourism impacting real-world tourism, though seemingly futuristic, is very realistic. In fact, VR tourism already exists. A company in Tokyo called First Airlines sells VR tourism experiences to destinations including Paris, New York, Rome and Hawaii. Customers don VR headsets while sitting in a replica airplane that is complete with flight attendants, 'in-flight' meals, and simulated take-off and landing. The business has proven popular since opening in 2016, with senior citizens comprising the majority of its customers (Ha, 2018; Kim, 2018).

Competing with Infinity: Implications of VR Tourism

The prospect of VR tourism providing an alternative to real travel has widespread implications, both positive and negative. To begin, most obviously, as VR comes to offer on-demand experiences to people anywhere in the world, and shifts real-world travel patterns, there would be positive and negative financial consequences for destinations and tourism suppliers (attractions, hotels, transport, etc.). The result would be negative for destinations and tourism suppliers if they were to lose customers to VR. Competition with VR would extend beyond the geographic boundaries that normally dictate tourism sector competition, and even would include fantasy experiences. For example, the San Diego Zoo would no longer be competing with other San Diego attractions, but with every VR zoo or animal-related experience. On the other hand, certain destinations and tourism experiences may become more popular with the emergence of VR tourism; for example, people may gravitate more towards active experiences less easily replicated in VR (e.g. volunteer tourism or kite sailing). Likewise, VR will provide a new medium for experiences to be sold more widely, without the usual limitations of space and geography. This potential has already been recognised in the concert industry, where companies including Live Nation are currently producing VR concerts (Coleman, 2017).

While the prospect of fewer people visiting a place in real life, and rather visiting it in VR, may have frightening financial implications, there would be social and environmental benefits to fewer tourists in some places. Myriad destinations, including Venice, Barcelona and Amsterdam, are struggling to remain desirable for residents in the face of surging waves of tourists, while myriad attractions, including Angkor Wat, Machu Picchu and the Taj Mahal, are not only losing character, but also sometimes suffering physical degradation due to being overrun by tourists. As the number of international travellers looks to continue increasing in the coming decades, VR could offer an antidote to overcrowding, while still providing virtual access to the masses. Likewise, native cultures and animals could be experienced in VR while avoiding concerns associated with cultural commodification or the ethics of zoos and aquariums. Nevertheless, particularly with regards to heritage sites, there may be a risk that as sites are preserved and regularly accessed in VR they may be neglected in real life, as the real-life versions are gradually perceived as less indispensable. This scenario somewhat parallels the setting for the film *Ready Player One*, in which the VR world of the popular game Oasis compensates for the misery of real life.

Tourism for the purpose of seeing other people, either for business or visiting friends and relatives (VFR), also will be impacted by VR. In fact, there already is some evidence that videoconferencing is sometimes used as a limited substituted for business air travel (Denstadli, 2004; Lu & Peeta, 2009), and videoconferencing can reduce grandchildren's visits to their grandparents (Harmon, 2008). As the social aspects of VR technology and the haptic feedback (ranging from virtual handshakes to virtual sex) further advance, it seems inevitable that the zero-cost and immediacy of VR will lead to it increasingly being accepted as a substitute for some business and VFR travel. On the one hand, the potential for highly realistic social interactions via VR may enrich social relationships, and together with emerging language translation technologies VR could facilitate the social benefits that sometimes result from intergroup contact via travel (e.g. reduced stereotypes). On the other hand, the artificiality of VR social interaction should not be discounted and Turkle (2011), for example, argued that technology-mediated interactions are paradoxically increasing connectedness but producing a new form of solitude.

This issue is further exacerbated by the potential for humanoid VR avatar bots that are programmed via AI and situated within VR tourism experiences. These avatars may be indistinguishable from human-controlled avatars, reminiscent of the 'replicants' in the film *Blade Runner*. Such a scenario raises important ethical questions as humans in VR potentially form emotional bonds with these AI avatars, just like tourists forge rich social bonds with people they meet on their travels, yet the VR users may be unaware they are interacting with an AI-directed bot. Moreover, concerns are raised by the possibility of people knowingly forming emotional bonds with AI avatars, somewhat like the basis of the film *Her*. This possibility is far from unimaginable, as it parallels the development of caregiving robots, which already are used in Japan (Tarantola, 2017). Also in Japan, some people even 'date' avatars via a dating simulation video game called *Love Plus+* , sometimes even going so far as to take their virtual girlfriends on real-world romantic trips (Lowry, 2015; Wakabayashi, 2010).

The rise of VR tourism also may lead to individuals regularly spending extended periods engaging in VR tourism (or other VR activities), and in turn neglecting their own physical and emotional well-being. This concern is underscored by modern internet, cell phone and video game addictions (Ng & Wiemer-Hastings, 2005); past data indicating Second Life users were averaging over six hours per day on the platform (Au, 2016), and disturbing anecdotes like people dying during video game marathons (e.g. Hunt & Ng, 2015), or a neglected baby who starved to death as her parents became obsessed with raising a virtual child in a video game (LeJacq, 2014). A more futuristic parallel to this concern is shown in the film *Surrogates*, in which humans spend much of their time reclined and wearing headsets that control robotic surrogates that engage in most daily activities.

In *Surrogates* people can have their surrogates engage in reckless behaviour because the human controlling a surrogate feels no pain if the surrogate is damaged. Likewise, VR tourism raises questions regarding users' behaviour in the VR world and, in turn, how this behaviour may impact people in real life. Being on vacation (Carr, 2002) and being online (Suler, 2004) both can reduce people's inhibitions, so combining the two may particularly alter people's behavioural tendencies. This possibility is noteworthy due to the previously described powerful impact of VR experiences, which could impact the virtual tourist or others with whom he/she is interacting. As an extreme example of such virtual behaviour, there have been numerous accounts of virtual sexual assaults in VR worlds that have led to real emotional harm for the victims (Cross, 2016; Huff *et al.*, 2003).

Incidents such as these raise questions regarding the ownership, management, policing and distribution of VR worlds and experiences. In the short-term, it seems likely that YouTube's status as a repository for 3D videos will lead it to becoming a chief distributor of VR tourism experiences (Guttentag *et al.*, 2018). Serving as a gatekeeper to such content will position YouTube (or whatever other company) well to also serve as an online travel agency, much like has happened with Google Search. In the long-term, social networks like Facebook may successfully pivot into becoming prominent VR platforms, yet existing peer-to-peer torrents and decentralised blockchains suggest the possibility of a more decentralised system. The plethora of pirated content currently accessible through torrents and elsewhere on the internet raises questions regarding 'experience ownership' and intellectual property of VR tourism. It would be very possible, for example, for programmers unassociated with the Louvre to profit by selling virtual visits to a VR Louvre, developed either on their own or by stealing the code of an 'official' VR Louvre (Guttentag, 2010). Even today, several famous landmarks, including Angkor Wat (Pheap, 2015) and the Hollywood sign (Osborne, 2015), cannot be photographed or filmed for commercial purposes, and such concerns would grow significantly if accessing a VR version of something were perceived as a potential substitute for actual paid visitation.

VR tourism platforms also will have immense tracking and surveillance capabilities, which raises questions regarding user privacy. The modern big data age already has ushered in an era of vast personal data tracking, as evidenced by the troves of personal data stored by companies like Facebook and Google, and by the expansive surveillance state being developed in China. VR will provide additional tracking opportunity, covering everything from body movements to eye movements to emotional responses (O'Brolcháin *et al.*, 2016). In fact, the HTC Vive's advertising platform already tracks a user's gaze to know whether a particular ad has been seen (Norman, 2017). Such data will be a boon for tourism marketers, but a threat to personal data privacy. Moreover, due to the power of VR experiences, personal VR data can be exploited to significantly impact and manipulate users, both in terms of what they believe and what they feel (Madary & Metzinger, 2016; O'Brolcháin *et al.*, 2016). As an example, O'Brolcháin *et al.* (2016: 16) explained, 'A virtual world could portray the city of New York as a den of vice and iniquity or as a vibrant and fun place'. The authors also noted an example of such manipulative power can be seen in Facebook's recent experiment that found manipulating users' newsfeeds by increasing the level of positive or negative content would in turn influence an individual's likelihood of posting positive or negative content. Concerns over manipulation and privacy are further exacerbated by the potential for hacking, which could affect both an individual's avatar or a whole virtual environment.

Concluding Arguments: Towards the Simulated Horizon

Attempting to predict the future centuries in advance is of course rife with challenges. What, for example, would someone in 1800 have predicted about the state of global tourism in 2000? It is important to keep in mind that, just as VR evolves, countless other significant changes will occur as well. Just to begin, lifespans may continue to increase, the geopolitical order may continue to shift, AI and robots may change the nature of work and the world economy, and transportation times may shrink. Nonetheless, by highlighting the VR technology already existing and being developed, this chapter has offered a futuristic glimpse of how VR tourism may emerge. Indeed, it seems inevitable that VR technology will continue advancing, and as the technology becomes better, more user-friendly and more closely woven into the fabric of everyday life, adoption levels by the general public and tourism firms should continually rise (Venkatesh *et al.*, 2003). The chapter has demonstrated that future technological advances in VR will mitigate the current limitations of VR tourism, whereas the benefits will remain clear. Therefore, it seems nearly certain that VR tourism eventually will compete with real-world travel and alter tourism purchase patterns more broadly, which in turn will have various impacts on the tourism sector. How the tourism sector responds to these impacts is far more difficult to predict. Nevertheless, to borrow the words of Neo from *The Matrix*, 'I didn't come here to tell you how this is going to end. I came here to tell you how it's going to begin'.

References

Allan, G. and Roy, G. (2008) Does television crowd out spectators? New evidence from the Scottish Premier League. *Journal of Sports Economics* 9 (6), 592–605.

Atherton, K. (2015) Neural implants let paralyzed woman fly plane with her mind, *Popular Science*, 3 March. See https://www.popsci.com/darpa-neural-implants-let-paralyzed-woman-fly-plane-simulator-her-mind (accessed 12 October 2018).

Au, W.J. (2016) VR will make life better – or just be an opiate for the masses. *Wired*, 25 February. See https://www.wired.com/2016/02/vr-moral-imperative-or-opiate-of-masses/ (accessed 12 October 2018).

Bailenson, J. (2018) *Experience on Demand: What Virtual Reality is, How it Works, and What it Can Do*. New York: W. W. Norton & Company.

Brooks, J. (2017) Swedish workers implanted with microchips to replace cash cards and ID passes. *Independent*, 6 April. See https://www.independent.co.uk/news/world/europe/sweden-workers-microchip-implant-cash-card-id-pass-replace-employee-hand-epicenter-rice-grain-size-a7670551.html (accessed 12 October 2018).

Buraimo, B., Paramio, J.L. and Campos, C. (2010) The impact of televised football on stadium attendances in English and Spanish league football. *Soccer & Society* 11 (4), 461–474.

Carr, N. (2002) The tourism–leisure behavioural continuum. *Annals of Tourism Research* 29 (4), 972–986.

Caughill, P. (2017) Nerve-stimulating implants may be the next level in virtual reality, *Futurism*, 26 September. See https://futurism.com/nerve-stimulating-implants-may-be-the-next-level-in-virtual-reality (accessed 12 October 2018).

Cheong, R. (1995) The virtual threat to travel and tourism. *Tourism Management* 16 (6), 417–422.

Chirico, A., Ferrise, F., Cordella, L. and Gaggioli, A. (2018) Designing awe in virtual reality: An experimental study. *Frontiers in Psychology* 8, 2351.

Christensen, C.M. (1997) *The Innovator's Dilemma: When New Technologies Cause Great Firms to Fail*. Boston, MA: Harvard Business School Press.

Cohen, O., Koppel, M., Malach, R. and Friedman, D. (2014) Controlling an avatar by thought using real-time fMRI. *Journal of Neural Engineering* 11 (3), 1–9.

Coleman, L.D. (2017) How the VR concert industry is boldly jockeying for a slice of a projected $660M pie. *Forbes*, 23 February. See https://www.forbes.com/sites/laurencoleman/2017/02/23/how-the-vr-concert-industry-is-boldly-jockeying-for-a-slice-of-a-projected-660m-pie/#313f014d1094 (accessed 12 October 2018).

Cross, K. (2016) Sexual assault enters virtual reality. *The Conversation*, 9 November. See http://theconversation.com/sexual-assault-enters-virtual-reality-67971 (accessed 12 October 2018).

Denstadli, J.M. (2004) Impacts of videoconferencing on business travel: The Norwegian experience. *Journal of Air Transport Management* 10 (6), 371–376.

Dewailly, J.M. (1999) Sustainable tourist space: From reality to virtual reality? *Tourism Geographies* 1 (1), 41–55.

Gallagher, B. (2018) The antidote to 'Black Mirror' virtual reality, *Nautilus* 12 January. See http://nautil.us/blog/the-antidote-to-black-mirror-virtual-reality (accessed 12 October 2018).

Gibson, W. (1984) *Neuromancer.* New York: Ace Books.

Guttentag, D.A. (2010) Virtual reality: Applications and implications for tourism. *Tourism Management* 31 (5), 637–651.

Guttentag, D., Griffin, T. and Lee, S.H. (2018) The future is now: How virtual reality and augmented reality are transforming tourism. In C. Cooper, S. Volo, W.C. Gartner and N. Scott (eds) *The SAGE Handbook of Tourism Management*. Washington, DC: SAGE.

Ha, A. (2013) Zynga's Pincus says FarmVille has passed $1b in total player purchases. *TechCrunch*, 5 February. See https://techcrunch.com/2013/02/05/farmville-1-billion/ (accessed 12 October 2018).

Ha, K. (2018) Boarding now for a flight from Tokyo to Paris that never takes off. *Reuters*, 15 February. See https://www.reuters.com/article/us-japan-restaurant-firstclass/boarding-now-for-a-flight-from-tokyo-to-paris-that-never-takes-off-idUSKCN1FZ0CE (accessed 12 October 2018).

Harmon, A. (2008) Grandma's on the computer screen. *The New York Times*, 26 November. See https://www.nytimes.com/2008/11/27/us/27minicam.html (accessed 12 October 2018).

Harpaz, B.J. (2016) Dali Museum adds virtual reality to new Disney exhibit. *Skift*, 24 January. See https://skift.com/2016/01/24/dali-museum-adds-virtual-reality-to-new-disneyexhibit (accessed 12 October 2018).

Hobson, J.S.P. and Williams, A.P. (1995) Virtual reality: a new horizon for the tourism industry. *Journal of Vacation Marketing* 1 (2), 124–135.

Horwitz, J. (2018) Pimax 8K headset will use Vaqso VR to emulate smells. *VentureBeat*, 18 June. See https://venturebeat.com/2018/06/18/pimax-8k-headset-will-use-vaqso-vr-to-emulate-smells/ (accessed 12 October 2018).

Hudson, L. (2016) 'It's not real, it's not real, it's not real'. *Slate*, 2 May. See http://www.slate.com/articles/arts/gaming/2016/05/virtual_reality_can_be_more_amazing_and_more_terrifying_than_a_traditional.html (accessed 12 October 2018).

Huff, C., Johnson, D.G. and Miller, K.W. (2003) Virtual harms and real responsibility. *IEEE Technology and Society Magazine*, (Summer), 12–19.

Hunt, K. and Ng, N. (2015) Man dies in Taiwan after 3-day online gaming binge. *CNN*, 19 January. See https://www.cnn.com/2015/01/19/world/taiwan-gamer-death/index.html (accessed 12 October 2018).

Kim, S. (2018) Armchair travel: The 'world's first virtual reality airline' launches'. *The Telegraph*, 21 February. See https://www.telegraph.co.uk/travel/destinations/asia/japan/tokyo/articles/worlds-first-virtual-reality-airline-vr-technology/ (accessed 12 October 2018).

Lai, R. (2016) Dexmo exoskeleton glove lets you touch and feel in VR. *Engadget*, 24 August. See https://www.engadget.com/2016/08/24/dexmo-exoskeleton-glove-force-feedback/ (accessed 12 October 2018).

Leary, K. (2017) The first mind-controlled virtual reality game has arrived. *Futurism*, 9 August 9. See https://futurism.com/the-first-mind-controlled-virtual-reality-game-has-arrived (accessed 12 October 2018).

LeJacq, Y. (2014) The story of a couple who played video games while their child died. *Kotaku*, 28 July. See https://kotaku.com/the-story-of-a-couple-who-played-video-games-while-thei-1611995782 (accessed 12 October 2018).

Lowry, R. (2015) Meet the lonely Japanese men in love with virtual girlfriends. *Time*, 25 September. See http://time.com/3998563/virtual-love-japan/ (accessed 12 October 2018).

Lu, J.L. and Peeta, S. (2009) Analysis of the factors that influence the relationship between business air travel and videoconferencing. *Transportation Research Part A: Policy and Practice* 43 (8), 709–721.

Madary, M. and Metzinger, T.K. (2016) Real virtuality: A code of ethical conduct. Recommendations for good scientific practice and the consumers of VR-technology. *Frontiers in Robotics and AI 3*. See https://www.frontiersin.org/articles/10.3389/frobt.2016.00003/full (accessed 12 October 2018).

Mest, E. (2016) Best Western invests in virtual reality training. *Hotel Management*, 3 May. See https://www.hotelmanagement.net/operate/best-western-invests-virtual-reality-training (accessed 12 October 2018).

Milk, C. (2015) How virtual reality can create the ultimate empathy machine, *TED*, March. See https://www.ted.com/talks/chris_milk_how_virtual_reality_can_create_the_ultimate_empathy_machine (accessed 12 October 2018).

Mokey, N. (2016) We have virtual reality. What's next is straight out of 'The Matrix'. *Digital Trends*, 19 December. See https://www.digitaltrends.com/features/dt10-we-have-virtual-reality-whats-next-is-straight-out-of-the-matrix/ (accessed 12 October 2018).

Moore, M. (2018) Fortnite: Battle Royale' crosses the billion-dollar threshold. *Fortune*, 20 July 20. See http://fortune.com/2018/07/20/fortnite-battle-royale-billion-dollars/ (accessed 12 October 2018).

Mura, P., Tavakoli, R. and Sharif, S.P. (2017) Authentic but not too much: Exploring perceptions of authenticity of virtual tourism. *Information Technology and Tourism* 17 (2), 145–59.

Murdoch, M. and Davies, J. (2017) Spiritual and affective responses to a physical church and corresponding virtual model. *Cyberpsychology, Behavior, and Social Networking* 20 (11), 702–708.

Ng, B.D. and Wiemer-Hastings, P. (2005) Addiction to the internet and online gaming. *Cyberpsychology & Behavior* 8 (2), 110–113.

Norman, A. (2017) New VR ads know if you're watching or ignoring commercials. *Futurism*, 29 March. See https://futurism.com/new-vr-ads-know-if-youre-watching-or-ignoring-commercials (accessed 12 October 2018).

O'Brolcháin, F., Jacquemard, T., Monaghan, D., O'Connor, N., Novitzky, P. and Gordijn, B. (2016) The convergence of virtual reality and social networks: threats to privacy and autonomy. *Science and Engineering Ethics* 22 (1), 1–29.

Ohanesian, L. (2016) Is the future of raves in virtual reality? *LA Weekly*, 15 September. See https://www.laweekly.com/music/is-the-future-of-raves-in-virtual-reality-7372388 (accessed 12 October 2018).

Osborne, S. (2015) Trademark law stops people filming Hollywood Sign. *The Independent*, 3 November. See https://www.independent.co.uk/news/world/americas/trademark-law-stops-people-filming-hollywood-sign-a6720311.html (accessed 12 October 2018).

Pangburn, D.J. (2016) This guy just spent 48 hours in virtual reality. *Vice*, 14 January. See https://www.vice.com/en_us/article/kzzgez/this-guy-just-spent-48-hours-in-virtual-reality (accessed 12 October 2018).

Pheap, A. (2015) Government imposes new rules for photos in Angkor Park. *The Cambodia Daily*, 2 December. See https://www.cambodiadaily.com/news/government-imposes-new-rules-for-photos-in-angkor-park-101693/ (accessed 12 October 2018).

Popper, B. (2016) Adding virtual reality to a roller coaster sounds dumb, but works amazingly well. *The Verge*, 15. See https://www.theverge.com/2016/6/15/11940194supermanvr-virtual-reality-roller-coaster-six-flags (accessed 12 October 2018).

Ranasinghe, N., Jain, P., Karwita, S. and Do, E.Y.L. (2017) Virtual lemonade: Let's teleport your lemonade! In *Proceedings of the 11th International Conference on Tangible,*

Embedded, and Embodied Interaction. ACM, 183–190. See https://dl.acm.org/doi/10.1145/3024969.3024977 (accessed 22 May 2019).

Reardon, S. (2017) AI-controlled brain implants for mood disorders tested in people. Nature, 22 November. See https://www.nature.com/news/ai-controlled-brain-implants-for-mood-disorders-tested-in-people-1.23031 (accessed 12 October 2018).

Revell, T. (2017) Virtual reality weather add-ons let you feel the sun and wind. New Scientist, 13 February. See https://www.newscientist.com/article/2121145-virtual-reality-weatheradd-ons-let-you-feel-the-sun-and-wind/ (accessed 12 October 2018).

Rubin, P. (2018) *Future Presence: How Virtual Reality is Changing Human Connection, Intimacy, and the Limits of Ordinary Life*. New York: HarperCollins.

Russon, C. (2018) Making a killing in virtual real estate. Bloomberg Businessweek, 12 June. See https://www.bloomberg.com/news/articles/2018-06-12/making-a-killing-in-virtual-real-estate (accessed 12 October 2018).

Statt, N. (2018) Google will help preserve endangered historical sites in virtual reality. The Verge, 16 April. See https://www.theverge.com/2018/4/16/17241710/google-cyark-partnership-vr-virtual-reality-preserve-historical-sites (accessed 12 October 2018).

Suler, J. (2004) The online disinhibition effect. Cyberpsychology & Behavior 7 (3), 321–326.

Sussmann, S. and Vanhegan, H. (2000) Virtual reality and the tourism product: Substitution or complement? *Proceedings of 8th European Conference on Information Systems, Trends in Information and Communication Systems for the 21st Century*, ECIS 2000, pp. 1077–1083. Vienna, Austria, 3–5 July.

Swerdloff, A. (2015) Eating the uncanny valley: Inside the virtual reality world of food. Vice, 13 April. See https://munchies.vice.com/en_us/article/ezq9zj/eating-the-uncanny-valley-inside-the-virtual-reality-world-of-food (accessed 12 October 2018).

Takahashi, D. (2010) Second Life's economy grows 65% to $567M. VentureBeat, 19 January. See https://venturebeat.com/2010/01/19/second-lifes-economy-grows-65-to-567m/ (accessed 12 October 2018).

Tangermann, V. (2018) Disney made a VR jacket that can simulate hugs, snakes crawling on your back. Futurism, 27 April. See https://futurism.com/disney-vr-jacket-simulate-hug-snakes (accessed 12 October 2018).

Tarantola, A. (2017) Robot caregivers are saving the elderly from lives of loneliness. Engadget, 29 August. See https://www.engadget.com/2017/08/29/robot-caregivers-are-saving-the-elderly-from-lives-of-loneliness/ (accessed 12 October 2018).

Turk, V. (2016) Face electrodes let you taste and chew in virtual reality. New Scientist, 4 November. See https://www.newscientist.com/article/2111371-face-electrodes-let-you-taste-and-chew-in-virtual-reality/ (accessed 12 October 2018).

Turkle, S. (2011) *Alone Together: Why We Expect More from Technology and Less from Each Other*. New York: Basic Books.

Vanderbilt, T. (2016) These tricks make virtual reality feel real realistic. Nautilus 7 January. See http://nautil.us/issue/32/space/these-tricks-make-virtual-reality-feel-real (accessed 12 October 2018).

Venkatesh, V., Morris, M.G., Davis, G.B. and Davis, F.D. (2003) User acceptance of information technology: Toward a unified view. MIS Quarterly 27 (3), 425–478.

Wakabayashi, D. (2010) Only in Japan, real men go to a hotel with virtual girlfriends. The Wall Street Journal, 31. See https://www.wsj.com/articles/SB10001424052748703632304575451414209658940 (accessed 12 October 2018).

Yung, R. and Khoo-Lattimore, C. (2017) New realities: A systematic literature review on virtual reality and augmented reality in tourism research. Current Issues in Tourism 22 (17), 2056–2081.

Zimmerman, K. (2015) Can virtual reality bring real tourists to B.C.? The Globe and Mail, 3 December. See https://www.theglobeandmail.com/life/travel/destinations/can-virtual-realitybring-real-tourists-to-bc/article27578414/ (accessed 12 October 2018).

13 The 'Safety Bubble' and the Future of Enclave Tourism

Cecilia de Bernardi

Chapter Highlights

- Perceived risk is an important factor in tourists' decision-making processes.
- Perceived risk is both subjective and objective, shaped by macro and micro factors.
- Concerns linked to perceived risk may steer tourists to enclave contexts.
- Two examples from the science fiction screen arts portray future no-risk bubbles.

Introduction

In tourism, the theme of the environmental bubble (also known as the tourist bubble) and enclave tourism are widely discussed topics (e.g. Jaakson, 2004; Lepp & Gibson, 2003; Saarinen, 2017). In these cases, the tourists travel to a place and remain mostly isolated from the local population. The research focus on enclave tourism usually concerns the economic disruption of the destinations (Freitag, 1994) and the flow of money – back to wealthy countries instead of contributing to the economic growth of the host country. The relation between risk perception and safety concerns and the potential choice of enclave settings has not yet received much attention in tourism research.

Safety concerns are important and may steer tourists towards enclave settings (Mossberg *et al.*, 2014). Examples of perceived risks are crime (Lorde & Jackman, 2013), terrorism (Rittichainuwat & Chakraborty, 2009; Sönmez & Graefe, 1998) and disease (Rittichainuwat & Chakraborty, 2009), among others. This behaviour is not only the case with European and North American travellers, but has been seen with Asian travellers as well (Lee & Wilkins, 2017). Since tourism demand can shift, it is relevant to reflect on the future consequences of tourists' perceived risk. The future possibility of enclave settings where risk is completely suppressed will also be discussed.

Yeoman (2012: 22) argues that using science fiction as a 'paradigm' for research is a way to predict the concrete future of tourism 'based on extrapolation from present-day trends.' This chapter first presents a sociological conceptualisation of different tourist categories and then discusses previous research from tourism and other disciplines. Future scenarios are conceptualised with the help of two examples from an animated series and a science fiction film. These examples are used to understand more thoroughly the potential future travel trends related to perceived risk and safety concerns.

The discussion considers relevant data from different disciplines to adhere to an interdisciplinary approach, as argued by Bhaskar *et al.* (2018: 1–6). Interdisciplinarity is necessary because the world in which we live is an open system with a multitude of different mechanisms acting at the same time. The investigation outlined in this study is ontologically and epistemologically based on the assumptions of critical realism, which means that reality is independent of how people conceptualise it, but there are also many aspects of reality that are mediated through human experience (Danermark *et al.*, 2002: 1–40). As discussed in the following sections, both objective and subjective factors contribute to the level of perceived risk and potential, related behaviour.

Tourism's Future: Science Fiction Scenarios

This section presents examples from the science fiction realm. The goal is to describe possible future scenarios of how people will act due to different factors related to perceived risk and fear of victimisation.

The ultimate tourist bubble

In an episode of the animated series *Rick and Morty* (Ridley *et al.*, 2017), the crazy and super-intelligent scientist Rick takes his son-in-law on vacation. Rick and his family live on Earth as we know it, but Rick has access to other dimensions and to alien worlds. Rick's access enables him to take his son-in-law Jerry to an alien resort where no one can die or be hurt. Rick takes Jerry there because he is weak and scared, and would probably end up getting hurt if they instead would go on an adventure as Rick usually does with his grandson. The resort is covered by an invisible protective shield called 'the immortality field.' Rick describes the resort as a place where people pay 'top dollar' to go, to avoid any kind of risk (Ridley *et al.*, 2017). In his words, 'you can't even get a heart attack' (Ridley *et al.*, 2017).

The ultimate 'space of containment'

In the futuristic movie *Elysium*, the rich and powerful have moved to a space station that they call 'Elysium' (Blomkamp, 2013). This space station is heavily guarded. It is full of gardens, and the residents have access to equipment that can heal disease. The rest of the human population lives on an overpopulated and crippled Earth, where everyone is interchangeable and expendable. The movie focuses on a coup by some of the people living on Earth to have access to Elysium. The people living on Elysium think that their lives are free of risk because of the security measures on the space station, because of their isolation from the rest of the population, and because of the healing stations that keep them free from disease.

A vacation free of risks

The scenarios described in *Rick and Morty* and *Elysium* present two different aspects of a similar situation. The people on vacation at the resort in *Rick and Morty* can let go of any fear related to physical damage, since there is no possibility of getting hurt. The same is true for the people living on Elysium, even though the difference is that they have healing equipment. At the alien resort, even the tourists most anxious about travelling would feel comfortable doing anything because there is no risk, neither perceived nor concrete. Tourists who go on vacation looking for novelty and adventure will also choose such a destination, as it still offers thrills (such as on the rollercoaster). There is simply no possibility to get physically hurt, which is a natural human concern (which will be discussed more in depth in the following section), true whether people are more adventurous or less so. In the case of *Elysium*, the people who would be attracted would be mostly the ones who prefer avoiding risks and isolating themselves from the rest of the population, even though the possibility to heal in Elysium should not be ignored.

The different preferences of tourists in terms of risk are exemplified in the categories elaborated by Cohen (1972: 166), who argued that many tourists can 'enjoy the experience of change and novelty only from a strong base of familiarity, which enables them to feel secure enough to enjoy the strangeness of what they experience.' This means that they would like to encounter a new strange place simultaneously from a macro perspective and from the micro perspective of a familiar environment. In this process, a sense of security is given by the familiarity of the so-called environmental bubble. The degree of familiarity is subjective to the tourists and depends on how the trip is organised (Cohen, 1972). Cohen (1972) created a continuum, with which he helped classify the tourist categories depending on their placement on said continuum. The two categories that are more likely to want to experience new places through an environmental bubble are the organised mass tourist and the individual mass tourist. The main difference between the two is that the organised mass tourist travels in a group package tour, while the individual one has more control over itineraries and free time (Cohen, 1972).

The other two tourist types in the continuum are the explorer and the drifter. The explorer still has a few ties to some familiar means of living and travelling, while the drifter separates completely from the home culture to become as much as possible part of the host culture. Karl (2018: 133) has argued that tourists who seek more familiarity in their travels 'tend to have significantly higher risk perception than novelty-seeking tourists.' In this case, it can then be argued that tourists that are more likely to seek a bubble will perceive risk negatively in a higher degree than the tourists that seek new places and new experiences. In the scenario described in the episode of *Rick and Morty*, even these categories of tourists would enjoy to be able to have a vacation in this resort because they can still enjoy the thrills that they seek when they go on vacation and explore, but the risks of actually getting hurt are completely eliminated.

Perceived Risk, Fear of Crime and Tourism

The focus of this chapter is in particular the fear of what could happen while people are on vacation. As exemplified in the two science fiction scenarios, in the future of tourism, there is a possibility of a risk-free vacation, but in order to understand the value of that; perceived risk and fear of crime have to be discussed.

Roehl and Fesenmaier (1992: 17) argue, 'a choice involves risk when the consequences associated with the decision are uncertain and some outcomes are more desirable than others.' Risk perception is bound to a specific situation and should be studied in close relation with the context (Roehl & Fesenmaier, 1992). To study behaviour related to risks and risk-taking, Roehl and Fesenmaier (1992) use an approach assuming that 'risk is subjective and that qualitative features of hazards will be linked to perceptions of risk' (Jenkin, 2006: 9). However, there are also objective aspects to assumptions of risk, which will be presented more in detail later. According to Ferraro (1995: 7), it is impossible to study fear of crime without first taking into consideration perceived risk (Ferraro, 1995).

The categories presented by Cohen (1972) show that people have different predisposition to search for adventure and therefore even different degrees of risk. The science fiction scenarios presented are meant to show that a vacation in which the risk of victimisation or physical damage is completely eliminated will appeal to all of the categories. The reason is that the factors that have been presented earlier are not only based on our interactions and subjective experiences, the objective factors are connected to our survival instincts. Decision-making influenced by interactions is governed by different organs and substances in the brain (Rilling *et al*., 2008). Response to threat has to be irrational in many cases for a person to react in the best way. These stimuli are a well-rooted response and not as easy to change compared to other kinds of stimuli. Many fear mechanisms are also not hereditary, but a learned behaviour. A better knowledge of threats, and how to respond to them, often corresponds to a higher chance of survival (Sidebottom & Tilley, 2008). Fear of victimisation is also different in males and females (Ferraro, 1995; Sidebottom & Tilley, 2008). These studies are important because they show that our responses to interactions and to threat or perceived threat are deeply rooted in how our brain is structured, especially related to threat, and that they are also deeply connected to our socialisation process.

There are also differences related to culture and levels of perceived risk (Karl, 2018; Quintal *et al.*, 2010; Reisinger & Mavondo, 2005). However, Reisinger and Mavondo (2005) argue that concerns that a trip to a destination may not be safe are present both in Australian and foreign surveyed groups. Another study on Finnish tourists showed that the major perceived threat was terrorism, but also that perceived risk sparks a search for information, which will be discussed in a separate section. Terrorism and the search for information were also important factors in the travelling decisions of people from the United States (Sönmez & Graefe, 1998). The findings of the study on the Finnish tourists confirmed Cohen's (1972) tourist typologies (Björk & Kauppinen-Räisänen, 2011).

Micro factors related to fear of victimisation (Ferraro, 1995) include one's relation to victimisation, both direct and indirect, and resources available to take care of a perceived threat. A neighbourhood cannot be cohesive if there is a high level of incivility, but perceptions always have a very important role. These impressions then affect perceived risk, and sometimes also result in changes in behaviour (Ferraro, 1995). Enclaves are often places in which tourists are shielded from seeing the reality of the places outside. As argued by Weaver (2005: 180) private beaches 'shield tourists from potentially unpleasant situations.' In a future in which tourism is based on enclaves, the surrounding of the tourists will be controlled down to the detail. The technological advancements will allow that. This is also shown in the contraposition of Earth to Elysium in which Elysium is beautiful and clean and Earth is the opposite: dusty, dirty and full of homemade sheds.

Macro factors include the characteristics of the community, the opportunities to commit crime, information from peers, and information received from the media. Ferraro (1995) has discussed the role of the media in influencing fear of crime, but suggested further studies on the topic. Hollis *et al.* (2017) have completed such a study and found a weak but significant relationship between media reporting, feelings of safety, and quality of life. This is connected to the information provided about a destination, which will be discussed in more detail later.

Perceived risk and destination choice

Aspects influencing decision-making on a destination include socio-economic factors and if the potential visitors have children or not (Karl,

2018). Destination choice is also influenced by personality, as is the propensity to take risks, as the decision to travel to a destination that is considered riskier than another (Lepp & Gibson, 2008). If crimes against tourists are perpetrated, the image of the destination suffers (Brunt et al., 2000; Milman et al., 1994). Furthermore, perceptions of safety can influence visitors' behaviour, such as if they will participate in certain activities when they go out. This behaviour may be present even in tourists who have not been victimised themselves (Barker et al., 2003). Insecurity and diminished sense of safety may also come from an unfamiliarity with the place and language constraints, and therefore a lack of confidence. Language constraints were found to be relevant in a study on a sample of Australian and foreign travellers (Reisinger & Mavondo, 2005). Perceptions of safety are also steered by different factors tied to the traveller herself such as sex, provenance and size of the group, among others. In general, an enclave setting and police presence convey a higher sense of safety for visitors (Milman et al., 1994).

When personal safety is perceived as lacking, the perception can be an obstacle to international travel. As an example, the United States warns its citizens about certain countries (Reisinger & Mavondo, 2005). Reisinger and Mavondo (2005) show that the perception of risk and safety deeply affects the decision to actually go through with vacation plans, change destination, modify behaviour, or acquire more information, as previously mentioned in the case of the Finnish tourists. Another consequence of safety concerns is that tourists may decide to vacation on a cruise ship. Weaver (2005) calls these ships 'spaces of containment.' The main point of Weaver's paper is that the ships create a containment of revenue. Weaver also mentions that during their cruise ship travel, tourists have access to private beaches 'that shield tourists from potentially unpleasant situations' (Weaver, 2005: 180). Weaver (2005) also theorizes that this kind of vacation will continue to be popular and may become even more contained. This phenomenon in tourism mirrors other developments, such as the rise of gated communities in many different countries (Clement & Grant, 2012; Rodgers, 2004). The space ship Elysium as well as the resort of Rick and Morty are expressions of this development. Elysium is the ultimate space of containment and also a very sophisticated gated community. As Weaver (2005) also theorises, this chapter see this development as increasing into new forms of risk-free isolations that in the future will pervade tourism and possibly be present also outside of our planet. This is especially true in the wake of the COVID-19 pandemic.

There are different factors that can influence the willingness to travel to a certain destination. The tourists are for instance afraid that they might contract a disease (Reisinger & Mavondo, 2005; Rittichainuwat & Chakraborty, 2009) and they are also afraid of crime and terrorism. The tourists will not need to fear any of these factors in the future scenarios presented here. Diseases cannot kill the tourists in the resort presented in

Rick and Morty, while the people living on Elysium have purposely isolated themselves from the crime and terrorism that they can experience on Earth.

Tourists can be overrepresented as crime victims (Milman *et al.*, 1994). A study by Maser and Weiermair (1998) contends that the concerns of the interviewed Austrian residents are related to crime, among other things. Fuchs and Reichel (2006) came to similar results in their interviews of tourists, tour guides and academics concerning Israel. The tourists were in fact mostly afraid of human-induced risk, which can relate to several types of victimisation. It is common to think that tourists are 'easy targets' for criminals (Lepp & Gibson, 2003). The respondents to a study by Lovelock (2004) were asked if they had ever felt unsafe at a destination and why. About half of the respondents indicated that they had felt unsafe, and the reason was potential crime victimisation. In an effort to avoid being victimised, British tourists indicated that they try to blend in and keep a low profile (Lepp & Gibson, 2003). However, it has also been shown that tourists can 'overstate the likelihood that they will experience crime' (Mawby, 2000: 119). Another factor that contributes to the choice of destination is terrorism, which has already been mentioned as an important risk factor, and the perception of a possible threat can have deep effects (Jenkin, 2006). The 9/11 attacks and the attack in Bali in 2002 influenced the perception of safety of travellers (Lovelock, 2004). Terrorism was also relevant in relation to perceived risk in the case of Australian and foreign tourists (Reisinger & Mavondo, 2005).

The role of information for perceived risk

Ferraro's (1995) study shows that people are quite realistic about all the information they receive regarding crime. Furthermore, the study does show that 'fear of crime is largely shaped by one's perceived risk of victimisation' (Ferraro, 1995: 120); fear of crime is not an irrational response. On the contrary, 'perceived risk is correlated with both official crime risk and fear; actors are more afraid when they sense a greater likelihood of potential criminal risk' (Ferraro, 1995: 120). This does not eliminate the different external and personal factors that can influence the fear levels (Ferraro, 1995). Jenkin (2006) and Rittichainuwat and Chakraborty (2009) argue that perceived risk and its consequences are not based on facts, but on perception. Ferraro (1995) acknowledges the fact that crime is related to emotions, but also considers it relevant that people do make decisions based on informed knowledge. Furthermore, Jenkin (2006: 4) also argues that 'fear of crime is an important predictor of defensive behaviors such as going out in groups, learning self-defense, carrying spray, or carrying a safety whistle.' One important factor is also trust for the source of the risk information (Jenkin, 2006). Especially nowadays, people travelling can gather a great amount of information through different channels, especially the internet (Lovelock, 2004). Some tourists may book their trip through a travel agent, another potential source of information; in a survey on travel agents and their advice regarding risky destinations, a high percentage of respondents indicated that advice is given regarding destinations that are politically unsafe (Lovelock, 2004). The respondents were also asked to rate destinations from very safe to very risky, and the ones considered the riskiest were the ones most mentioned in media coverage (Lovelock, 2004). The potential influence from the media has been previously mentioned as an important factor (Ferraro, 1995; Rittichainuwat & Chakraborty, 2009).

Ferraro's (1995) study was completed when social media did not exist, and there were not as many different kinds of true crime program on TV. More recently, Schroeder and Pennington-Gray (2014) completed a study on the effect of reading printed news stories on the perception that risk of crime was higher at a destination during a mega sport event. Callanan (2012) found that there are differences between ethnic groups, but that certain TV programs affected all of the groups. These results show that even though there is still a rational component to perceived risk and fear of crime, different media channels can nowadays influence the irrational side. The study by Ferraro (1995) found that people had a perception that was in line with statistics. However, because of all of the information and interaction stimuli to which people are now exposed, it can be argued that people may ignore instances when risk is actually higher or have a heightened but unjustified risk perception level, even though more research is needed on the subject. One example is a study carried out using Facebook, which showed that trust for the information source is an important factor, but the multiplicity of sources makes it difficult to know which to trust (Turcotte *et al.*, 2015). As argued earlier in this chapter, the use of social media and the presence of many TV channels creates a situation in which potential travellers can be bombarded by information on the potential dangers of travelling and they will want a guaranteed safe familiar enclave. Considering that the use of social media has only increased since their invention, in the future the use of social media will be even more extensive and enhanced, gaining a more demarcated role in shaping the decision-making of potential tourists. As in the case of terrorism, the tourists could also end up choosing enclaves because deemed the only really safe option when the information on destinations does not provide a sure answer on where to travel. This is supported by research on the so-called 'fake news' and the different factors that can shape people's ideas. One factor is related to social media filtering algorithms, even though active choices by the users are also involved (Spohr, 2017). Without going into details into this line of study, what is argued here is the potential that the social media and other information channels have in shaping potential tourists' perceived risk and destination choice.

The Future of Enclaves

A recent study on tourists' behaviour has found that in the end, all respondents seem to choose similar locations when they finally decide where to travel (Karl, 2018). The possible explanation is that some destinations are excluded in the beginning of the decision-making process because of safety and security concerns, but are later reconsidered because of cheaper alternatives (Karl, 2018). These kinds of constraints are considered 'the most common' when the tourists proceed to the actual booking (Karl, 2018: 143). The results by Karl (2018) directly contradict the study of Rittichainuwat and Chakraborty (2009), who argue that the respondents would not forego their personal safety even if the cost were low. However, Rittichainuwat and Chakraborty performed their study before the financial crisis of 2008, which might be why the surveyed tourists were still open to paying more to ensure their safety.

Despite these results, which should not be ignored, it is argued that, in the future, enclaves and spaces of containment will be growing in number. The reason is the connection between the role of information and perceived risk. The growing number of platforms can make tourists feel unsure about which information can be trusted, as shown in a study by Turcotte *et al.* (2015). Furthermore, media reporting has been linked to safety feelings (Hollis *et al.*, 2017). Ferraro (1995) argued that people base their perceived risk on facts, while Rittichainuwat and Chakraborty (2009) argue that it is based on perception. A growing number of platforms and recurrent reports about a certain country can steer the tourists towards a risk-free enclave. Cruise ships are already an example of these spaces of containment in which language and other potential problems are not an issue. Gated communities are another example and some people are willing to pay for both cruise trips and a residence in a gated community already at this time. Cohen's (1972) categories show that different types of tourists are more or less willing to take risks. In this chapter, it is argued that places in which tourists can completely isolate themselves from all kinds of risks (be that disease, terrorism or crime) will be very popular in the future, especially for those who can afford them.

Considering the growing role of media in people's lives, and the media's influence on risk perceptions, tourism involving no-risk bubbles will grow as fast as technology allows. For instance, these enclaves can be equipped with robots speaking the tourists' language. In a book chapter about the future of food tourism, Yeoman and McMahon-Beattie (2015: 29) talk about the food replicator of the series *Star Trek* and its capability to 'rearrange subatomic particles' to produce any kind of food. This technology would make risk-free enclaves very appealing for tourists seeking this kind of commodities. The tourist typologies described by Cohen (1972) prefer certain destinations depending on their desire to have a more or less adventurous vacation and different factors (language, perceived risk and food) are involved in the preferences. These scenarios show a future of tourism in which tourists can be steered to risk-free destinations, which they will be happy to pay for in order to avoid risks all over or to make sure that even though the vacation is adventurous, there will be no danger.

Concluding Arguments

This chapter has discussed the potential of perceived risk related to the choice of destination by the tourists as well as the role of information in perceived risk especially related to destination choice in the future. As mentioned throughout the chapter, the research on the consequences of perceived risk and fear of crime does not agree on how perceived risk works and on the consequences on tourists' behaviour. However, the studies that have been presented throughout the discussion are meant to support a strong argument on the fact that there are different factors influencing people's perceived risk and that they will play an important role in future destination development. The source of the information has been found to be important (Jenkin, 2006), but when information comes from many different sources the tourists will not know which source to trust. However, the same information coming from many different sources has the potential to influence decision-making. This can also induce them to pay more money for a completely safe destination, as mentioned by Rick (Ridley *et al.*, 2017), which allows them to go on both risk-free adventures as well as to be sure that absolutely nothing will happen. This also makes it a good destination choice for all of Cohen's (1972) categories. Considering the factors that have been compiled in this chapter as well as the growing number of media platforms, it is important for future destination development to take into consideration the potential positive and negative aspects of the creation of risk-free environmental bubbles.

References

Barker, M., Page, S.J. and Meyer, D. (2003) Urban visitor perceptions of safety during a special event. *Journal of Travel Research* 41 (4), 355–361.
Bhaskar, R., Danermark, B. and Price, L. (2018) *Interdisciplinarity and Wellbeing: A Critical Realist General Theory of Interdisciplinarity*. Abingdon: Routledge.
Björk, P. and Kauppinen-Räisänen, H. (2011) The impact of perceived risk on information search: A study of Finnish tourists. *Scandinavian Journal of Hospitality and Tourism* 11 (3), 306–323.
Blomkamp, N. (Writer and Director) (2013) *Elysium* [Motion Picture]. United States: Sony.
Brunt, P., Mawby, R. and Hambly, Z. (2000) Tourist victimisation and the fear of crime on holiday. *Tourism Management* 21 (4), 417–424.
Callanan, V.J. (2012) Media consumption, perceptions of crime risk and fear of crime: Examining race/ethnic differences. *Sociological Perspectives* 55 (1), 93–115.

Clement, R. and Grant, J.L. (2012) Enclosing paradise: The design of gated communities. *Journal of Urban Design* 17 (October), 37–41.
Cohen, E. (1972) Toward a sociology of international tourism. *Social Research* 39 (1), 164–182.
Danermark, B., Ekstrom, M., Jakobsen, L., Karlsson, J.C. and Ch. Karlsson, J. (2002) *Explaining Society: Critical Realism in the Social Sciences*. London: Routledge.
Ferraro, K.F. (1995) *Fear of Crime: Interpreting Victimization Risk*. Albany: State University of New York Press.
Freitag, T.G. (1994) Enclave tourism development for whom the benefits roll? *Annals of Tourism Research* 21 (3), 538–554.
Fuchs, G. and Reichel, A. (2006) Tourist destination risk perception: The case of Israel. *Journal of Hospitality & Leisure Marketing* 14 (2), 83–108.
Hollis, M.E., Downey, S., del Carmen, A. and Dobbs, R.R. (2017) The relationship between media portrayals and crime: Perceptions of fear of crime among citizens. *Crime Prevention and Community Safety* 19 (1), 46–60.
Jaakson, R. (2004) Beyond the tourist bubble? Cruise ship passengers in port. *Annals of Tourism Research* 31 (1), 44–60.
Jenkin, C.M. (2006) Risk perception and terrorism: Applying the psychometric paradigm. *Homeland Security Affairs* 2 (2), 1–14.
Karl, M. (2018) Risk and uncertainty in travel decision-making: Tourist and destination perspective. *Journal of Travel Research* 57 (1), 129–146.
Lee, H.J. and Wilkins, H. (2017) Mass tourists and destination interaction avoidance. *Journal of Vacation Marketing* 23 (1), 3–19.
Lepp, A. and Gibson, H. (2003) Tourist roles, perceived risk and international tourism. *Annals of Tourism Research* 30 (3), 606–624.
Lepp, A. and Gibson, H. (2008) Sensation seeking and tourism: Tourist role, perception of risk and destination choice. *Tourism Management* 29 (4), 740–750.
Lorde, T. and Jackman, M. (2013) Evaluating the impact of crime on tourism in Barbados: A transfer function approach. *Tourism Analysis* 18 (2), 183–191.
Lovelock, B. (2004) New Zealand travel agent practice in the provision of advice for travel to risky destinations. *Journal of Travel & Tourism Marketing* 15 (4), 259–279.
Maser, B. and Weiermair, K. (1998) Travel decision-making: From the vantage point of perceived risk and information preferences. *Journal of Travel & Tourism Marketing* 7 (4), 107–121.
Mawby, R.I. (2000) Tourists' perceptions of security: The risk-fear paradox. *Tourism Economics* 6 (2), 109–121.
Milman, A., Jones, F. and Bach, S. (1994) The impact of security devices on tourists' perceived safety: The central Florida example. *Journal of Hospitality & Tourism Research* 23 (4), 371–386.
Mossberg, L., Hanefors, M. and Hansen, A. H. (2014) Guide performance: Co-created experiences for tourist immersion. In N.K. Prebensen, J.S. Chen and M. Uysal (eds) *Creating Experience Value in Tourism* (pp. 234–247). CABI: Wallingford.
Quintal, V.A., Lee, J.A. and Soutar, G.N. (2010) Risk, uncertainty and the theory of planned behavior: A tourism example. *Tourism Management* 31 (6), 797–805.
Reisinger, Y. and Mavondo, F. (2005) Travel anxiety and intentions to travel internationally: Implications of travel risk perception. *Journal of Travel Research* 43 (3), 212–225.
Ridley, R. (Writer), Meza-Leon, J.J. (Director) and Archer, W. (Director) (2017) *The Whirly Dirly Conspiracy* [Television series episode]. In J. Roiland and D. Harmon (Creators), Rick and Morty. Atlanta, Georgia: Cartoon Network.
Rilling, J.K., King-Casas, B. and Sanfey, A.G. (2008) The neurobiology of social decision-making. *Current Opinion in Neurobiology* 18 (2), 159–165.
Rittichainuwat, B.N. and Chakraborty, G. (2009) Perceived travel risks regarding terrorism and disease: The case of Thailand. *Tourism Management* 30 (3), 410–418.

Rodgers, D. (2004) 'Disembedding' the city: Crime, insecurity and spatial organization in Managua, Nicaragua. *Environment and Urbanization* 16 (2), 113–124.

Roehl, W.S. and Fesenmaier, D.R. (1992) Risk perceptions and pleasure travel: An exploratory analysis. *Journal of Travel Research* 30 (4), 17–26.

Saarinen, J. (2017) Enclavic tourism spaces: Territorialization and bordering in tourism destination development and planning. *Tourism Geographies* 19 (3), 425–437.

Schroeder, A. and Pennington-Gray, L. (2014) Perceptions of crime at the Olympic Games: What role does media, travel advisories, and social media play? *Journal of Vacation Marketing* 20 (3), 225–237.

Sidebottom, A. and Tilley, N. (2008) Evolutionary psychology and fear of crime. *Policing: A Journal of Policy and Practice* 2 (2), 167–174.

Spohr, D. (2017) Fake news and ideological polarization: Filter bubbles and selective exposure on social media. *Business Information Review* 34 (3), 150–160.

Sönmez, S.F. and Graefe, A.R. (1998) Influence of terrorism risk on foreign tourism decisions. *Annals of Tourism Research* 25 (1), 112–144.

Turcotte, J., York, C., Irving, J., Scholl, R.M. and Pingree, R.J. (2015) News recommendations from social media opinion leaders: Effects on media trust and information seeking. *Journal of Computer-Mediated Communication* 20 (5), 520–535.

Weaver, A. (2005) Spaces of containment and revenue capture: 'Super-sized' cruise ships as mobile tourism enclaves. *Tourism Geographies* 7 (2), 165–184.

Yeoman, I. (2012) *2050 – Tomorrow's Tourism*. Bristol: Channel View Publications.

Yeoman, I. and McMahon-Beattie, U. (2015) The future of food tourism: The Star Trek replicator and exclusivity. In I. Yeoman, U. McMahon-Beattie, K. Fields, J.N. Albrecht and K. Meethan (eds) *The Future of Food Tourism: Foodies, Experiences, Exclusivity, Visions and Political Capital* (pp. 23–45). Bristol: Channel View Publications.

Part 4
Dystopia

14 The Coming of the Fugue and the Blind Tourist?

Stuart Reid and Richard Ek

Chapter Highlights

- Implications of mass tourism: mobility, overtourism, unsustainability.
- Capitalism, conspicuous consumption, self-image, social status.
- Dystopic trajectories: tourism as social pathology, digitalisation and pathology.
- Mapping alternative futures: localism and staycations; virtual reality and hyper-reality.

Introduction

> Among travellers we may distinguish five grades.
>
> The first and lowest grade is of those who travel and are seen –
>
> they become really travelled and are, as it were, blind.
>
> (Nietzsche 2006/1879, thesis 228: 321)

Already in 1879, Friedrich Nietzsche characterised the blind tourist as a hypermobile subject that remained oblivious to the places he or she travelled through, not capable of really seeing, experiencing and living a life based on his or her travelling experiences. Nietzsche's conservatism clearly shines through here, but nevertheless we argue that his dismissive verdict works as a cipher to address two related societal converging trajectories that find empirical expression most clearly in tourism: traveller fugue and a tourist increasingly immersed in self. The purpose of this chapter is to ask if mass tourism, morphing into overtourism, can be conceptualised as an emerging plague of 'walking tourists', and what kinds of tourism futures might come of it.

The chapter will explore topical thematic implications of mass tourism: the passing of thresholds of overtourism, unsustainability and the role of social media and mobile digital devices in the creation of a plague of blind tourists, auto-communicating rather than mindfully interacting with the places visited. Working from currently visible trends, we will relate to the seemingly entrenched pathology of mass tourism (travel to places in order to stay blind) in the contemporary cultural imaginary, as humans witnessing an escalating plague of travel obsession.

To do this, the chapter essentially presents a dystopian tale depicting the onset of the pathological fugue condition (inspired by the critical theory of the Frankfurt School), whereby natural progression of current technological and social developments impels a (worsening) pathology of travel fugue. The result? The visual similarities of zombies invading still-human strongholds in the TV series *The Walking Dead* and cruise tourists entering touristic cities like Dubrovnik or Barcelona, but also smaller towns like Hobart, Tasmania, cannot be denied. Both scenes metaphorically depict the possible demise of civilisation. As a counterweight to this dystopic tale, we contemplate alternative future scenarios that not only challenge the established dominant understanding of tourism, but also the bleak narrative of ours. Here, we propose that the deleterious effects of the plague can be expected to limit overtourism through the imposition of assorted policy restrictions (e.g. higher visas costs, lotteries to enter certain places, seasonal restrictions and closures); inviting movement toward alternative forms of tourism. This dystopic tale then serves as a springboard for contemplation of alternative tourism futures, and two future possibilities are presented in a single bifurcating scenario.

First, we draw inspiration from *We Can Remember it for You Wholesale*, the short story by Phillip K. Dick (1997), originally published in 1966, and made famous through the movie *Total Recall*. In this short story, one alternative tourism future entails development of implantable vacation memories or virtual reality (VR) holidays. It is a future of digital tourism without travel, disconnected from real places and beings, but one offering limitless possibilities for virtual experiences. Second, we are encouraged by the rise of 'staycations', 'holidays at home or nearby' (Fox, 2009; Merriam-Webster, 2018, Oxford Dictionary, 2019); here, the future is one of intimate connection to, and mindful appreciation of, local areas and beings. It portrays a slower, geographically limited but, perhaps, existentially richer kind of local tourism.

Through this bifurcating scenario, we seek to provoke contemplation of the meanings and logics of travel and tourism in a wider context. Notably, tourism is redefined from the current and unsustainable practices of travelling to far-flung places (in vain?) to collect memorable experiences, and we invite contemplation of a future in which people variously create valued and memorable experiences without the need for distant travel. The final contemplative question is which of our imagined futures is utopian or dystopian?

The Pathology of the Blind Zombie Tourist

Overtourism is about to be a keyword in tourism studies. It describes a situation in which citizens of popular destinations not only figuratively but literally apprehend that their city and their quality of everyday life is threatened by a relentless inflow of tourists. Lately, vulnerable destinations like Venice (Seraphin *et al.*, 2018) and extremely popular destinations like Barcelona (Martins, 2018) have become hotspots of attention. It is ironic that one of the destinations with problems of overtourism is Dubrovnik; the city that acts as the fantasy capital King's Landing in the TV series *Game of Thrones*, in which cities are regularly sacked by invading forces. It is even more telling that the TV series has even increased the inflow of film tourists (Tkalec *et al.*, 2017). The cultural imaginary is thus manifested – life imitates art.

As repeatedly stated in tourism studies and beyond, tourism is one of the largest and fastest-growing sectors of consumption and nowadays even a paramount ordering force not only in, but of, the societal (Franklin, 2004). The hotspots of overtourism are not (only) extreme exception, but emblematic instances indicating a future for all landscapes, cities and destinations. Overtourism is a systematic result of a stratified but nevertheless increasing hypermobility on a global scale. Tourism is not only business opportunities for private and public stakeholders, it is not only that corporations and destinations profit on increased tourism and to the best of their ability order, design and encourage increased tourism and leisure consumption. Overtourism is also a logical outcome of a well-functioning capitalistic system.

Tourism fulfils an inherent function in the capitalistic economic system of late modernity. From a Marxist perspective, mass tourism as mobile consumption practices serves the purpose of a safety valve that at least temporarily suppresses the alienation the worker experiences in a system in which his or her means of production is in another hands. Getting away from work sometimes is enjoyable and this makes it less likely the masses will rise to challenge the system itself. The problem is, as MacCannell (1999) points out, that as tourism itself reifies and commodifies the tourist as well as the touristic place, tourism just reinforces estrangement and adds new layers of alienation (see also Wang, 2000).

Tourism can be regarded as an upholder of the social pathologies that are inherent in modern capitalist society. To the first generation of the Frankfurt School (Horkheimer, Adorno, Marcuse), social pathologies are an outcome of a pathological society per se, sustained by reifying and alienating powers inherent in capitalist social and material relations of hierarchies, stratifications and precarious conditions of inclusion and belonging. Social pathologies are connected to the particular logic of economic life in a capitalist society, serving to integrate each individual subject into its own functional interests (in a Foucauldian parlour we could see this as self-inclined subjectification in the biopolitical contemporary). The social pathologies inherent in capitalist society literally shape how the individual understands the world around him or her, his or her ontology, with the distortion of individualisation into heteronomous and alienated forms of individual consciousness as the inexorable end. In other words, it becomes increasingly difficult for the members of the pathological society to grasp that they are living in a sick society (Thompson, 2016, 2018).

To summarise, tourism as an alienating and commodifying consumption practice works as an imperative to travel more and visit destinations and other touristic places with increased frequency, promising an escape from the bleak sedimentary life of work and everyday routine. However, as tourism only increases the alienation, it only enhances the pathological state in contemporary capitalism. Besides that, tourism in itself creates and reproduces hierarchies and patterns of stratification and segmentation. The uneven possibility to travel, to be able to be a tourist subject in the first place, is inherent in the global tourist mobility apparatus. From this vantage, tourism is a key capitalist practice in the production of the uneven politics of mobility (Cresswell, 2010; Edensor, 2007), and has become a dominant sign system on the planet of consumerism (Hall, 2011).

An additional recent societal development that, we argue, increases the pathological state of the blind tourist, metaphorically turning him or her into a blind zombie tourist, is the advent of tourist social media and mobile digital devices. Social media describes internet applications that carry consumer-generated content (CGC) (Kietzmann *et al.*, 2011), enabling users to become the 'media' themselves (Leung *et al.*, 2013). Due to the advances in mobile digital technology, there has been an explosive growth in CGC on the internet (Parra-López *et al.*, 2011). Various digital platforms have gained increasing popularity among travellers, with social media tools assisting tourists in posting and sharing their travel-related comments, opinions and personal experiences widely with others (Xiang & Gretzel, 2010). With over 70% of travellers now using the internet for travel information (Mendes-Filho *et al.*, 2018), social media platforms are a dominant source of travel information. The confluence of the hypermobile travel society and digital society has changed the way people choose to travel, communicate their travel, and in the continuation, how they relate to themselves as travellers and to the places they visit (Tussyadiah & Fesenmaier, 2009).

The perceived trustworthiness of CGC has played a crucial role in the success of social media as a travel information source (Marchi, 2012). The trust owes much to a perceived lack of commercial self-interest (Casaló *et al.*, 2011). Travel and tourist information is considered trustworthy when it seems to describe real experiences by real people (Burgess *et al.*, 2009). The tourist trusts in his or her co-tourists. To advocates of the digital development, this implies a 'user democracy' culture and information sharing that reduces the information asymmetry of markets and shifts more power toward the consumer (Leung *et al.*, 2013). Social media has ostensibly turned destination branding into a two-way competition (Lim *et al.*, 2012). However, as consumers post content of their travel experiences in social media they also become marketers, doing the marketing work for free. Also, the empowerment of social media is not a one-way street; it also offers possibilities for marketers to nudge travellers toward desired consumption ends, by using analytic strategies (Munar, 2011). Successful market intervention hinges upon cultivation (and manipulation?) of trust (Kozinet et al., 2010).

The dimension of trust is thus crucial: trust offers an imagination of a consumer as free to express and represent their experiences and images of tourist destinations and attractions; while that same trust also hides that the consumer acts in a techno-social ecology that is designed and structured to create a specific behaviour among its users and ensures that markets are created, unfolded, expanded and capitalised upon (Munar & Ek, 2015). As a business idea it is a marvellous one: make the users, for free, fill the digital platforms that are designed to create and sustain a certain vocabulary, style of thinking and representations of the (touristic) world with content, then exploit that content by selling digital space for advertising and other marketing messages to other economic actors (Fuchs, 2014). From this critical theoretical perspective, tourist social media becomes just another instrument of control. Albeit offering democratising potential, in this specific context it works as a disciplining force in maintaining the status quo of a capitalist cultural totality, designing and shaping the tourist into a tech-savvy zombie embedded in a digital hyper-reality of the places and destinations he or she is blindly passing through. The erratic pedestrian walks of tourists immersed in the screen arguably offers a mundane expression of this commercialised embeddedness.

An outstanding trait of tourist social media is that it is increasingly centred on visual depictions of the tourist self in the tourist place. Social media allow travellers to visually portray, reconstruct and re-live their trips (Tussyadiah & Fesenmaier, 2009) as well as re-create and share their travel experiences to a global public (Lo et al., 2011). According to Lo et al. (2011) some 89% of pleasure travellers take photographs and 41% of them post photographs online. Many are selfies. Following a set of cultural codes of image design and conduct, the selfie now transforms ephemeral touristic experience into a digital reality, providing the ultimate proof of 'being there' (Stylianou-Lambert, 2012). The 'selfie' is now the emblematic tourist social media picture (Lo & McKercher, 2015).

Despite apparent individual freedom, how we relate to, depict and represent the world is not at all separate from commercial interests. Kodak transformed photography from a specialised activity to an activity integral to everyday life, fostering the memorialisation of travel in holiday pictures. By doing so, Kodak has played its part in the transformation of travel into a superficial tourist activity, with exploration and experience supplanted by more of a checklist approach (Munir & Phillips, 2005), a change in traveling practices that Nietzsche for sure should have disapproved of. But at least 'old-fashioned' tourism photography practice put the place of visitation in focus. The selfie now puts the touristic self in focus. This has ontological implications as photography creates a certain sense of the self, shaping how the self relates to the world and the travels to different places (Lo & McKercher, 2015). The long queues of three hours or more, following an arduous nine-hour long hike, to take just the right selfie photograph at the remote site of Trolltunga in Norway serves as a potent example (Corderoy, 2017).

Image-driven identity-work through travel selfies is central in the selective presentation of the tourist self; it is the digital equivalent of Goffman's (1956) impression management (Lo & McKercher, 2015; Lyu, 2016); the social reward is found in the 'likes' and 'shares' (Lyu, 2016), constituting the addicting sign systems of commercial digital social platforms. Indeed, Fox and Rooney (2015) describe selfies as a kind of pathology; part of a 'Dark Triad' of narcissism, psychopathology and Machiavellianism (see also Fox *et al.*, 2018). Maddox (2017) argues that exhibitionism rather than narcissism is behind the selfie culture. In popular media selfies are increasingly reported as a cause of injury and death, typically implicating states of inattention, exhibition and narcissism (Fantz & Nottingham, 2015; Graham, 2018; Lovittt, 2016). Albeit the media reports are probably sensationalised (Maddox, 2017), physical harm associated with selfies is increasing in tourism and travel (Flaherty & Choi, 2016) as in life in general (Bhogesha *et al.*, 2016). In existential search for the perfect self-image photography, the blind zombie tourist runs a real risk of ending up dead.

Toward a Future of Tourism Without Travel: A Slower Richer Tourism?

Looking back, it all seemed so obvious. Although Thomas Cook's tours popularised travel, bringing the leisure travel formerly only within reach of the elite – Veblen's leisure class (Veblen & Banta, 2008) – into the aspirational reach of the masses (Sezgin & Yolal, 2012), it was the memorialisation of the photograph that ultimately propelled Cook's mass tourism to its logically unsustainable end. Ironically, just as one capitalist (Cook) had seeded the idea of travel for leisure as a basic human right, another (Kodak) inadvertently sowed the seeds of its demise. As surely as the coming of the Night King's winter in *Game of Thrones*, one thing inexorably led to another until the cumulative weight of developments undid the whole capitalist tourism project.

The trouble started when people began travelling to collect photographic evidence of travel. Travel photographs soon became valued tokens of cultural capital, parleyed for social standing (Bourdieu, 1990). Capitalising on this development, Kodak and other enterprises did their bit to foster consumption of travel as an existential end, coincidentally contributing to the wider good of growing the tourism economy. However, as travel became increasingly affordable, the value of these emblematic tokens predictably declined. The consumer response was to collect more tokens, by travelling more and, especially, further. Soon billions of trips were being taken by millions of people travelling around the world, ironically travelling to each other's places of origin in an existential search for photographs of exotic sights. It was but a taste of things to come.

When the internet came along, people started sharing these photographic tokens online. Soon most were sharing stories and images online (Xiang & Gretzel, 2010). The battleground for status differentiation then shifted: it was no longer sufficient to collect tokens documenting visitation to a place and the photographs now needed to emphasise the uniqueness of the person against the backdrop of place (e.g. Saltzman, 2017). The travel selfie quickly became the new photographic evidentiary norm. The manipulated self-images fanned a growing demand for travel (Lyu, 2016). As the new competition for cultural status symbols intensified, so too did the requirements for exhibition of self – both in terms of personal appearance, with 'selfie face' becoming an increasingly elaborate art form (Saltzman, 2017), and in terms of dramatic posturing, even at cost of injury and death (e.g. Flaherty & Choi, 2016; Lovittt, 2016). As the destination remained a necessary backdrop, travel to popular sites escalated.

Tapping fears of social and existential irrelevancy, humanity soon found itself in the grip of a fugue-like condition of pathological travel frenzy. The effects of environmental damage and climate change were, in the main, typically dismissed as too remote and indirect to be of personal concern. The essential problem was the 'market's need for selfish escapism into hedonism and consumption' (Lane, 2009: 35). Holiday making was considered a basic human right. After decades of uninterrupted growth, the tourist system was addicted to growth (Lane, 2009); and the addicted members were simply unwilling to countenance any change to their increasingly unsustainable behaviour (Becken & Hay, 2007; Cohen *et al.*, 2011).

Ultimately, it was the more immediate human effects that brought the whole tourism system undone. In ever more places, the swelling tourist hordes began to resemble a zombie invasion, sparking local backlash (Seraphin *et al.*, 2018). Threatened with local uprisings, policy-makers had little choice but to act: travel taxes, quotas, seasonal restrictions and other policy measures limited, and then severely restricted, travel to the most popular sites. Pressured cities, such as Venice, Rome, Paris and Dubrovnik, were the first to fall. Travel demand, and consequential problems, shifted to other destinations, sparking imposition of similar travel controls elsewhere. Travel controls soon spread right across the world. For the first time since World War II travel volumes began a precipitous decline. Faced with unfamiliar notions of travel restriction, business groups and consumers protested vigorously, but to no avail. For the first time in decades, public order and community welfare took policy precedence. It became increasingly clear that travel would never be the same again.

Seeing an extant business opportunity, astute entrepreneurs dutifully furnished innovations (Schumpeter, 1934) providing new ways for stymied travel consumers to acquire the essential cultural tokens that were now no longer so readily available. For a not-inconsiderable fee, people could visit virtual reality (VR) clinics where holidays were digitally experienced and the memories simply, and quickly, implanted. It was literally the stuff of science fiction made real (Dick, 1997). These clinics also generated multimedia imagery packages providing an enduring memento of the trip, providing the vital cultural tokens of a tourist visit. The fantastic technology offered many practical advantages too – the effort of planning a trip and visiting a place was removed, saving time and effort; the experiences were never marred by the kinds of annoyances and hazards that plagued travel in real life; moreover, the photos were always amazing and could be readily remastered to provide the intended angles and effects. The technology was such that the experience was no longer a facsimile (Dewailly, 1999) but seemed entirely real. In time, it was even more real than the physical reality, a physical reality that was pushed more and more into the background of people's awareness, finally verifying Jean Baudrillard's (1983) thesis on the hyperreality. It was, for all intents and purposes just like taking the perfect holiday, and one could, for a price, visit anywhere anytime. Premium destinations were, of course, more expensive, providing scope for status differentiation or digital stratification, enabling the continuation of the market based on the foundational capitalist premise of aspirational consumption (Veblen & Banta, 2008).

In reality, the virtual implants of a certain destination were even more costly than actually travelling to the destination, something that just increased the appeal of the digital experiences, not only because they were so much more aesthetically appealing than the real world and because they were impossible to forget (as is the case with 'real' memories), but because they were more expensive per se. Eventually, the customers became tired of having these static, eternal, memories and cunning entrepreneurs quickly came up with a new device, a tele-transmitter wired in the skull that changed the implanted memories seamlessly, thus with the same high standard of unreal aesthetics. This specific technological device fell out of fashion when it was revealed that it could be used to monitor affect and state-of-mind, implying a new level of possible invasion of private integrity. Quite soon, however, people did not see a problem with this as they 'did not have anything to hide'. In a few cases, the implanted memories were creating cognitive and psychological havoc as layers of memory blurred into each other in destructive ways. This phenomenon was, however, not that considered among the public, after all, there were car accidents still (even since the driver-less car has become the norm). Also, haphazardly, or at least as an unintended side effect, the external environment was an inadvertent beneficiary since the new technology reduced physical travel.

As the new technology was expensive, as it usually is at least initially, the rest of the population had to find a different way to satisfy the culturally ingrained need for travel. Taking inspiration from the slow tourism movement, and the notion of mindfulness, for these folks a new local form of tourism emerged – the staycation (Fox, 2009; Merriam-Webster, 2018). Instead of travelling more and more, to further and further places, people started to 'holiday at or near home' (Merriam-Webster, 2018; Oxford Dictionary, 2019); either taking day trips from their own homes or, via a burgeoning sharing-economy, visiting the nearby homes of others. The growth of different sharing-economy digital hospitality platforms rocketed, but never managed to threaten the omnipresent Airbnb.inc. The growth of the sharing economy implied reinforcement of the neoliberal economic paradigm and in some senses, there remained a lack of concern about environmental sustainability (Martin, 2016), but at least the possibilities for staycation practices increased.

Following the ethos of slow tourism and mindfulness, this new kind of travel was grounded in carefully noticing one's surroundings. The new photographic tokens of cultural capital were thus found, not in the idea of exoticism, but in the self-images that depicted people mindfully attending to their immediate surrounds. It eventually became almost a genre of its own, 'mindfulness-selfies' (sometimes also called 'consciousness-selfies'). This selfie genre was primarily characterised through the attempts to express and represent an increased access to the consciousness and a mindful but highly existential awareness. In these pictorial representations, certainly still very much arranged and commercially embedded, subjects tried to express a blissful relationship between the self and the immediate physical environment of the 'staycationer'. The differentiation standard was now the degree to which people were able to show, in their circulated self-images, their ability to convey their mindful appreciation of the detailed essence of a place, and the more intricate it was, the better. The writings of Nietzsche and Heidegger had a renaissance, and one particularly popular subgenre was to illustrate ideas and statements stemming out of their work. The travel was considerably less arduous, and, since it occurred within a relatively familiar local context, it removed a great deal of the attendant risks. Not that the carbon emissions of travel had ever really been a central concern of travellers (Cohen *et al.*, 2011), but the reduction of travel did also entail an unintended side effect of environmental benefit.

Concluding Arguments

The question arises as to which imagined tourism future is utopian or dystopian, and for whom? A planet dominated by zombies, walking dead or walking tourists, sounds very much like a dystopian future; but for capitalism, a planet full of zombies is a utopian scenario. The zombie has for decades been a trope or metaphor for the consumer in popular culture and in critical social science and humanities research as the zombie has an insatiable appetite, a drive to consume and is all-absorbing with respect to its environment (Webb & Byrnand, 2008). Utopia, meaning both a 'good place' and 'non-place', indicates that for the zombie tourist, the Earth is a good place as well as non-place. For the zombie tourist, the imperative to indulge in hedonistic enjoyment crystallises into a passive nihilism; the world of touristic places is there for enjoyment, indulgence, consumption. These are 'good places'. Simultaneously they are just there as a background for the ego of the tourist, so heavily mediated through mobile technological devices, social media platforms and virtual worlds that they disappear so far in the ontological background that they cannot be found; they belong to a world that cannot be found. It is thus fitting to conclude with the question: can a future tourism, even if less based on unsustainable travelling practices, be anything than dystopic as long as it is embedded in the logic of a pathologic capitalism?

References

Baudrillard, J. (1983) *Simulations*. New York: Semiotext(e).
Becken, S. and Hay, J.E. (2007) *Tourism and Climate Change: Risks and Opportunities*. Clevedon: Channel View Publications.
Bhogesha, S., John, J.R. and Tripathy, S. (2016) Death in a flash: Selfie and the lack of self-awareness. *Journal of Travel Medicine* 23 (4), taw033.
Bourdieu, P. (1990) *The Logic of Practice*. Redwood City, CA: Stanford University Press.
Burgess, S., Sellitto, C., Cox, C. and Buultjens, J. (2009) User-Generated Content (UGC) in tourism: Benefits and concerns of online consumers. *Proceedings of European Conference on Information Systems (ECIS)*. Association for Information Systems (pp. 417–429).
Casaló, L.V., Flavián, C. and Guinalíu, M. (2011) Understanding the intention to follow the advice obtained in an online travel community. *Computers in Human Behavior* 27 (2), 622–633.
Cohen, S.A., Higham, J.E.S. and Cavaliere, C.T. (2011) Binge flying: Behavioural addiction and climate change. *Annals of Tourism Research* 38 (3), 1070–1089.
Corderoy, J. (2017) Harsh reality behind one of world's most popular tourist attractions. *News.com.au*, 6 November. See https://www.news.com.au/travel/world-travel/europe/harsh-reality-behind-one-of-worlds-most-popular-tourist-attractions/news-story/6cb1ea05272bdf97fa73d882a9d24392 (accessed 20 July 2018).
Cresswell, T. (2010) Towards a politics of mobility. *Environment and Planning D: Society and Space* 28 (1), 17–31.
Dewailly, J.M. (1999) Sustainable tourist space: From reality to virtual reality? *Tourism Geographies* 1 (1), 41–55.
Dick, P.K. (1997) We can remember it for you wholesale. In *The Phillip K. Dick Reader*. Secaucus, NJ: Citadel Twilight (original work published 1966).
Edensor, T. (2007) Mundane mobilities, performances and spaces of tourism. *Social and Cultural Geography* 8 (2), 199–215.
Fantz, A. and Nottingham, S. (2015) NTSB: Taking selfies likely caused fatal Colorado plane crash, *CNN*. See https://edition.cnn.com/2015/02/03/us/selfie-plane-crash/index.html (accessed 10 September 2018).

Flaherty, G.T. and Choi, J. (2016) The 'selfie' phenomenon: Reducing the risk of harm while using smartphones during international travel. *Journal of Travel Medicine* 23 (2), tav026.

Fox, A.K., Bacile, T.J., Nakhata, C. and Weible, A. (2018) Selfie-marketing: Exploring narcissism and self-concept in visual user-generated content on social media. *Journal of Consumer Marketing* 35 (1), 11–21.

Fox, J. and Rooney, M.C. (2015) The Dark Triad and trait self-objectification as predictors of men's use and self-presentation behaviors on social networking sites. *Personality and Individual Differences* 76, 161–165.

Fox, S. (2009) Vacation or staycation See http://citeseerx.ist.psu.edu/viewdoc/download?doi=10.1.1.474.7210&rep=rep1&type=pdf (accessed 18 January 2019).

Franklin, A. (2004) Tourism as an ordering: Towards a new ontology of tourism. *Tourist Studies* 4 (3), 277–301.

Fuchs, C. (2014) *Social Media: A Critical Introduction*. London: SAGE.

Goffman, E. (1956) *The Presentation Of self in Everyday Life*. Edinburgh: University of Edinburgh.

Graham, B. (2018) Teen struck by train was distracted by phone. *News.com.au*, 28 April 2018. See https://www.news.com.au/lifestyle/parenting/teens/teen-struck-by-train-was-distracted-by-phone/news-story/7df10acf5b9768922057faabffca054e (accessed 2 July 2018).

Hall, C.M. (2011) Consumerism, tourism and voluntary simplicity: We all have to consume, but do we really have to travel so much to be happy? *Tourism Recreation Research* 36 (3), 298–303.

Kietzmann, J.H., Hermkens, K., McCarthy, I.P. and Silvestre, B.S. (2011) Social media? Get serious! Understanding the functional building blocks of social media. *Business Horizons* 54 (3), 241–251.

Kozinet, R.V., Valck, K.D., Wojnicki, A.C. and Wilner, S.J.S. (2010) Networked narratives: understanding word-of-mouth marketing in online communities. *Journal of Marketing* 74 (2), 71–89.

Lane, B. (2009) Thirty years of sustainable tourism. In S. Gossling, C.M. Hall and D.B. Weaver (eds) *Sustainable Tourism Futures*. New York: Routledge.

Leung, D., Law, R., Van Hoof, H. and Buhalis, D. (2013) Social media in tourism and hospitality: A literature review. *Journal of Travel & Tourism Marketing* 30 (1–2), 3–22.

Lim, Y., Chung, Y. and Weaver, P.A. (2012) The impact of social media on destination branding: Consumer-generated videos versus destination marketer-generated videos. *Journal of Vacation Marketing* 18 (3), 197–206.

Lo, I.S. and Mckercher, B. (2015) Ideal image in process: Online tourist photography and impression management. *Annals of Tourism Research* 52, 104–116.

Lo, I.S., Mckercher, B., Lo, A., Cheung, C. and Law, R. (2011) Tourism and online photography. *Tourism Management* 32 (4), 725–731.

Lovitt, B. (2016) Death by selfie: 11 disturbing stories of social media pics gone wrong: Fatal poses with walruses, bulls and even a live grenade. *Rolling Stone*, 14 July. See https://www.rollingstone.com/culture/culture-lists/death-by-selfie-11-disturbing-stories-of-social-media-pics-gone-wrong-15091/ (accessed 10 September 2018).

Lyu, S.O. (2016) Travel selfies on social media as objectified self-presentation. *Tourism Management* 54 (C), 185–195.

MacCannell, D. (1999) *The Tourist. A New Theory of the Leisure Class*. Berkeley: The University of California Press.

Maddox, J. (2017) 'Guns don't kill people ... selfies do': Rethinking narcissism as exhibitionism in selfie-related deaths. *Critical Studies in Media Communication* 34 (3), 193–205.

Marchi, R. (2012) With Facebook, blogs, and fake news, teens reject journalistic 'objectivity'. *Journal of Communication Inquiry* 36 (3), 246–262.

Martin, C.J. (2016) The sharing economy: A pathway to sustainability or a nightmarish form of neoliberal capitalism? *Ecological Economics* 121, 149–159.

Martins, M. (2018) Tourism planning and tourismphobia: An analysis of the strategic tourism plan of Barcelona 2010–2015. *Journal of Tourism, Heritage & Services Marketing* 4 (1), 3–7.

Mendes-Filho, L., Mills, A.M., Tan, F.B. and Milne, S. (2018) Empowering the traveler: An examination of the impact of user-generated content on travel planning. *Journal of Travel & Tourism Marketing* 35 (4), 425–436.

Merriam-Webster (2018) Staycation. In: The Open Dictionary. [online] Springfield, MA: Merriam-Webster. See https://www.merriam-webster.com/dictionary/staycation (accessed 10 September 2018).

Munar, A.M. (2011) Tourist-created content: Rethinking destination branding. *International Journal of Culture, Tourism and Hospitality Research* 5 (3), 291–305.

Munar, A.M. and Ek, R. (2015) Relationsbits: You, me and the other. In T. Miller (ed.) *The Routledge Companion to Global Popular Culture*. New York & London: Routledge.

Munir, K.A. and Phillips, N. (2005) The birth of the 'Kodak moment': Institutional entrepreneurship and the adoption of new technologies. *Organization Studies* 26 (11), 1665–1687.

Nietzsche, F. (2006) *Human, All-Too-Human. Parts One and Two*. Mineova, NY: Dover Publications (original work published 1879).

Oxford Dictionary (2019) Staycation. In: Oxford Living Dictionaries. [online] Oxford, UK: Oxford University Press. See https://en.oxforddictionaries.com/definition/staycation (accessed 2 July 2018).

Parra-López, E., Bulchand-Gidumal, J., Gutiérrez-Taño, D. and Díaz-Armas, R. (2011) Intentions to use social media in organizing and taking vacation trips. *Computers in Human Behavior* 27 (2), 640–654.

Saltzman, S. (2017) How to take a good selfie: 12 selfie tips to consider. *Allure*. See https://www.allure.com/story/how-to-take-good-selfies (accessed 10 September 2018).

Schumpeter, J.A. (1934) *The Theory of Economic development: An Inquiry Into Profits, Capital, Credit, Interest and the Business Cycle*. Cambridge: Harvard University Press.

Seraphin, H., Sheeran, P. and Pilato, M. (2018) Over-tourism and the fall of Venice as a destination. *Journal of Destination Marketing & Management* 9 (February), 374–376.

Sezgin, E. and Yolal, M. (2012) Golden age of mass tourism: Its history and development. In M. Kasimoglu (ed.) *Visions for Global Tourism Industry – Creating and Sustaining Competitive Strategies*. London: In Tech Open.

Stylianou-Lambert, T. (2012) Tourists with cameras: Reproducing or producing? *Annals of Tourism Research* 39 (4), 1817–1838.

Thompson, M.J. (2016) *The Domestication of Criticalt*. London: Rowman & Littlefield.

Thompson, M.J. (2018) Hierarchy, social pathology and the failure of recognition theory. *European Journal of Social Theory*, 1–17.

Tkalec, M., Zilic, I. and Recher, V. (2017) The effect of film industry on tourism: Game of Thrones and Dubrovnik. *International Journal of Tourism Research* 19, 705–714.

Tussyadiah, I.P. and Fesenmaier, D.R. (2009) Mediating tourist experiences: Access to places via shared videos. *Annals of Tourism Research* 36 (1), 24–40.

Veblen, T. and Banta, M. (2008) *The Theory of the Leisure Class*. Oxford: Oxford University Press.

Webb, J. and Byrnand, S. (2008) Some kind of virus: The zombie as body and as trope. *Body & Society* 14 (2), 83–98.

Wang, N. (2000) *Tourism and Modernity: A Sociological Analysis*. Amsterdam: Pergamon.

Xiang, Z. and Gretzel, U. (2010) Role of social media in online travel information search. *Tourism Management* 31 (2), 179–188.

15 Technological Frontiers: From the Wild West Myth to the Dystopia of the Westworld's Post-human Theme Park

Jane Lovell and Sam Hitchmough

Chapter Highlights

- The technological frontier will employ hyper-staged authenticity, furthering MacCannell's (1973) concepts of 'staged authenticity' by using technology to more intricately produce nostalgic, original, reconstructing period detail from authentic sources.
- The progressive, technological frontier is ironically driven by nostalgia for the 19th-century past, which can be tailored in a bespoke experience for future tourism.
- Westworld's android host–guest relationships are a metaphor for tourism Othering; the theme park in *Westworld* resembled a slave plantation in its brutal oppression of the android 'hosts'.
- Fear of the post-human and 'becoming machine' will haunt future tourism, by threatening security and employment with the desire for the experience of the mass-produced machine 'hosts' in Westworld gaining consciousness.

Introduction

This chapter deconstructs the narrative of the frontier Science Fiction television show, *Westworld* (2016–) in order to examine future dystopic trends in tourism. Westworld is an HBO television series based on the book by Michael Crichton and the 1973 film of the same name. The series is highly relevant when considering the future of tourism because it is set in a hyperreal, futuristic, android-hosted theme park celebrating the 19th-century American frontier. Science fiction and westerns both draw on popular myths (Handley & Lewis, 2007; Stoeltje, 1987) and the myth of the frontier has been inextricably associated with narratives of pioneering individualism and adventure (Spurgeon, 2005). According to HBO, viewing figures for the first series eventually averaged 13.2 million per episode across different platforms and season 2, 10 million (Ottersen, 2018). Its viewers are 63.77% male and it is most popular in the 25–34 (31.11%) and 35–44 (25.86%) year old age brackets; it is aimed at the 'nerds' (Statsocial, 2016). According to Nielson, *Westworld*, whose cast is drawn from multiple ethnic groups, draws an audience that is 76% white, 8% African-American, 7% Hispanic, 7% Asian and 2% other (Lawler, 2018). The audience profile complies with the white, male audience of Westerns and science fiction fans, the 'nerds' but differs from the family-oriented frontier attraction and theme park tourist audience (Maher, 2016). The series depicts the type of experiential travel which may become available to the third and fourth generation of the 'experience economy' (Pine & Gilmore, 1999). This chapter makes an original contribution by examining hyper-staged authenticity (adapted from MacCannell, 1973) on the 'technological frontier', examining how period detail such as costumes is produced by android technology, in forms of nostalgic-futurism.

Western Tourism

It is highly significant that the series *Westworld* is set on the 19th-century American frontier, replaying the battle for superiority and colonisation in the same territory. When Turner (1893) stated that the frontier had closed, at the time it was a source of anxiety and nostalgia, as the connection to a rugged, national sense of freedom was lost and those associations continue. Describing how science fiction novels were inspired by a sense of loss at the closing of the frontier, Stoeltje (1987: 48) states '…recalling the early science fiction novels that served as a scientific variant of the myth, we can view the myth of the Western frontier and the myth of the scientific frontier as contiguous at their birth. Each one assumed prominence at the time when the setting of the socio-political process matched the setting of the myth: 1900 and 1960'. Western tourism flourished during the Romantic Movement, which 'inspired travel to the American West' from 1830–70 (Brégent-Heald, 2007: 50). A utopic frontier idyll was born, inspired by characters such as William Cody, whose Buffalo Bill's Wild West shows toured extensively. The Arcadian landscape was reimagined by the romantics as transcendental and sublime; re-enactments and celebratory events such as Frontier Day celebrated western skills.

According to Maher (2016: 14–17) western tourism boomed from 1920–1980 in what he described as the 'frontier era of recreation' when tourist attractions proliferated. Maher outlined different elements of frontier tourism supply: 'This vast array of pulp fiction, western films and television shows, theme parks, entire towns catering to tourists in search of the Wild West (Dodge City, Deadwood and Tombstone in particular) with national parks and state parks reconstructing 19th-century forts...' (2016: 17). The popularity of western tourism then began to recede from 1980, which Maher (2016: 21) attributes to the lack of constant television exposure and the enculturation policies of America, moving frontier tourism from entertainment into the realm of cultural heritage.

Westworld reopened the frontier, reframing the iconic scenery and skyscapes within a theme park narrative. Western tourism has long mingled with film tourism (Beeton, 2015; Handley & Lewis, 2007; Laing & Frost, 2015), adding to the atmosphere of the cinematic environment. As Tkalec *et al.* (2017: 705) argue, based upon the positive effect of *Game of Thrones* on visitor numbers to Dubrovnik from 2012, associations with fantasy television series provokes interest in destinations. Film-induced tourism has attracted tourists to visit classic western locations such as Monument Valley, where John Ford's film *The Searchers* and other classic Westerns were shot. As films become more sophisticated, the implication for tourism is that they will attract a more knowing type of visitor, familiar with the effects of CGI and editing, who visits the location not for an exact imprint of a cinematic place, but for an impression which forms a patchwork with virtual versions.

The series of *Westworld* perpetuates and revives the frontier tropes, recreating a dystopia in contrast to an idyll, by repeating the aggressive suppression of Native American peoples and Ghost Dancer resistance, echoed in the alienation, and oppression of the android hosts. The drama re-plays the wider colonisation of places by tourists, also presenting a form of dystopic future dreaming, which demonstrates the consequences of a tourist-host relationship inversion after a rebellion by androids. Many historians point to the fact the United States took its search for expanding new frontiers of technology and modernity that took the country, ultimately, into space: the 'final Frontier' of expansionism. That the technological frontier would return the tourist gaze (Urry, 2002) to the Western was inevitable, as was the dystopic perspective it would project. This paper addresses a vision of the next phase of Western tourism, when, with President Trump's cuts to national park funding (National Parks Conservation Association, ND), it is likely that investment will increasingly come from private sources and the experiential sensationalism of the theme park West will be heightened to compete with other fantasy-heritage attractions.

Fast Authenticity

It is said that in the globalised economy, future tourists will increasingly search for authenticity (Yeoman *et al.*, 2012). The quest for the real, 'true West' or 'Old West', symbolised by the frontier, seems to be an enduring tourism trope (Delyser, 1999; Handley Lewis, 2007). Knudsen *et al.* (2016) and Vidon and Rickly (2018) argue that tourists seek a fantasy to escape the alienation of their daily lives in their pursuit of authentic experiences. It is probable that tourists will continue to both incorporate immersive technologies such as virtual reality and seek to escape their online connectivity to explore a more authentic 'nonline' leisure existence. The believability of the Westworld premise is echoed in the increasing popularity of the pursuit of alternate realities, as some tourists undergo transformational experiences at living history sites, re-enactment festivals, cos-play events and at fantasy film locations (Beeton, 2015; Everett & Parakoottathil, 2018).

Theme parks specialise in immersive technology designed to counter reality and simulate other places. In the case of frontier attractions, the illusion of 'elsewhereness' (Butler, 1991), also leads to 'elsewhenness' as they revive the past. Visiting frontier theme parks indicates 'postmodern authenticity' (Wang, 1999), predicated on the dominance of hyperreality (Baudrillard, 1981; Eco, 1986; Fjellman, 1992) and acceptance of the copy and simulated visitor experience. Lovell and Bull (2017: 8) examined authentic and inauthentic tourist places, exploring the attributes of Fjellman's (1992) heuristic categories of 'fake fake, fake real, real fake and fake fake' and added categories including 'hyperreal'. They assessed the ways in which tourists authenticated 'real fake' theme parks and their hyperreal sensations of dissonance and 'the wobble, when reality is inverted and seems to be a copy' (Fjellman, 1992). Inverted reality is apparent when studying western theme parks, which use different and sliding scales of 'rhetorics of authenticity' for example at the theme park *Knotts Berry Farm* by recreating an old schoolhouse from documentary evidence, but adding a red bell tower (Hsu, 2007). Cowboy and Wild West iconography have inspired 'fake real' places based on myths; Knotts Berry Farm claims: 'The difference is real' drawing attention to its distinctiveness in comparison with attractions such as Disney's *Frontierland*, which are based on filmic references. Westworld moves the themed environment into the realm of the hyperreal, by providing a backdrop of a terraformed copy of the western frontier as a theme park, a simulacrum of a simulacrum (Baudrillard, 1981), which is depicted as attracting tomorrow's Western tourists. The series takes MacCannell's (1973) concepts of 'staged authenticity' further, generating 'hyper-staged authenticity' where technology is used to more intricately reconstruct period detail from authentic sources. Lovell and Bull (2017: 171) discuss the real-fakery of theme parks, which place the emphasis on the staging, rather than the authenticity. They posit that theme parks are 'hotly inauthenticated' (adapted from Cohen & Cohen, 2012) using 'fast authenticity' by visitors searching for the artificial, for example fast food, souvenirs as signifiers of themed sense of place. The premise of *Westworld* is to simulate and augment the illusion of fast experiences (death, sex) and to provide reprogrammable hosts for the shorter attention spans of the third and fourth generations of the experience economy (Yeoman, ND), who, as the series suggests, will require their gaming enacted in an embodied arena to enhance its experiential authenticity.

Hyper-staged authenticity in theme parks

The pursuit of alternative realities by tourists means that forms of fantasy theming are increasing, for example, 'Harry Potterisation', the recent take on Bryman's (2004) Disneyisation, which has inspired theme parks in America, London and Japan and visitor attractions at King's Cross Station and York. Different variations of Western theme parks, such as *Knotts Berry Farm* and *Frontierland*, exhibit the Disneyfication of the West, using what could be termed hyper-nostalgic framing. While theme parks seem to embrace Ritzer's concepts of McDonaldisation, Bryman (2004:13) differentiates between the two terms, arguing: 'Disneyization's affinities are with a post-Fordist world of variety and consumer choice'. As Migacz and Petrick (2018) state in a study of Millennial tourism motivations, Millennials vary in their taste and requirements. Their typology of 'Young and free Millennials' require educational sites which also incorporate adventure and fun. Urry's (2002) theory of 'edutainment' therefore seems likely to continue in popularity. Eco terms Knotts Berry Farm a 'city of robots' (1986: 42) and goes on to define it as a hallucinatory 'ghost town' filled with waxworks and models. He argues (1986: 43) that 'Disneyland is more hyperrealistic than the wax museum, precisely because the latter still tries to make us believe that what we are seeing reproduces reality absolutely, whereas Disneyland makes it clear that within its magic enclosure it is fantasy that is absolutely reproduced'. Gottdeiner (2001: 152) suggests that theme parks offer a utopian space, for example, Disneyland's main street USA. While Baudrillard (1981) has argued that the Disney fantasy exists to make the rest of America look real, Eco contends that 'Disneyland tells us that technology can give us more reality than nature can' (Eco, 1986: 44) and Fjellman (1992) discusses Walt Disney World's 'vinyl leaves on plastic trees'.

The pursuit of the authentic past by tourists is a chimera which, like other forms of heritage-fantasy Wild West tourism, could also be said to be partially motivated by nostalgia (Timothy, 2011). Thus, the future of tourism attraction design is indicated in what Katerberg terms the 'Western paradox' (2008: 31) which is the interweaving of experience and design with the two strands of progression and nostalgia. While Western films tended to look backward, science fiction is future-facing (Katerberg, 2008) and this contradiction is reflected by the vertiginously nostalgic–futuristic theme park of *Westworld*. That *Westworld* celebrates the imagined future with postmodern heritage techniques is substantiated by the comments of Zack Grobler, *Westworld*'s production designer, when outlining how showrunner Jonathan Nolan, instructed him to create a 'unique aesthetic' (Zacharin, 2017):

> Westworld has to be both nostalgic and futuristic while also seeming ever so slightly artificial. Because the theme park costs the character more money to visit than their life insurance is likely to pay out, it also can't be completely wild'. As Grobler points out, 'We tried to give a much higher quality and much more realistic setting. Disneyland is not bad now, but a few hundred years from now, the technology will be so much better that you will not even realise you're in the park'.

Hyper-staged authenticity is fully realised in *Westworld*, which depicts nostalgic-futurist methods of enlivening a heritage theme park. Lovell and Bull (2017: 176) suggest that attractions such as the Wizarding World of Harry Potter are often meticulously researched, offering parallels to the efforts made in traditional skills and research to maintain artefacts in historic sites. In future, tourism may rely increasingly on using hyper-staging processes, which create, not only the impression of elsewhere, but also 'elsewhen-ness'. The likelihood of this approach is substantiated by *Westworld* costume designer Ane Crabtree, who stated how she struggled to replicate authentic 19th-century fabrics and hired 3D printers to help with the work, mirroring the visuals of androids constructed by what appear to be organic 3D printers in the series title sequence. 'Costume designers found vintage fabrics, hand-painted and distressed them, and then replicated the textiles using 3D printing...fabrics today just aren't as intricate – they're not made the same way' (Minton, 2016). Costume enables tourist to inhabit simulacra and have transformational experiences (Everett & Parakoottathil, 2018) at cos-play events; crafting 3D copies of vintage fabrics typifies the hyper-staged authenticity of the simulacrum where reproductions are produced from sources loser to the originals.

Westworld incorporates a number of future trends in tourist; the theme park offers increasingly bespoke tourism to the wealthy guests. Theme parks are predicated on the concepts of deeper immersion, 'riskless risk' (Hannigan, 2007) and control, drawing upon the operational hallmarks of regulation and staging designed by Disneyland. *Knotts Berry Farm* recreates the tropes of saloon shows, an Old Wild West Ghost Town and a Wild West stunt show which are tightly scripted in a safe space, with guns shooting blanks, as are the set pieces staged in the main street of Westworld's Sweetwater and Mariposa saloon. Katerberg (2008: 34) pointed out that Frontierland and Tomorrowland are also ideologically safe: 'Neither threatens the settled frontier mythology or comfortable lifestyle that most visitors bring to the park'. Rather than providing thrill rides, Westworld revives the vanishing, closed frontier, restoring the lost freedoms of the lawless past in the theme park ethical vortex to provide peak experiences. Imitating a future-real experience, the official HBO series *Website* includes an ethical questionnaire which ascribes you a role as a potential Westworld guest to test the visitor experience of 'Life without limits'. However, Westworld's future customers aren't the passive observers of the staged, formulaic cheroot-chewing gunfights of *Knotts Berry Farm* or deadwood, they are abusers, indulging their desire for violence. Yet, even in 'black hat' mode, participants can safely and invincibly indulge their fantasies, even as they range further away from the more controlled township to have a peripherally authentic dangerous experience. While the peripheral areas of the park ramp up the terror for guests with wilder-west scenarios in the edgy town of Pariah and encounters with the fictional Ghost Nation, they are also controlled. In episode 2:05 Acane no Mai, a park hierarchy is revealed; Shogun World offering Ronin and Ninja challenges for those guests who find Westworld too tame. The future experience economy takes the thrill of gaming from a virtual space into embodiment, while preserving the unreality.

Another aspect of future tourism is explored in *Westworld* episode 2:1 The Passenger. In order to secure an investment, the protagonist William reveals to James Delos that it is not the hosts who are being copied but the tourist guests, whose behaviour is recorded, in order to both explore the intricacies of human decision-making and ultimately, to copy human minds into host bodies, offering life after death. Bryman (2004) discussed the focus on control through surveillance in theme parks and this plotline uncovers the increasing monitoring of data, in which theme parks seem to specialise. The attractions share the characteristics of 'non-places' (Augé, 1995) such as airports; people are known by wristbands; the interaction of guests tends to be with technical appliances. It seems likely that future theme parks will not simply further the fantasy experience, but also that convenience technology masks the collection of big data. While this may lead to bespoke and personalised experiences, the implication is that the guests' identity can be duplicated for manipulation and meta-control. The plot arc also reveals the commercial potential of the increasing desire for immortality, which is likely to be reflected in future tourism trends such as wellness. It is important to note that when the Westworld engineers craft a cyborg James Delos by copying his mind into an android replica of his body after his death, he continually rejects his reality, experiencing paralysing dissonance.

The imagination of visitors in Disney theme parks is controlled by 'imagineers', as Bryman (2004: 135) indicates; according to *Knotts Berry Farm's* website the experience 'Ghost Town Alive allows people to play a role in the re-enactment of the West:'

> Guests can visit the beloved town of Calico and experience an authentic western adventure by becoming a star in this summer's story of the Wild West, where the power to unlock adventures is in the hands of every guest. Ghost Town Alive! offers guests a first-hand experience to play an important part in the unfolding story of the Wild West, with specific tasks and activities and an essential role as an honorary citizen of Calico.

Familiar theme park conventions are followed in series 1 of *Westworld*, where the guests also participate in scripted plotlines. These narratives seem tailored but contain tropes and set-pieces such as the inevitable gunfight and robbery, which are repeated on a loop, recalling Edensor's (2001) analysis of choreographed guided tours. The script 'loops' move theme park manipulation into a fresh domain; different robots act the same storylines interchangeably. For example, when the android Peter Abernathy, the father of Delores, finds a photograph which reveals a different reality and triggers an existential crisis, he is replaced by a different robot the following day. His disposability only becomes deeply sinister when his host daughter, Delores, initially does not notice the switch. In season 2, the androids meet Japanese versions of themselves, following similar scripts in Shogun World, creating an existentially hyperreal wobble when the simulation recognises itself in another simulation, a Benjaminian nightmare of infinite reproduction and loss of self.

Becoming machine

The most striking aspect of future tourism depicted in *Westworld* is that the central theme park has taken immersive experiences to the level of producing disposable android 'hosts'. Yeoman (2012: 206) suggests that by 2050 some hotels may be staffed by robots and that people may consider androids as long-term partners. Each host rises dripping on the wheel as an android version of post-Vitruvian man, which is a drawing by Leonardo da Vinci which shows man in two positions with his arms and legs apart within a circle and square. Baudrillard and Eco have both discussed the value of cybernetics as a hyperreal phenomenon. Baudrillard (1981: 124) states that:

> It is no longer possible to fabricate the unreal from the real, the imaginary from the givens of the real. The process will, rather, be the opposite: it will be put to decentred situations, models of simulation in place and to contrive to give them the feeling of the real, of the banal, of lived experience, to reinvent the real as fiction because it has disappeared from our lives.

Automata have always attracted visitors, from the creations of Robert-Houdin onwards, for example his writing automaton, which was built in 1844 for the Universal Exposition. Androids, illusion and special effects tend to inspire two reactions in visitors. Firstly, tourists have long been drawn to androids or even suggestions of androids, for example the highly popular holograms projected on to mannequins which appeared to talk sing and wink in Jean Paul Gaultier's recent touring Blockbuster retrospective (2011–2015) and dominated the reviews of the exhibits on show. As Baudrillard (1981: 5) argues, androids seem to be 'perfect simulacra, forever radiant with their own fascination'. The Westworld androids seem so real that the character Logan Delos, cannot identify them when he first meets them at a sales pitch in episode 2:2 Reunion declaring: 'We are not here yet!' His meaning is that the future has arrived faster than he had supposed.

The second reaction to androids is that they also inspire 'uncanny valley phenomenon' summarised by Saygin *et al.* (2011: 414) as 'as an agent's appearance is made more human-like, people's disposition toward it becomes more positive, until a point at which increasing human-likeness leads to the agent being considered strange, unfamiliar and disconcerting'. Robots and androids form a central branch of Science Fiction film and literature, dating from Fritz Lang's *Metropolis*. Perhaps the most famous examples are the computer Hal in 2001 and the dystopian novel *Do Androids Dream of Electric Sheep* by Philip K. Dick (1966) which was developed further in the movie *Blade Runner* (1982). The atavistic fear of awakening, superior machines is accompanied by the dread of economic redundancy, which was explored in the television series *Humans* (2015–17), in which the modern workforce is being replaced by robots and machinery. This was also a central psychological and real effect of 19th-century industrialisation, which helped to idealise the frontier idyll of the past (Lovell & Bull, 2017). Fear of androids is predicated on the suspicion of the increasingly organic nature of artificial intelligence, and 'hyper-real-fake hosts' becoming 'real'. *Westworld* explores the tensions between these two forces of fascination and atavistic fear; the response to 'unhappy valley' is to abuse the android 'hosts'.

Identity politics are becoming increasingly important within tourism studies in the age of globalisation. It is important to note that androids – the Other – are not post-racial, or post-gender (Braidotti, 2013), they carry markers of race, which, in the case of the Ghost Nation Lakotan, Akecheta, may have significance for their programming. Their position as hosts symbolises those many races who have been oppressed: the technological frontier becomes a metaphor for a colonised reservation, portraying a group of people who have resources, which are being exploited and controlled. The theme park also resembles a slave plantation, where the enslaved have to do whatever they're told to serve the visitors. There is a further parallel with the reaction by residents to overtourism, the protests against the exploitation of resources and vicious cycle of tourism at its peak in cities such as Venice, Barcelona and Amsterdam (Seraphim *et al.*, 2018). For a central android character, Delores Abernathy, the West is certainly the edge and boundary of the theme park, outside which she can no longer operate (a familiar narrative used to control slaves who feared an unknown brutality outside the boundaries of their owner's jurisdiction where they would be hunted down or preyed upon). As discussed earlier, Baudrillard stated that theme parks create a simulated world, which makes the rest of existence seem real in contrast and this is referenced by the Man in Black, when he barks at a guest who mentioned his 'real life' that he is 'on holiday' and desires escape. It is this glimpse into his alternative reality as a philanthropist which seems unreal as he inhabits the avenging character of the Man in Black.

The technological frontier creates a dividing line between host and tourist. It is mixed by their architect as he enables the hosts to remember the abuse they suffered from guests and theme park workers. Our sympathy is with the plight of Delores, who is positioned as the exploited heroine (before she evolves into the avenging character of Wyatt). The scripted control of tomorrow's theme parks raises ethical questions in the host-slave motif, where the memories of disposable, reprogrammable robot hosts are 'rolled back' or they are 'put down' and placed into cold storage (the stillness and storage of the basement resembling the morgue) when they malfunction. Artificial intelligence is no longer a copy when it becomes self-aware; Delores assumes an authenticity of her own. In the episode 1:05 Contrapasso where all characters start to rebel against their roles, first their more authentic selves seem to appear, and then performances are dropped, breaking prescribed behaviours. When hosts recall the violence dealt to them by tourists, or other hosts, for example, the character Maeve reflecting on how both she and her 'child' were murdered, their present reality 'wobbles' in a hyperreal moment and retribution for the guests is inevitable, as the tourist becomes the Other and the host a new, superior species.

The counterpoint to theme park control is the small acts of defiance by workers and visitors according to Bryman (2004: 151); the danger which tourists seek can only truly be explored outside the loop of surveillance. The slow awakening of resistance by artificial intelligence is traced in episode 1.1 The Original, from the inability to blink an eye in reaction to a fly, to the swatting of the insect as a tremor of coming resistance. The future is foreshadowed in that simple motion; differences are blurring between host and guest are blurring. The action substantiates the thoughts of Braidotti (2013: 91) who states: 'This machinic vitality is not so much about determinism, inbuilt purpose or finality, but rather about becoming and transformation'. The deus ex machina is explored in both Science Fiction and Frontier Science Fiction such as Leslie Marmo Silko's *Almanac of the Dead*. The deus ex machina is made explicit in Season 2 episode 2:01 Journey into Night; after death, Robert Ford inhabits the cyberworld which he has constructed, building a meta-game for the Man in Black/William. Ford's omniscience resembles Walt Disney, whose design, rules and personality seeped into every aspect of his theme parks until man, place and philosophy were merged. The plot traced the ultimate dystopian fear of 'becoming machine' discussed by Deleuze and Guettari (1988) where humanity is not simply conquered, but subsumed.

The building and eventual rebellion of the hosts is significant for this paper, because Westworld plays out a fear of the post-Anthropocene which is reflected in a proliferation of recent, multi-platform dystopic science fiction series, such as *The Walking Dead* (2010–). While, as Katerberg (2008:

39) argues, the frontier marries the nostalgic and progressive, there is a warning in the conceptual reinvention of another 'New World:' 'If American progress required the destruction of the Indians, will progress of the future require the destruction of America?' Simulated dangers which tourists seek in video gaming and theme parks are inverted, as William states in episode 2:01 Journey Into Night when he re-authenticates his experience: 'The stakes are real in this place now. Real consequences'. Ultimately, the tourist-host relationship is overturned, the loop revolving into a cycle of retribution for hyper-consumption, the post-Anthropocene encapsulated in the sentence: 'These violent delights will have violent ends'. Yet, in episode 8:1: Trace Decay, when Delores finds a church buried under the sand with only the steeple visible, we are reminded that even the technological frontier will pass.

Concluding Arguments

The chapter suggests that while visitors seek pseudo 19th-century dangers on the frontier, adopting android technology poses significant risks and ethical challenges on the hyper-frontier of the future. The theme parks in Westworld act as metaphors for change, demonstrating how, as inventions become more sophisticated, future tourism trends will reflect the fear of a post-human and post-Anthropocentric universe. While *Westworld* appears to offer a utopic future consumer dream of 'fast authenticity', it forges a new dystopia; the out of control theme park, where the visitors are locked into a more authentically brutal, truly 'wild' west nightmare. Authentic period details intended to enhance the immersive experience are hyper-staged, new technologies resurrecting the past in *Jurassic Park* scenarios, moving ever closer to original sources. *Westworld*'s treatment of a new Other, the post-human, has been compared to colonisation processes. The hosts act as signifiers for past oppression and mirror the protests about overtourism in cities such as Amsterdam, where residents feel exploited; a movement which is likely to increase as access to travel accelerates.

According to *Westworld*, future tourists will still pursue unethical hyperreal versions of authentic experiences. Offered 'life without limits', guests choose to repeat historical mistakes and re-enact past injustices on the frontier, potentially fuelling a dystopian themed playground where tourism is used to cloak 'violent delights'. Simultaneously, while dealing death, the park is used to further dreams of immortality. Life after death is evident in the performances of actors such as Carrie Fisher and Peter Cushing in the *Star Wars* series. As the android host Angela states: 'We're here'. It is the digitally de-aged face of Anthony Hopkins, edited onto the body of an actor in episode 1:3 The Stray, which gives the most unsettling effect. This provokes the most interesting thoughts about nostalgia-driven future tourism: in Westworld, our decisions, behaviours, experiences and memories will be recorded and recreated and the past brought back to life.

As the younger Robert Ford speaks, the effect is startling, life on a loop, cyber-immortality achieved, individuality blended, becoming machine.

References

Augé, M. (1995) *Non-places: Introduction to an Anthropology of Supermodernity*. London: London: Verso.
Blade Runner (1992) *Directed by Ridley Scott [Film]*. USA: Warner Brothers Pictures.
Baudrillard, J. (1981) *Selected Writings: Simulacra and Simulation*. Cambridge: Polity Press.
Beeton, S. (2015) *Travel, Tourism and the Moving Image*. Bristol: Channel View Publications.
Braidotti R. (2013) *The Posthuman*. Cambridge: Polity.
Brégent-Heald, D. (2007) Primitive encounters: Film and tourism in the North American West. *Western Historical Quarterly* 38 (1), 47–67.
Bryman, A. (2004) *The Disneyization of Society*. London: Sage.
Butler, R.W. (1991) West Edmonton Mall as a tourist attraction. *The Canadian Geographer/Le Géographe Canadie* 35 (3), 287–295.
Cohen, E. and Cohen, S.A. (2012) Authentication: Hot and cool. *Annals of Tourism Research* 39 (3), 1295–1314.
Deleuze, G. and Guattari, F. (1988) *A Thousand Plateaus: Capitalism and Schizophrenia*. London: Bloomsbury Publishing.
DeLyser, D. (1999) Authenticity on the ground: Engaging the past in a California ghost town. *Annals of the Association of American Geographers* 89 (4), 602–632.
Dick, P.K. (1966) *Do Androids Dream of Electric Sheep*. New York: Doubleday.
Eco, U. (1986, 2014) *Faith in Fakes*. London: Random House.
Edensor, T. (2001) Performing tourism, staging tourism: (Re) producing tourist space and practice. *Tourist Studies* 1 (1), 59–80.
Everett, S. and Parakoottathil, D. (2018) Transformation, meaning-making and identity creation through folklore tourism: The case of the Robin Hood Festival. *Journal of Heritage Tourism* 13 (1), 30–45.
Fjellman, S.M. (1992) *Vinyl Leaves: Walt Disney World and America (Institutional Structures of Feeling)*. Boulder: Westview Press.
Gottdiener, M. (2001) *The Theming of America: Dreams, Media Fantasies, and Themed Environments*. Cambridge, MA: Westview Press.
Knudsen, D., Rickly, J. and Vidon, E. (2016) The fantasy of authenticity: Touring with Lacan. *Annals of Tourism Research*, 58, 33–45.
Handley, W. and Lewis, N. (eds) (2007) *True West: Authenticity and the American West*. Lincoln, NE: University of Nebraska Press.
Hannigan, K. (2007) From Fantasy City to Creative City. In G. Richards and J. Wilson (eds) *Tourism Creativity and Development* (pp. 48–56). Abingdon: Routledge.
Hsu, H. (2007) Authentic re-creations: Ideology, practice, and regional history along Buena Park's entertainment corridor. In W.R. Handley and N. Lewis (eds) *True West: Authenticity and the American West*. Lincoln, NE: University of Nebraska Press.
Katerberg, W.H. (2008) *Future West: Utopia and Apocalypse in Frontier Science Fiction*. Lawrence: University Press of Kansas.
Knotts Berry Farm (ND) See https://www.knotts.com/explore/ghost-town-alive (accessed 29 August 2018).
Lawler (2018) Study: How diverse are the audiences for 'This Is Us', 'Empire', 'Westworld'? See https://eu.usatoday.com/story/life/tv/2018/06/28/how-diverse-audiences-us-empire-westworld-nielsen/738216002/ (accessed 30 January 2018).
Laing, J. and Frost, W. (2015) *Imagining the American West through Film and Tourism*. Abingdon: Routledge.
Lovell, J. and Bull, C. (2017) *Authentic and Inauthentic Places in Tourism: From Heritage Sites to Theme Parks*. Abingdon: Routledge.

MacCannell, D. (1973) Staged authenticity: Arrangements of social space in tourist settings. *American Sociological Review* 79 (3), 589–603.
Maher, D.R. (2016) *Mythic Frontiers: Remembering, Forgetting, and Profiting with Cultural Heritage Tourism*. Gainesville, FL: University Press of Florida.
Migacz, S.J. and Petrick, J.F. (2018) Millennials: America's cash cow is not necessarily a herd. *Journal of Tourism Futures* 4 (1), 16–30.
Minton (2016) HBO's Westworld 3D printed its period costumes. *Architectural Digest*. See https://www.architecturaldigest.com/story/hbo-Westworld-3d-printed-costumes (accessed 31 July 2018).
National Parks Conservation Association (ND) Trump proposals fail National Parks. See https://www.npca.org/articles/1750-trump-proposals-fail-national-parks (accessed 31 August 2–18).
Otterson, J. (2018) Westworld' season 2 premiere ratings steady with season 1, *Variety*. See https://variety.com/2018/tv/news/Westworld-season-2-premiere-ratings-1202782514/ (accessed 31 July 2018).
Pine, B. and Gilmore, H. (1999) *The Experience Economy: Work is Theatre and Every Business a Stage*. Boston MA: Harvard University Press.
Saygin, A.P., Chaminade, T., Ishiguro, H., Driver, J. and Frith, C. (2011) The thing that should not be: Predictive coding and the uncanny valley in perceiving human and humanoid robot actions. *Social Cognitive and Affective Neuroscience* 7 (4), 413–422.
Seraphin, H., Sheeran, P. and Pilato, M. (2018) Over-tourism and the fall of Venice as a destination. *Journal of Destination Marketing & Management Access* 9 (000), 374–376.
Spurgeon, S. (2005) *Exploding the Frontier: Myths of the Postmodern Frontier*. Texas: A&M University Press.
Slotkin, R. (1992) *Gunfighter Nation: The Myth of the Frontier in Twentieth-Century America*. Oklahoma City, OKC: University of Oklahoma Press.
Statsocial.com (2018) Who Exactly Is Watching HBO's New Hit Westworld? https://blog.statsocial.com/who-exactly-is-watching-hbos-new-hit-westworld-72d1b4a3c608 (accessed 17 February, 2018).
Stoeltje, B. (1987) Making the frontier myth: Folklore process in a modern nation. *Western Folklore* 46 (4), 235–253.
Timothy, D. (2011) *Cultural Heritage and Tourism: An Introduction* (1st edn). Bristol: Channel View Publications.
Tkalec, M., Zilic, I. and Recher, V. (2017) The effect of the film industry on tourism: Game of Thrones and Dubrovnik. *International Journal of Tourism Research* 19 (6), 705–714.
Urry, J. (2002) *The Tourist Gaze* (2nd edn). London: Sage.
Urry, J. and Larsen, J. (2011) *The Tourist Gaze 3.0*. London: Sage.
Vidon, E. and Rickly, J. (2018) Alienation and anxiety in tourism motivation. *Annals of Tourism Research* 69 (1), 65–75.
Wang N. (1999) Rethinking authenticity in tourism experience. *Annals of Tourism Research* 26 (2), 349–370.
Yeoman, I. (2012) *2050 – Tomorrow's Tourism*. Bristol: Channel View Publications.
Yeoman, I. (ND) The future of experiential travel: Call for Papers See https://www.etfi.nl/en/blog/future-experiential-travel-call-papers (accessed 31 August 2018).
Yeoman, I., Robertson, M. and Smith, K. (2012) A futurist's view on the future of events. In S. Page and J. Connell (eds) *The Routledge Handbook of Events* (pp. 507–525). Abingdon: Routledge.
Zacharin, J. (2016) Inside the Westworld set's luxury dystopian cowboy Disneyland. *Inverse Magazine*. See https://www.inverse.com/article/21830-design-of-Westworlds-old-western-town-set-hbo-melody-ranch (accessed 25 June 2017).

16 The Future of Music Concerts and Tourism in Dystopian Times

Daniel Wright

Chapter Highlights

- This chapter applies a futurist approach to consider the role of music events and tourism in dystopian times.
- The role of music as an influencer on society, cultural identity and global communities are examined.
- The chapter explores how music can impact on individual and community emotions and behaviour.
- Tourism and benefit concerts are explored as a mechanism to support future societies and or communities living in dystopian realities.

Introduction

> The true beauty of music is that it connects people. It carries a message, and we, the musicians, are the messengers.
>
> (Roy Ayers see Harmonious Assembler, ND)

Society is constantly being exposed and overwhelmed with stories, images, movies and music that express fear. Our world is saturated with fictional ideas of what a dystopian future would look like. For some people and communities, a dystopian world is their non-fictional day-to-day reality. Just like our ancestors were exposed to great horrors and tragedies, our world is not free of terror and pain. The future has the potential to also be littered with pockets of dystopian realities throughout our globe. So, what if at our disposal, humans had a force so powerful that it could bring people together in the darkest of times. In times of natural disaster, war, human atrocity, a time and place that could be likened to a dystopian reality, be it for a few people, a community, a nation, continent or even our global society. What if we could set in motion a movement that would help those in need in the darkest of periods. What if society had a tangible influencer, something that unites people. Society does have such a tangible force, we can establish a movement in which victims of a dystopian world can be supported. This movement can be found within the music tourism industry, in what is often termed benefit concerts. Society has already shown its capabilities to come together through music to support victims of tragedy, a popular example being Live Aid. However, current academic literature provides limited consideration and research exploring benefit concerts and importantly, how they could be influential in the future as a support mechanism for communities living in dystopia. This chapter, applying a futurist approach, considers how future foresight can provide an ability to look ahead, to consider what potential realities await humanity. As noted by Slaughter (2003: xxii) a central task of futures studies is the consideration and depiction of future landscapes. With such knowledge, insights and information gained, people, companies, nations, can make decisions, strategise and plan for change. Society is susceptible to unexpected and volatile changes and events, be they environmental phenomena or changes in the geo-political landscape that can result in human conflict or environmental disasters. Consequently, societies and communities can be willingly or not pulled into chaos and suffering. In the event of such scenarios, crisis and disaster management, and contingency plans are highly important. This chapter offers insights into how music tourism through events such as benefits concerts should be considered as a contingency plan to support communities living in dystopia. Thus, this chapter offers original and novel research and discussions on the future of benefit concerts and tourism as a means of supporting people living in dystopian realities. It considers how tourism and benefit concerts can ensure global exposure and support to communities' suffering. To explore this, the chapter will initially consider the role of music in society, the potential of looming dystopian realities and subsequently, how future benefit concerts and tourism could have the ability to bring about relief to communities.

Music's Influence on Culture and Society

Ever since different cultures started to form, there has always been a place for rhythmic sounds that can communicate our feelings. In every age and civilisation this particular form of expression has existed, and in so many varied styles.

(Exploring Your Mind, 2018)

Throughout medieval times culture retained a more constant form, much to the dominance of the Catholic Church as an undeniable influential force. Today, culture is more complex, fluid and interchangeable for individuals and collectives. In the past (in Europe and the Middle Ages) music might have remained similar for hundreds of years (Duke, 2014). As society moves away from the dominant cultural stabiliser approach (provided by institutions such as the Catholic Church), our world and its people are more likely to pick and choose different aspects of cultural from around the world and establish their own preferences, ideas, beliefs and practices. The ability to share information instantly across social platforms has magnified this to phenomenal levels. Such sensory overload of digital information and real-life experiences (often attributed to the growth of international travel) ensures people are disposed to continually change (Duke, 2014; Huang, 2015). Music has been a significant pillar throughout human history. Today, popular music of our time is said to reflect our present culture. Accordingly, if one is to observe the transition of music throughout time, 'we can see the fingerprints of a certain generation in the lyrics and sound of that time' (Huang, 2015). With that in mind Huang (2015) suggests that 'In other words, culture and music flow together. What our parents used to dig, kids of today would deem as lame. And in a few years, the music we think is cool now will probably be outdated. It's nothing against the music. It's just a representation, a manifestation of what's constantly changing around us. With that said, we need to be very aware of our modern day culture, but more importantly, we need to be intentional about the cultures we want to create and cultivate with our music'. The culture and identify of people and communities can change because of the music being produced and consumed.

From this perspective it could be suggested that music and importantly musicians have an important role in society as they and their music are not only influential on people but can change and even establish new cultures. Such a premise is noted by Huang (2015), 'As musicians, we are carriers of influence, whether or not we are aware of it and whether or not we intend to be. The sound and messages we release through our art form directly impact our listeners in powerful ways. This is especially true of the youth and adolescents of our society, who are still extremely malleable to the world around them'. Ludwig van Beethoven said, 'music is a higher revelation than all wisdom and philosophy' and Bono noted that 'music can change the world because it can change people'. Music is a powerful cultural tool within society and the role of musicians is also of note, as fans embrace their musical celebrities as role models. Consequently, musicians have an important role to play in society. There are many examples in which the power of music has impacted people, cultures and has operated as a support mechanism during difficult times. Examples are discussed by Nuñez (2015) and include, Bob Dylan: 'Times They Are A-Changin', which was intended for a young generation of the 60s who felt that segregation and oppression were outdated practices and were looking for change. Widely regarded as John Lennon's signature song, 'Imagine' conveys his wish for world peace. The song was originally inspired by a poem written by Yoko Ono. The song is poignant, encouraging us to look to the future and work towards a world without extreme poverty.

Music and Tourism

> I travel the garden of music, thru inspiration. It's a large, very large garden, seen?
>
> (Peter Tosh, see IMDb, 1990–2018)

According to Gibson and Connell (2003), music tourism is an undervalued sector of the cultural economy. Furthermore, it is suggested that there are no 'typical music tourists' especially considering the number of participants and experiences, and the ability for people to move between them (Gibson & Connell, 2007).

> 'Music tourism constitutes evolving clusters of tourists, activities, locations, attractions, workers and events that utilise musical resources for tourist purposes. Related sites exist within sets of networks of transport and tourist infrastructures, social relations, business linkages and cultural performances that support certain activities and economies. Music tourism can be seen as a range of practices where sites of music production and expression (whether in past or present "scenes") become the points of attraction for tourists, and may also become central to strategies employed by the local state, tourist promotion boards and companies to market musical heritage and a musical environment'. (Gibson & Connell, 2007: 167)

The diversity of music tourism is evident, with opera houses, global music events and festivals, youth beach parties, and importantly, these reflect the current times, desires and demands of society. Much to the varied and diverse nature of the music tourism industry and the fluid nature in which people can interact with the different forms in which it can manifest, 'It is thus difficult to define music tourism and tricky to conduct straightforward economic or social impact analysis' (Gibson & Connell, 2007: 167).

The music sector closely related to the theme is this chapter, benefit and relief concerts, could be aligned closely to the festivals and events sector. Festivals and events have throughout history played a role in communities. And as noted by Getz (2008), events are an important motivator of tourism, and in recent years, developing and establishing events has become a prominent figure in the international tourism arena. Academic attention has been provided to the different types of motivations for implementing events for benefits such as enhancing economy (profit-making), social welfare (i.e. celebrating culture, history, religion etc.) and politically motivated (such as the staging of mega events) perspectives (Richards & Palmers, 2010; Van der Wagen & White, 2010). From a music event perspective, an article in *The Guardian* by Dean (2016) suggested that the number of festivals listed on festival website eFestivals jumped from 496 in 2007 to 1070 in 2015. According to the IFPI Global Music Report (2017: 3) music is 'being enjoyed by more people is more ways than ever before'. Taking data from the UK music sector, the core music industry in 2016 made an estimated economic contribution (Gross Value Added or GVA) of £4.4bn to the UK economy, supported by 142,208 jobs (CIC, 2017). Additionally, UK live music audience numbers in 2016 were 30.9m with 27m attending concerts and 3.9m going to music festivals (CIC, 2017). Music and tourism events and festivals play a significant role in the leisure economies around the globe. Tourists travel purposefully to national and international destinations to engage with music concerts, be it motivated by music culture, popular music genres, or to watch artists in action. This is not new to society, and current consumption and staging trends suggest that the popularity of music events will continue to grow in the foreseeable future. A potential reason for this is the impact music has on human emotion.

Music's Influence on Emotions

More profoundly, humans experience emotional reactions to music.

(Wood & Smith, 2004)

Music has the potential to change our moods, to impact our emotions and humans apply music throughout various aspects of life to purposely influence us. As Huang (2015) suggests, 'music has the potential to change a mood, to shift an atmosphere, and to encourage a different behaviour'. As humans, our behaviour can be manipulated by the sound of music and as humans our behaviour is influenced by the way we feel, by our emotions. It is not only the language or the lyrics that are poignant, it's the sounds that can engage and impact people. 'Music is a universal language capable of awakening emotions and unique sensations. Sometimes, even though we hear someone singing in an unknown language, we're still able to feel what they're trying to convey, even if we don't know what the lyrics mean. What we do know is that they are expressing something cheerful, sad, or dramatic, etc. The influence of music on our soul is so far ranging' (Exploring Your Mind, 2018). Duke (2014) stresses, 'without a doubt music affects the way we feel and our bodies respond to the sounds that we hear'. See research by Professor Roberto Valderrama Hernández at the Psychology Faculty of the BUAP University in Mexico as an example of how music can impact on emotions. The research explored the effects of heavy metal music and anxiety (Exploring Your Mind, 2018). Through music, humans can attain emotional feelings and experiences, and rhythms can influence human emotions. 'The key is to find the rhythm that can benefit us in any given situation. In this way the influence of music can be a real positive influence on our lives' (Exploring Your Mind, 2018). Understanding this is very important. While music is often recognised and applied throughout health and medical fields, spiritual development, marketing and promotion and encouraging consumers, our understanding and application in tourism deserves more attention. As noted by Gibson and Connell (2007: 165), music 'evokes feelings of nostalgia, elation, energy and melancholy, and that this is so is both a lure for tourists and an opportunity for tourism promoters'. Fundamentally, music is an invisible and ephemeral sensory experience (Gibson & Connell, 2007). Suggesting that music has often had its visual and more permanent artefacts (performers, instruments, stages etc.), but fundamentally, without sound, music would not exist. While the essence of tourism, the 'tourist gaze' (Urry, 1990) relies on that which is concrete, visible and (semi-)permanent. Music festivals, concerts and events are a place where the essence of tourism and the essence of music come together, allowing people with shared interests to express themselves, where their emotions will be carried by the sound of music.

Benefit Concerts and Tourism

> When tragedy strikes, celebrities take center stage to bring attention to worthwhile causes.
>
> (Jimison, 2017)

In times of humanitarian crisis, natural, human-inflicted disasters, or other forms of pain and suffering, a benefit concert or charity concert is an opportunity to exhibit empathetic and financial support for victims and people in need. Such concerts offer musical performances with their main purpose being of a charitable nature, aiming to raise cash donations and influencing legislation (concrete objectives) and raising awareness (subjective objective) towards a specific, immediate or ongoing humanitarian crisis and or event. As noted by Jimison (2017) 'billions of dollars have been raised since benefit concerts gained popularity and thousands of fans have followed the example of their favourite actor, comedian or musician and become supporters of causes they might not have otherwise'. There are many examples of past benefit concerts that have raised awareness and funds for people living dystopian lives: The Concert for Bangladesh in 1971; Live Aid in 1985; A Concert for Life: The Freddie Mercury Tribute Concert for AIDS Awareness in 1992; A Concert for Hurricane Relief 2005; Live Earth in 2007 (Swertlow, 2017) and One Love Manchester 2017 (Jimison, 2017). Benefit concerts are often recognised as an effective method of establishing support, awareness and raising funds due to the large media coverage they receive. They are often internationally broadcasted to ensure mass coverage and attention. Dayan and Katz (1992) categorise benefit concerts as 'media events'. Here suggesting that media events can be used as a tool to encourage shared experiences with the aim of uniting viewers and spectators with one another in and across societies. In as much, international concerts with high levels of interpersonal communication (where the viewer becomes more interactive with the purpose of the event) have the potential to create a 'fellow feeling' among viewers. The interactive nature is often due to the unpredictable nature of the charitable event. For example, in the early stages of a disaster, the events interruption of normal media coverage and television programming and the act of donating money encourages engagement (Dayan & Katz, 1992).

Benefit concerts often involve popular musicians (and actors, actresses, comedians, popular social figures, etc.) to further encourage wider audiences to engage (Pomerantz, 2010). While there are criticisms of celebrities participating in benefit concerts, partaking for motives such as public image and promotion, the potential of their participation to engage a wider audience is likely to be more beneficial. As noted by Hague *et al.* (2008) celebrities have the potential to produce a social effect termed 'Geldofism', the meaning of which, is the grouping of pop stars and their fans behind a cause. The power and quality of the celebrity is important. For people to be engaged, interested, and to support, be it through discussing, participation and or through cash donations, the entertainment and music should be of high entertainment quality. At its most effective, benefits concerts will create a space where the audience come together, and when people sing in the presence of other people (often common at benefit events) a collective effervesce can be achieved, individuals become connected and thus, more likely to work towards a common goal (Bennett, 2001).

Defining Dystopian

Your dystopia is my utopia – your utopia is my dystopia! A utopia is a paradise while a dystopia could be a paradise lost. Before utopias and dystopias became imagined futures, they were imagined pasts, or imagined places, like the Garden of Eden (Lepore, 2017). Dystopia can be defined as an 'imagined state or society in which there is great suffering or injustice, typically one that is totalitarian or post-apocalyptic' (Oxford Dictionaries, 2018a). Contrary, utopia can be defined as an 'imagined place or state of things in which everything is perfect' (Oxford Dictionaries, 2018b). *Brave New Words*, a dictionary of science fiction, defines dystopia as 'an imagined society or state of affairs in which conditions are extremely bad, especially in which these conditions result from the continuation of some current trend to an extreme' (see Liptak, 2013). However, as noted by Magid (2015) if one is to consider such definitions as undisputed conditions of the two, then one is automatically presuming that such places and or situations of dystopia could be considered good and bad to every individual from the society, and or alternative societies. Cultures and societies do not share the same beliefs and practise, and what is deemed good and bad is not the same for all. The following are said to be characteristics of a dystopian future and can be a useful list in which to consider the range of different types and levels of dystopia:

- Propaganda is used to control the citizens of society,
- Information, independent thought, and freedom are restricted,
- A figurehead or concept is worshipped by the citizens of the society,
- Citizens are perceived to be under constant surveillance,
- Citizens have a fear of the outside world,
- Citizens live in a dehumanized state,
- The natural world is banished and distrusted,
- Citizens conform to uniform expectations. Individuality and dissent are bad,
- The society is an illusion of a perfect utopian world.

(Read, Write, Think, 2006)

Dystopia can exist on different levels and can be influenced by varying factors. Many of the definitions of dystopia could be applied to current communities – from conflicts in Syria or Myanmar, to natural disasters that frequently strike communities. Additionally, not all people will define a dystopian reality in the same manner and thus, dystopian ideologies and realities can exist throughout our global community while other communities could be said to live in peace, comfort and what is deemed a more utopian existence. So, a dystopian reality is personal to the individual or to a collective group of people and their view of what a dystopian world looks like will be shaped by their real-life experiences and culturally relevant. However, this is not to say that people across cultures cannot appreciate and understand suffering other communities could be experiencing. The point is that culture across time and space is different, and we do not all share the same life views. Through the application of the above characteristics of dystopia, some common ground could be realised.

Future Visions of Dystopia

Dystopian reality is often portrayed, but not limited to fiction in the form of novels and movies. As noted by Slaughter (2003: xxi–xxii), 'In fiction, however, we can allow ourselves a glimpse at the truth without directly challenging the prevailing social order. We can experience our anxiety and fear in the safe confines of a book, a movie theatre or TV screen'. The music industry is not short of dark dystopian visions.

> Melodramatic types that they are, musicians love a bit of doom and gloom set in far-flung futures where society's crumbled and sits on the brink of collapse. From the Rolling Stones and Kate Bush to St Vincent and Rage Against the Machine, loads of acts have ended up with killer songs after imagining the end is nigh. (Barker, 2015)

Is our global home on a path of self-destruction? Are future societies heading for dystopian realities? There are no simple answers to such questions, just observation and consideration of current trends and then as noted, opinions vary depending of cultural perspectives of what constitutes dystopian and utopian realities. As noted, discussions surrounding dystopian futures are not limited to fiction. Applying the above characteristics of dystopia, the following are all relevant current examinations suggesting future dystopian worlds are possible. Media outlets provide explorations of possible dystopian futures, far too many to provide full coverage here. However, some examples include, Haynes (2013) who asks the question, is humanity heading for utopia or dystopia, focusing on nuclear explosions, global warming and solar flares as the forces at work. Writing in the *Financial Times*, Foroohar (2017) explores the idea of military enforcers of world order filled with corrupt and even crazy political leaders. The author makes links to themes in emerging markets such as Turkey and Argentina. Card (2018), on *USNews*, explores similar themes such as big government and media dominating American society and free speech. Scientific research also alludes to evidence that we are entering dangerous times, especially when related to the natural environment. Research by Ceballos *et al.* (2015) in *Science Advances* argues that earth is entering a sixth-mass extinction due to loss and extinction of animal species, suggesting that the world in the next 50 years could be completely different to what we know today. Ban Ki-moon (Un Secretary-General 24 January 2008) suggests more conflicts will continue to terrorise our societies if water shortage issues are not taken seriously. Gössling *et al.* (2015) also suggest global conflicts could arise in the future due to water shortages. An international team of scientists published a study stressing that even if the carbon emission reductions called for in the Paris Agreement are met, there is a strong risk of Earth entering what they call 'Hothouse Earth' conditions. The researchers considered 10 natural feedback processes, many of which are 'tipping elements' (Steffen *et al.*, 2018). If the tipping elements are breached, there is great potential that it will lead to abrupt changes in our natural environment with the potential to cause further devastation to communities. As for human conflict, the likelihood of it being a continued reality in the future is expressed by Pearce (2004), who suggests that violence is a part of humans and a result of our failure to transcend. Pearce suggests that humans are shaped by the culture we created, one in which violence is part of our innate need to survive. If we are unable to transcend culture, then we are unlikely to transcend violence. Consequently, the future, is likely to continue to whiteness violence and thus dystopia as a result of war and conflict.

Benefit Concerts and Tourism: Future Considerations

Music is a powerful tool for change. It can influence people's emotions, but can it change our minds? This is a somewhat more complicated question. According to Huang (2015) researchers cannot categorically link cause-and-effect between listening to certain songs and how it can encourage a certain type of behaviour. However, many researchers and people agree that certain music surely encourages certain behaviour (here reference was given to rap and hip-hop music and the excess of songs with lyrics glorifying sex, drugs, and violence and the potential influence it can have on people). In line with this Huang (2015) states the following:

> I believe that morals and behavior, especially in teens, aren't completely steered by the lyrics they're listening to, because there are so many factors to building a moral compass. However, music can definitely play a significant role in determining what seems to be right or wrong, okay or not okay, and good or bad. Because of this, we need to become wary about the messages that we are putting out with our songs, but to take it a step further, what if the songs we wrote intentionally carried positive messages? What if they became anthems that declared hope and joy, triumphs over weaknesses, courage and love? We would have the influence to empower the hearts and minds of the next generation, and that is something to truly take hold of.

It could be argued that a global projection of either a dystopian or utopian future limits the human imagination, it confines culture and difference across societies. Thus, as noted above, similarly to present day, a more comprehensive appreciation to the future is taken, one where dystopia and utopia exist across the globe, a more fluid recognition of the hardship that some societies experience while others live in relative comfort, and or luxury. As highlighted in this chapter, research predictions suggest that future societies will continue to experience and live in worlds that hold dystopian characteristics. Consequently, there is significant scope for the growth of the benefit concert to be more widely applied within the tourism market. As noted, music has the power to impact culture, our emotions, it can unite communities. Likewise, tourism has the power to influence emotions, bring about cultural change and unite people. Music and tourism together can play a significant role in bringing about social change. Thus, benefit concerts should continue to be recognised and embraced not only as a contingency action in times of difficulty, but premeditated events that support dystopian worlds. Be it on a local level or global one, as not all benefit concerts need to be on the large scale like those identified above. In a time where tourism can be recognised for its negative impacts across the globe (i.e. overtourism in destinations, mass tourism consumption, negative cultural impacts, lack of sustainable approaches), benefit concerts and tourism have the potential to bring about positive social change. As identified by Mattern (1998), popular music has the ability to bring people together with the intention of establishing more effective political communities.

Communities should come together and consider how music events with a benefit concert approach can bring about positive change for themselves and others. However, their success could be dependent on various key players involved. Organisers have the responsibility to manage the social engagement of tourists and how this is reflected to the societies living in dystopia. The media need to consider their role in promoting and presenting benefit concerts to the wider society. Significantly, musicians have a responsibility to consider the types of music they are sharing and the impact it can have on culture and people. Society and tourists need to ensure they are aware of the potential role they play in engaging with benefit concerts and that the ultimate aim is to provide support for people living in dystopia, with the eventual aim of alleviating pain and suffering for people. As noted by Taleb Rifai (2015) UNWTO Secretary-General, 'Every time we travel, we become part of a global movement that has the power to drive positive change for our planet and all people'. And alongside music, which as suggested by musician Albert Ayler, 'music is the healing force of the universe'. Brining the forces of music and tourism together, society has the potential to support communities in the greatest of need.

Concluding Arguments

Past and current societies have and continue to experience great suffering and pain that hold at their core, the characteristics of what a dystopian world is like, and evidence suggests that future societies will continue to experience dystopian realities. This chapter thus explored the role of music in society. It presented evidence of music power in influencing behaviour in people and more significantly, how music can change culture. Music and tourism operate together on local and global levels, and examples were presented to highlight the power of music to bring people together to support communities in time of natural and human disasters. People come together to support victims, a show of empathetic unity, and as a means of raising financial support. Music events of this nature are often identified as benefit concerts, purposely staged in times of difficulty. However, the world continues to witness great suffering and pain for communities, and consequently, the question being asked is, can benefit concerts become a more common scene in the tourism market as a method to support communities? If this is to be the case, there is great responsibility on the part of organisers, musicians, media and tourists in the way they promote, engage and ultimately support communities living in dystopia.

He who can listen to the music in the midst of noise can achieve great things.

(Vikram Sarabhai)

References

Attali, J. (1992) *Noise: The Political Economy of Music*. Minneapolis: University of Minnesota Press.
Ban. K-M. (2008) At world economic forum, ban ki-moon pledges action on water resources. UN News Centre, 24 January. See https://news.un.org/en/story/2008/01/246802-world-economic-forum-ban-ki-moon-pledges-action-water-resources (accessed 31 May 2018).
Barker, E. (2010) 40 Bone-chilling tracks that predict a dark dystopian future. See http://www.nme.com/photos/40-bone-chilling-tracks-that-predict-a-dark-dystopian-future-1404708 (accessed 29 May 2018).
Bennett, J. (2001) Ethical energetics. In J. Bennett (ed.) *The Enchantment of Modern Life: Attachments, Crossings, and Ethics* (pp. 131–158). Princeton, New Jersey: Princeton University Press.
Card, J. (2018) America the dystopia? See https://www.usnews.com/opinion/articles/2016-05-13/could-america-and-the-world-become-a-real-version-of-dystopian-fiction#close-modal (accessed 31 May 2018).
Ceballos, C., Ehrlich, P.R., Barnosky, A.D., Garcia, A., Pringle, R.M. and Palmer, T.M. (2015) Accelerated modern human-induced species losses: Entering the sixth mass extinction. *Science Advances* 1 (5). DOI: 10.1126/sciadv.1400253.
CIC (2017) Music sector size and value. See http://www.thecreativeindustries.co.uk/industries/music/music-facts-and-figures/uk-music-market-size-and-value (accessed 6 August 2018).
Dayan, D. and Katz, E. (1992) *Media Events: The Live Broadcasting of History*. Cambridge, MA: Harvard University Press.
Dean, S. (2016) Do the growing number of music festivals actually make any money? See https://www.telegraph.co.uk/business/2016/07/02/do-the-growing-number-of-music-festivals-actually-make-any-money/ (accessed 6 August 2018).
Duke, S. (2014) Influential beats: The cultural impact of music. See https://www.thenewamerican.com/culture/item/17311-influential-beats-the-cultural-impact-of-music (accessed 30 May 2018).
Exploring Your Mind (2018) The influence of music on our lives. See https://exploringyourmind.com/influence-music-lives/ (accessed 30 May 2018).
Foroohar, R. (2017) Dystopian America: How far are we from Gilead? See https://www.ft.com/content/c40e11e8-928a-11e7-a9e6-11d2f0ebb7f0 (accessed 31 May 2018).
Getz, D. (2008) Event tourism: Definition, evolution, and research. *Tourism Management* 29 (3), 403–428.
Gibson, C. and Connell, J. (2003) Bongo Fury: Tourism, music and cultural economy at Byron Bay, Australia. *Tijdschrift voor Economische en Sociale Geografie* 94 (2), 164–187.
Gibson C. and Connell, J. (2007) Music, tourism and the transformation of Memphis. *Tourism Geographies* 9 (2), 160–190.
Gössling, S., Hall, C.M. and Scott, D. (2015) *Tourism and Water*. Bristol: Channel View Publications.
IMDb (1990–2018) Peter Tosh quotes. See https://m.imdb.com/name/nm0869167/quotes?ref_=m_nm_trv_trv (accessed 19 December 2018).
Hague, S., Street, J. and Savigny, H. (2008) The voice of the people? Musicians as political actors. *Cultural Politics* 4 (1) 5–23.
Harmonious Assembler (ND) The ubiquitous Roy Ayers. See http://www.revive-music.com/2011/01/26/the-ubiquitous-roy-ayers/ (accessed 19 December 2019).
Haynes, G. (2013) Is humanity headed for utopia or dystopia? See https://www.vice.com/en_uk/article/4w8zmb/things-that-may-destroy-the-world-but-may-lead-to-nirvana (accessed 31 May 2018).
Huang, B. (2015) What kind of impact does our music really make on society? See https://www.thenewamerican.com/culture/item/17311-influential-beats-the-cultural-impact-of-music (accessed 30 May 2018).

IFPI Global Music Report (2017) Global music report 2017: Annual state of the industry. See http://www.ifpi.org/downloads/GMR2017.pdf (accessed 6 August 2018).

Jimison, R. (2017) From Farm Aid to Manchester: The biggest celebrity benefit concerts. See https://edition.cnn.com/2017/06/09/health/champions-for-change-biggest-celebrity-driven-fundraiser-events/index.html (accessed 6 August 2018).

Lepore, J. (2017) A golden age for dystopian fiction. See https://www.newyorker.com/magazine/2017/06/05/a-golden-age-for-dystopian-fiction (accessed 31 May 2018).

Liptak, A. (2013) A brief history of the dystopian novel. See https://www.kirkusreviews.com/features/brief-history-dystopian-novel/ (accessed 28 June 2018).

Magid, A.M. (2015) *Apocalyptic Projections: A Study of Past Predictions, Current Trends and Future Intimations as Related to Film and Literature*. Newcastle: Cambridge Scholars.

Mattern, M. (1998) *Acting in Concert: Music, Community, and Political Action*. New Brunswick: Rutgers University Press.

Nuñez, C. (2015) Music that has changed the world. See https://www.globalcitizen.org/en/content/music-that-has-changed-the-world/ (accessed 31 May 2018).

Oxford Dictionaries (2018a) Dystopia. See https://en.oxforddictionaries.com/definition/dystopia (accessed 29 May 2018).

Oxford Dictionaries (2018b) Utopia. See https://en.oxforddictionaries.com/definition/utopia (accessed 29 May 2018).

Pearce, J.C. (2004) *The Biology of Transcendence: A Blueprint of the Human Spirit*. Vermont: Park Street Press.

Pomerantz, D. (2010) The truth About celebrity benefit concerts. See https://www.forbes.com/2010/01/26/haiti-clooney-wyclef-business-entertainment-charitable-celebs.html#528d901a4c46 (accessed 6 August 2018).

Read, Write, Think (2006) Dystopias: Definition and characteristics. See http://myfisd.com/jh/wp-content/uploads/sites/13/2016/08/Dystopias.pdf (accessed 31 May 2018).

Richards, G. and Palmer, R. (2010) *The Eventful City: Cultural Management and Urban Revitalisation*. Oxford: Butterworth Heinemann.

Rifai, T. (2015) Message for World Tourism Day 2015. See http://wtd.unwto.org/content/official-messages-world-tourism-day (accessed 6 August 2018).

Slaughter, R.A. (2003) *Futures beyond Dystopia: Creating Social Foresight*. London: Routledge.

Steffen, W., Rockström, J., Richardson, K., Lenton, T.M., Folke, C., Liverman, D., Summerhayes, C.P., Barnosky, A.D., Cornell, S.E., Crucifix, M., Donges, J.F., Fetzer, J., Lade, S.J., Scheffer, M., Winkelmann, R. and Schellnhuber, Hans, J. (2018) Trajectories of the Earth System in the Anthropocene. Proceedings of *the National Academy of Sciences of the United States of America (PNAS)*. See https://doi.org/10.1073/pnas.1810141115 (accessed 31 August 2018).

Swertlow, M. (2017) 8 of the biggest benefit concerts of all time and how much they really made. See https://www.eonline.com/uk/news/858226/8-of-the-biggest-benefit-concerts-of-all-time-and-how-much-they-really-made (accessed 6 August 2018).

Urry, J. (1990) *The Tourist Gaze*. London: Sage.

Van der Wagen, L. and White, L. (2010) *Events Management for Tourism, Culture, Business and Sporting Events*. Australia: Pearson.

Wood, N. and Smith, S.J. (2004) Instrumental routes to emotional geographies. *Social and Cultural Geography* 5 (4), 533–548.

17 Exclusion Tourism: Sci-Fi Stalkers and Subjunctive Plays in Apocalyptic Destinations from Chernobyl to Plymouth, Montserrat

Magdalena Banaszkiewicz and Jonathan Skinner

Chapter Highlights

- Exclusion tourism is a form of niche tourism where the tourist visits Exclusion Zones.
- These exclusion zones are places where the tourist projects utopic or dystopic futures from films and books.
- There is an apocalyptic dimension to this tourism found in the Chernobyl Exclusion Zone and the Montserrat volcano exclusion zone examined in this chapter.

Overview

This chapter examines venues that had been destroyed or abandoned but in their apocalyptic ruin have emerged as conflicted tourist attractions associated with science fiction. There is no doubt that post-apocalyptic landscapes are particularly attractive for science fiction texts (books, films and video games). As tourist destinations, they elicit reactions of emptiness, unevenness, loss and desolation that are counterbalanced by awe, fascination and a sublime enjoyment that Manjikian (2012) refers to as 'the romance of the end'. The apocalyptic – whether 'man-made' or 'natural' – are dystopic locations where the future subjunctive can be played out by temporary visitors who court the uniqueness of their exclusion.

Specifically, the chapter engages anthropologically through participant observation with two apocalyptic destinations: the Chernobyl Exclusion Zone – CEZ (Ukraine) and the Plymouth, Montserrat Exclusion Zone (Eastern Caribbean), both destinations for tourists seeking to engage with their paradoxical aesthetics. The CEZ, a case of exclusion through radioactive leak, of modernity's ills, is frequented by 'stalkers' – a community of illegal urban explorers whose archetype was presented by Andrei Tarkovsky in *Stalker* (1979), a film loosely based on the novel *Roadside Picnic* by Boris and Arkady Strugatsky (1977). This contrasts with visitors to the island of Montserrat that is riven down the middle by an on-going volcanic eruption. Tourists to Montserrat contrast 'the green' of the north with 'the gritty' of the south and the ruin of Plymouth, 'a modern-day Pompeii', exploring the former capital and imagining by reference to films such as *Jumanji* (Johnston, 1995; Kasdan, 2017) a new politics from the hybridity of its nature-wrought debris. Together, these case studies of the apocalyptic show how science fiction texts have established a new form of tourist practice: 'exclusion tourism'.

Introduction

We are attracted to what we cannot or should not reach for. This desire or impulse is most apparent in the liminal, in the in-between. In post-colonial discourses, the fantasy of the Other is an Orientalist flame. We suggest, in this chapter, that the embers of this colonial desiring identified by Young (1995: 167) from Edward Said as a trope of repression in a capitalist society, is also apparent in hyper-modern tourist settings, particularly places of exclusion: empty spaces that are no-go areas, prohibited unsafe spaces, neither civilised nor entirely natural. They defy us as the demilitarised No Man's Land running between the North and the South, ironically less safe for its lack of hardware. In the radioactive lifelessness of Chernobyl, northern Ukraine, they become radio-attractive for the adventurous tourist. Similarly in the barrenness of ruined capital Plymouth, in the south of the small British Caribbean colony Montserrat, the contrast between tropical island and apocalyptic wasteland is humbling. In these two examples, we hope as anthropologists to flesh out the post-colonial theory with empirical example, with ethnography of visits to the locations ourselves, interviews with tour guides and tourists to these empty places. This will enable us to test Young's literary post-structural conjectures, to illustrate his theorisation on desire in far-flung locations, both threatened not just by disaster (natural and unnatural) but also by the machinations of colonialism from mother Russia and motherland Britain. Abandoned, by necessity, these spaces are frozen in time and span a temporality with the past (1986 for Chernobyl, 1997 for Plymouth) but also allude to a future, to a dystopic proto-future. The warning signs are 'larger than life', to use the expression in a full sense of the words.

They reference not to the abandonment of signification but a series of futuristic Anthropocenes depending upon the visitors.

Brown *et al.* (1996) open their 'end of the era' marketing book with a chapter titled *Apocaholics Anonymous*. In the initial section, *Apocalypse Then*, they write:

> We are, so it seems, transfixed by terminal visions, mesmerised by the millennium, entranced by eschatological expectations, consumed by chiliastic conjecture and, thanks to the protection afforded by the well-lubricated prophylactic of prophesy, ever eager to embrace and fructify the end of time. (Brown *et al.*, 1996: 1)

The authors point to a public addiction to 'imagined endings' such as Armageddon from the Bible, to which we can add the millenarianism or the capitalisation in recent popular culture of Doomsday scenarios that have shifted from nuclear meltdown to zombie face-off and human termination. Technically, an apocalypse is a revelation, a sense of ending when only the few struggle to survive. In this Apoca-scape, the ruins of civilisation are what is left. This is a warning to the present. Repent now in thy ways to preserve thy present whether it be preservation from modern-day famine, plague, disaster. The intimation is that these end of times are man-made through our self-destructive traits whether irreligious or negligent. Brown *et al.* (1996) invoke Baudrillard's 'illusion of the end' when fin de millennium social thought is plaited and distorted. For us, in a tourism context, the hell-bent is more than post-ideological (cf. the end of history [Fukuyama 1992] or neoliberal disaster capitalism [Klein 2007]), or Brown *et al.*'s (1996) rather lovely phrased 'not for prophet' marketing. We take their use of the term – 'the word "apocalypse" is widely associated with death, destruction, chaos and carnage' (Brown *et al.*, 1996: 4) – and link it to our dark tourism destinations that have become increasingly popular tourist attractions for all their exclusivity.

Chernobyl Exclusion Zone Tours

In the late night, 26 April 1986, in the reactor No. 4 during a safety test which simulated a station blackout power-failure, uncontrolled reaction conditions happened – water flashed into steam generating a destructive steam explosion and a subsequent open-air graphite fire. That was the first time when such an accident happened. 'The battle of Chernobyl' lasted for the next half year and led to innumerable consequences (Yablokov *et al.*, 2009; Yaroshinskaya, 2017): not only physiological and psychological health disorders (Petryna, 2002) and environmental pollution (Mycio, 2005), but also a political crisis and economical crush connected with the collapse of the USSR (Sekuła, 2014). As a direct result of the disaster, approximately 350 thousand people were evacuated from the contaminated areas of Ukraine, Belarus and Russia – including approximately 163,000 from Ukraine alone. However, it is not possible to calculate the exact number of evacuations: it ranges from 116,000 to over 163,000 even though the evacuees were counted three times (Sekula, 2014: 235). The CEZ has been established in an area with a radius of approximately 30 km from the power plant, the territory most affected by radioactive waste. It is an area where the population must not live, where economic activity must not be carried out, and where food must not be produced.

In the following years, when the level of radiation was decreasing and the space of the Zone visits to it began to be organised more or less officially (Banaszkiewicz *et al.*, 2017), it became possible to explore the Zone not only as a memorial site but also as a space of post-apocalyptic aesthetics, representing the fascination of decay (Edensor, 2005; Pora & Olsen 2014). Until 2010, former residents of the Zone, delegations and individuals who adequately argued for the need to visit the CEZ were granted special permits to enter the CEZ – most often they were scientists and journalists. The area has been open for official visitors since 2011. Data obtained from the Ukrainian State Agency on Exclusion Zone Management show a dynamic increase in the number of visitors to the CEZ: 8000 tourists visited the Zone in 2010, almost 18,000 in 2013 and 36,000 in 2016 during the 30th anniversary of the disaster (almost 25,000 of whom were non-nationals). In 2017, the number of visitors to the Zone reached 50,000, including approximately 34,000 foreigners. In 2018 it is estimated that the number of visitors will grow to 80,000. This is twice as high as the 2016 level.

In recent years, a number of research projects focusing on various aspects of visiting the CEZ have been published. The unique status of exclusiveness of the Zone's space attracted the attention of Philip Stone (2013), who explored the problem through the lens of Foucault's concept of heterotopia. According to Stone (2013):

> Chernobyl is now an-Other place. It exists alongside ordinary spaces of the everyday, yet it is a place where disaster has been captured and suspended. It is a place of crisis, of deviation, of serious reflection. […] A surreal place to juxtapose our apocalyptic nightmares, Chernobyl is both real and imagined. […] Indeed, tourists now ritually consume the place as a site of environmental disaster, failed technology and political collapse.
>
> (Stone, 2013: 90–91)

Dobraszczyk (2010) argues that Pripyat has become a model example of a contemporary ruin and the target of urban explorers, for whom the 'ghost town' Pripyat is what Paris is for cultural tourists. Goatcher and Brundsen (2011) focus on the emotional encounters of visits and propose to use the term sublime in order to depict the special psychological state of a person being in the Zone. However, none of the above-mentioned authors conducted prolonged fieldwork research among the visiting tourists. The study presented here is based on a dozen visits to the Zone with different tourist groups in the period between 2016 and 2018 with a focus on gathering material on strategies of tour-guiding. In addition to this participant observation, 20 in-depth and semi-structured interviews were conducted with guides, stalkers and tour-organisers during this period.

Yankowska and Hannam (2014, 932) define visits to the CEZ both as dark tourism and toxic tourism, adopting the term introduced by Pezzullo (2007) who used it to describe organised tours to places of environmental degradation. They agree with Di Chiro (2000) that toxic tourism 'can provide a strong educational experience, raising awareness about the current environmental issues and the polluted environmental conditions around us' (Yankowska & Hannam, 2016: 937). Just as Sather-Wagstaff (2011) concluded in her study of visitors to the site of 9/11, neither dark tourism nor toxic tourism exhaust the heterogenous character of recent visits to the Zone. As the fieldwork proved, people wish to visit the Zone due to a variety of reasons. Some tourists are interested in atomic energy; some would like to know more about the Zone's functioning so many years after the disaster; others want to enter this very specific sanctuary where wildlife prevails. For the majority of them, the main motive is not to pay tribute to the victims of the disaster or pay attention of environmental issues, but to see the 'emptiness' of the post-apocalyptic landscape. They are also thrilled by a risk resulting from a visit to a contaminated space. Although the level of radiation should not be hazardous to human health (for such short exposure) on the so-called tourist route, the proximity of radiation is an additional element of enhancing the experience. Many visitors consider the CEZ as the quintessence of 'the radical experience' that oversees everything that they had experienced before. This is deliberate risk-taking in an increasingly safe world characterised by sociologist Stephen Lyng (2005a) as a form of 'edgework'. Here the edges are quite literally the edges of the Exclusion Zone, a place of contrast with the everyday and an opportunity to express a disenchantment with typical leisure consumption (cf. Lyng, 2005b: 21). However, probably the most interesting factor depicted by the respondents was the chance to make comparison between real, embodied experience of being in the Zone with the images of that site known from books, films, games as well as social media or purposefully designed virtual tours

The CEZ, despite its long-term unavailability, or perhaps thanks to it, is a space where visual representations are particularly intriguing. The first pictures appeared with the disaster itself, and the more vaguely the facts were presented, the more the pictures fed the societies (mostly in the Ukraine, but also in other countries of the region) engulfed in atomic horror. The presence of Chernobyl in mass culture from the very early moments of the disaster has been playing a significant role in the popularisation of the Zone. Tourism imaginaries in the context of the Zone have been fed particularly by three science fiction works. The first one is the science fiction novel *Roadside Picnic* (Russian: Пикник на обочине) written by the brothers Arkadiy and Boris Strugatsky, published in the Soviet Union in 1972. The novel describes the situation on Earth after the arrival and departure of aliens from space. It is not known who the aliens were, but they left a lot of traces of themselves in the six Landing Zones. These areas are called Zones, and these traces, in the form of various mysterious objects, are collected by single explorers who then sell them on the black market. The novel regained fame after the release of STALKER (https://www.stalker-game.com/), a game that linked literary themes with space of the CEZ. Actually, it is a series of three games: 'Shadow of Chernobyl – 2007', 'Clear Sky – 2008', 'Call of Pripyat' – 2009). It is a first-person shooter game with some elements of role playing. The series was created by Ukrainian studio GSC Game World and turned out to be a spectacular success. 'STALKER' sold more than 5 million copies (more than half a million copies were purchased in the former USSR alone during the first two weeks) and the revenue was more than 100 million dollars (http://www.gsc-game.com/). The game received numerous awards and very positive reviews from players (Степанец 2017, 379). The plot of the game refers to the disaster of 1986. Twenty years later, in 2006, there was an explosion of the sarcophagus, the construction built around the damaged reactor of the Chernobyl nuclear power plant. Radioactive contamination is repeated, resulting in natural anomalies (mutant animals and plants) occurring in the 30 km Exclusion Zone. What is more, the area of the Zone expands spontaneously, and the mysterious force literally tears the living organisms apart or causes non-healing wounds. Access to the Zone is closely guarded by the army. A few years have passed before a few brave men called stalkers appear; they go to the Zone to get the remaining artefacts, strange plants and other objects there and then sell them. The player is one of these stalkers who searches the Zone.

The last work is Andrei Tarkovsky's film *Stalker* (1979). The film was based on a script written in cooperation with the Strugacki brothers, loosely based on the motifs of their novel. Contrary to the expectations of the Soviet authorities, A. Tarkovsky's next film, after *Solaris*, is not a traditional science fiction type – the plot serves only as a pretext for a slow reflection on the image, far removed from the visual effects. Seweryn Kuśmierczyk, a film critic and expert on A. Tarkovsky's work argues that a stalker is not a hero seeking adventure, but God's madman, a spiritual guide, who with his paradoxical behaviour awakens transcendental sensitivity in others (Kuśmierczyk, 2012; 295–322). In this sense, the story of the journey to the Zone can also be read as a parable of a spiritual journey, a journey into oneself. After the Chernobyl catastrophe, A. Tarkovsky's film began to be interpreted as a prophetic announcement of the tragedy, and the term 'stalker' came into common use to describe people illegally crossing the border of the CEZ.

A pivotal role in the popularisation of the Zone has been its virtuality. Chernobyl portrayed in virtual space has enriched the repertoire of experiences related to travelling, expanding it with simulacric experience of hyperreality – the stage of physical travelling has been extended with pre- and post-stages by consuming and sharing travel stories about the place. Since the CEZ appeared within cellphone transmitter reach, visitors to the region have been able to immediately send their messages (both textual and visual) and be connected to the internet for most of the time of their visit in the Zone. Consequently, there has been a drastic growth in the number of easily accessible visual representations of the Zone in virtual reality. Currently, pictures of Pripyat or Chernobyl cities made by tourists can be traced on google.maps; Tripadvisor is filled with reviews and galleries commenting visits to Zone (https://www.tripadvisor.com/Attraction_Review-g294474-d3370334-Reviews-CHERNOBYL_TOUR-Kiev.html) and one can easily find photos from the CEZ on Instagram and on Facebook (https://www.facebook.com/groups/1564934223757227/). Not only tourists create virtual, visual representations of exploring the Zone, also professionals willing to use the virtual space as an environment that supports their commercial or educational aims (or both of them combined) contribute to what McDaniel labels as virtual dark tourism (2018: 4). To give an example, the Chernobyl VR Project made by The Farm 51 combines the world of computer games with the world of educational applications and film narration. It is the first ever virtual tour around Chernobyl and Pripyat for virtual reality (VR) devices (Oculus, PlayStation VR, HTC Vive) and mobile devices (Samsung Gear VR). Its creators call it a 'virtual Chernobyl museum' (http://www.chernobylvrproject.com/en/o-projekcie.html). Experiencing journeys in hyperreality becomes a great challenge both for tourism industry as well as for researchers since the very idea of travel is confronted with the need of redefinition.

Virtual tours even if they strive for offering a 'real' experience are deprived of things that constitute a physical exploration: social interactions (like small talks with other visitors or guides), full sensory exposure (smelling or feeling temperature) (MacDaniel, 2018). Moreover, virtual visits do not raise many ethical concerns that appear while being in the Zone in reality, i.e. with the ability to physically enter every building, one loses the issue of respect paid for the victims by not interrupting the abandoned privacy. Therefore, virtual tours constitute a legitimate experience on its own terms and extend physical journeys rather than substitutes them. Yet, they are not simply a 'lesser version' of reality, but a different form of experiencing sites performatively encouraging one to immerse into mental, intellectual or even spiritual liminality.

In addition, the active WiFi connection enhances the tourist encounter in the Zone by combining real, embodied experiences with the virtual engagement. Popular tourist applications offer activities such as hunting for Pokemons or searching for hidden items (geocaching). In the nearest future it would be probably possible to combine virtual exploration with the physical one simultaneously. The Farm 51 has started cooperation with tourist companies organising tours to the CEZ, in order to create a version of their tour as a simple application that would enable comparison of what is now with what was in the past. Amateur applications tracking between the past and the present are already in operation: in 2017, during the Chernobyling Festival, participants of a trip had the opportunity to use a special version of the 'Here we go' application (with a GPS comparable to Google maps). During that trip, its participants could download special maps showing the most important places of Pripyat in the past and today.

Montserrat Exclusion Zone Tours

The island of Montserrat in the Eastern Caribbean is a related case of exclusion zone tourism. It too is an example of ruin and decay. Where Chernobyl has the atomic dimension as a technological 'cultural' warning to the future, the Montserrat case study is 'natural', a grey wasteland that is presented as a modern-day Pompeii for the tourists, a working illustration of volcanology for the modern geography student. It too is mediated by popular culture with interpretative descriptions and film parallels or dramatic apocalyptic association: *Silent Hill* meets *Hawaii Five-0* for urbex explorer Shane Thoms (2016), or *STALKER* with aspects of *The Shining* and *The Andromeda Strain* (Skinner, 2008a; 2018) or 'the set of a 1960s episode of *The Twilight Zone*' for journalist Mark Rogers (2016). Keri Jones, producer and presenter of the *Great Destinations Radio Show*, went on a tour of the former capital Plymouth and was shocked when he crossed into the exclusion zone with his tour guide for a strictly timed visit, describing the view before him as one resembling 'the aftermath of a nuclear attack' (Jones, 2016). And features writer for *The Telegraph*, Neil Tweedie (2006) makes similar simile approximations:

> Plymouth is often compared with Pompeii but feels post-nuclear. The Barclays bank, Texaco petrol station, Seventh Day Adventist church and courthouse are half buried in rubble and ash, as if overwhelmed by the shock wave of a hydrogen bomb.

The gritty side of Montserrat is not yet a 'film-induced' (Beeton, 2016) tourist attraction destination. It is 'the other side' of the Emerald Isle of the Caribbean where a pyroclastic volcanic eruption has been venting and creeping its way across the capital since 1995. The unexpected eruption of a volcano in the south of the island, where it overlooked the island capital Plymouth, initially led to an evacuation of the capital and a mass exodus of the island's population, a loss in tourism visit revenue and a relocation of business and government works to the north of the island, often by commandeering expatriate villas (Skinner, 2003a). The island suffered catastrophic eruptions in 1997, literally swamping the capital and south east sides of the island in pyroclastic mudflow, tragically killing 19 Montserratians in the process and turning much of the south of the island into an uninhabitable apocalyptic landscape (Skinner, 2018). The Montserrat Government were led by the British government-aided Montserrat Volcano Observatory (MVO) to initially institute an Alert Scheme of microzonation across the island. Following the June 1997 eruption and fatalities, this was simplified into a risk map of northern zone (low risk), central zone (residential heightened state of alert) and Exclusion Zone (no admittance except for scientific monitoring and National Security Matters) (Aspinall *et al.*, 2002). The boundaries to these zones shifted according to volcanic activity and 'probabilistic forecasting' involving controversial expert elicitation (see Skinner, 2008b). This was also subject to liaison with and filtration through the Montserrat Government and island administration, the Governor, and the decisions of the National Disaster Preparedness and Response Advisory Committee (NDPRAC). Not without its political machinations, Donovan and Oppenheimer (2014: 154) refer to this as a complexity of 'blurred local and colonial geographies of science and governance', with some of the exasperated locals and expatriates disputing and lampooning the zoning of the island into acceptable spaces of risk with the printing of T-shirts showing the island divided between an 'Exclusion Zone' in the south and a 'Logic Free Zone' in the north (Skinner, 2003a: 112).

This second example of Exclusion tourism in this chapter is more about visits to the area than its 'policy-oriented delineation' and the evolving spatial delineation of social order and shift from alert-based to hazard-based management systems which is the focus of Donovan, Oppenheimer and Bravo's (2012) paper 'Contested boundaries: Delineating the "safe zone" on Montserrat'. The current system of governance of this area is a hazard level system linked with zones. Hicks and Few (2015: 5–6) describe the shifts as the island map (Figure 17.1: i) is partitioned/'microzoned' into seven hazard zones with fluctuating levels of alert from September 1996 (Figure 17.1: ii). Zones F and G were occupied by the general population and Zones A and B were generally inaccessible bar short approved visits. This microzoning was replaced by a simple set of three broad zones in September 1997 (and finally revised by April 1999) – a clear response to the tragic loss of life on 25 June by 19 Montserratians (Figure 17.1: iii); Northern Zone for living in; Daytime Entry Zone for limited access; and the Exclusion Zone prohibited to the public. As defined by the MVO, the zones consider access and activity for the general public (see Table 17.1). The final iteration of the map occurred between August 2008 and November 2011 and linked a hazard level system with new hazard zones. This traffic light system reinforced the safe entry parts of the island, accompanied by a numbered system based upon scientific perceptions of hazard and safety. It divided the southern

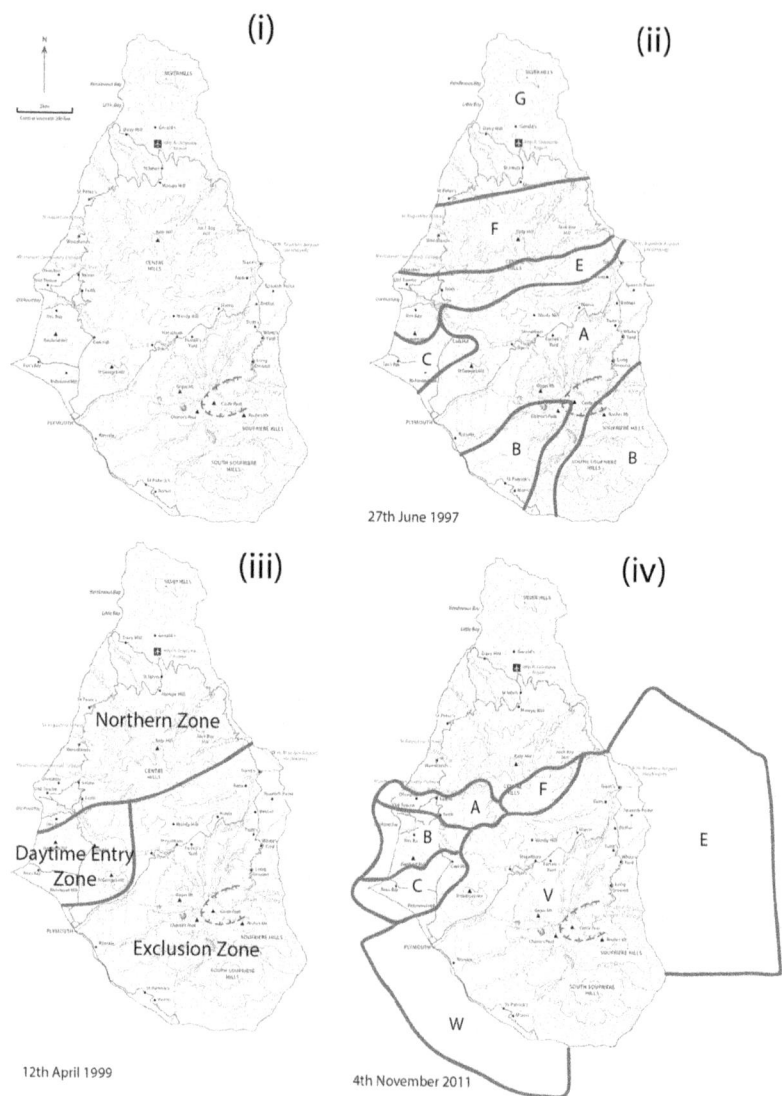

Figure 17.1 Montserrat Exclusion Zones from 1997 to 2011 (from Hicks & Frew, 2015: 3)

Table 17.1 Exclusion Zone definitions (from Michigan Tech Archives, 1997)

Exclusion Zone	No admittance except for scientific monitoring and National Security Matters.
Central Zone	Residential area only, all residents on heightened state of alert. All residents to have rapid means of exit 24 hours per day. Hard hat area all residents to have hard hats and dust masks.
Northern Zone	Area with significantly lower risk, suitable for residential and commercial occupation.

two-thirds of the island into five zones (A, B, C, F and V) and included two maritime exclusion zones (W and E) (Figure 17.1: iv); Zones A, B and C broadly approximate the former Daytime Entry Zone and V the former Exclusion Zone. This underwent minor amendments by 1 August 2014 to include new growth areas in the east of the island and is currently the latest version of the Montserrat Hazard Level System Zones (see Figure 17.2). It is summarised for tourists on the UK government's foreign

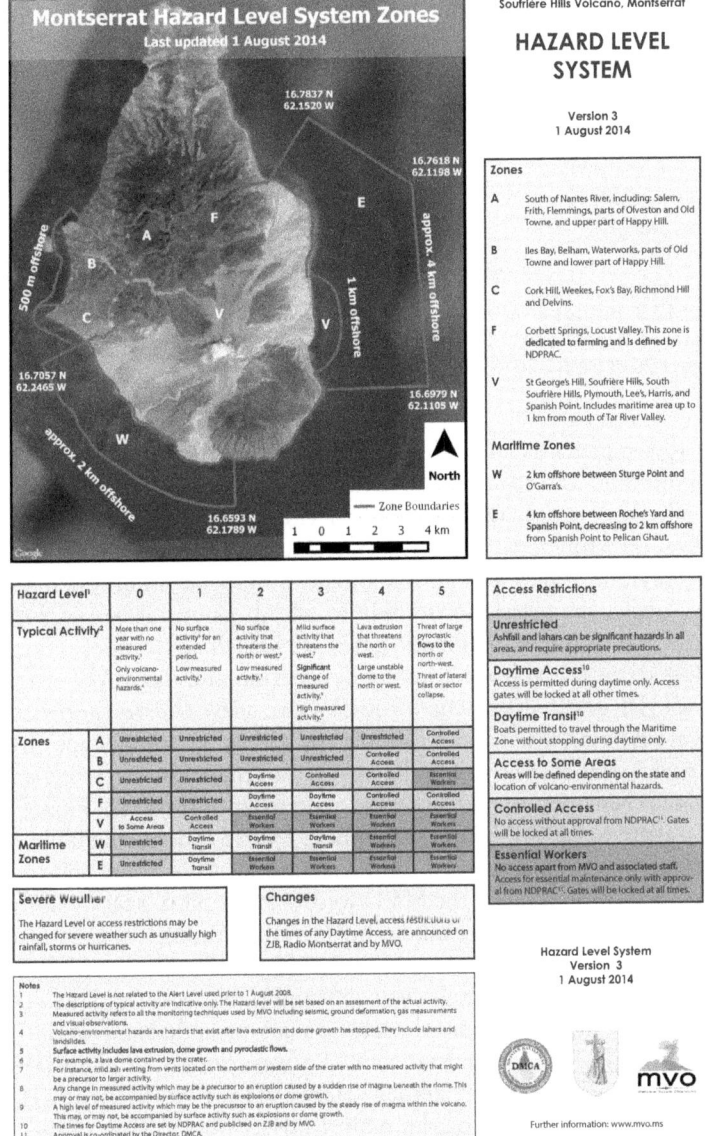

Figure 17.2 Montserrat Hazard Level System Zones, 1 August 2014 (MVO, 2019)

Montserrat Volcano Observatory

Hazard Level

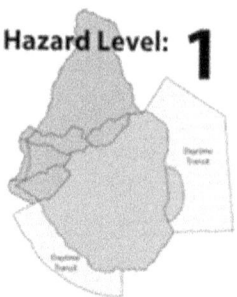

Figure 17.3 Branded MVO and Hazard Alerts as of 21 December 2018 (MVO, 2018)

Figure 17.4 Monitoring the Soufriere Hills volcano from the MVO (Skinner, 2005)

travel advice pages as a 0 to 5 hazard scale with 5 as the highest. Level 1 gives unlimited access to zones A, B, C and F by vehicle, and Zone V is 'controlled access only with permission' (FCO, 2019). At Level 1, the current level on Montserrat, the maritime exclusion zone is accessible only during daylight hours and no anchorage is allowed. In an escalation of hazard levels, the restrictions apply to all bar essential workers (MVO staff and associated workers) and gates are locked at all times.

Montserrat is a volcano tourism destination and the MVO is a workplace tourist destination where tourists can view the volcano from an observation point, learn about the volcano and its on-going eruptions and, in the early years of the work of the MVO, visit the scientists' work stations and watch them monitor the live volcano in front of them (see Figure 17.4). This is 'alienated leisure' (cf. MacCannell, 1989: 57) at an alienation zone. The island has been at Level 1 on the hazard alert for some considerable time, a status promoted on the MVO web and Facebook pages. Figure 17.3 shows the MVO brand: an island relief logo illustrating 'the green and the gritty' Montserrat Tourist Board marketing slogan (Skinner, 2015) – an island divided between the relatively safe and stable north and the destruction and devastation of the south – above a colour-coded map of the island with its three key areas (the Maritime east/west sides in bright yellow, the Exclusion Zone in the centre golden yellow, and the Northern and interim zones in green). It is one of the main tourist spots on the island, with another viewing facility on the east of the island at the more remote Jack Boy Hill. In the former, tourists gaze from afar at the peaks, unthreatened

Figure 17.5 Viewing the Soufriere Hills volcano from the MVO and the Montserrat Springs Hotel (Skinner, 2017)

and often taking selfies (see Figure 17.5). In the latter, tourists are closer to the volcano, not far from its eruptive pyroclastic flows, under it and looking up at the gases and flames through binoculars.

The contrast – the extremes – are dramatic. The casual attire of the visitor contrasts with the work overalls of the MVO staff in an ashen environment, just as the white disposable coveralls and Geiger counters were de rigueur costume of reporters on early visits to the 'humanless' but very much alive Chernobyl alienation zone. On Montserrat, the split is between tropical, musical, lively Caribbean island life – often visited during the week-long St Patrick's Day carnival – and the silent, greyscape of a dangerous and deserted space. Once you cross Belham Valley, disorientating mounds of mud surround the vehicles. Attention has to be paid to the weather, to the forecast, and to the ZJB government radio station as you drive past expatriate apartment blocks now reclaimed by nature, navigate mud-cleared trafficless traffic junctions, and drive towards the former capital avoiding debris on the roads and gawping at the tropical undergrowth suffocating the residential buildings and the mudflows reabsorbing the built-up environment. Unescorted excursions can drive up to the perimeter of the Central Exclusion Zone: up Richmond Hill to the former windmill that stood as the National Museum, surrounded by expatriate houses next to the Montserrat Springs Hotel. From these locations one can see the ghauts that the mudflows followed as they swept down the flanks of the Soufriere Hills to entomb Plymouth. The city shape used to be visible from the first floor up. More recently only certain landmark buildings exist as the city has been swallowed whole. The view from the Springs veranda is from the Exclusion Zone boundary. It is closer and more immediate than from the MVO. The both terrible and sublime nature of the volcano is in front of you with the buried capital and abandoned harbour. And around you is the debris of a ruined hotel where tourists visit but can't stay, view rooms frozen in time, dilapidated, ash buried, abandoned with graffiti on the remaining window panes. 'Weekend Roady' Phil Sites visited the hotel and likened it to a scene from *Jumanji*:

> We visited buildings on the very edge of the exclusion zone (a forbidden area which includes all of what's left of Plymouth) and got some amazing views of this modern day Pompeii from the old Montserrat Springs Hotel.

> The 'hotel' is now being overtaken by the earth itself, with about two feet worth of ash building up around the floors of the lobby area. Some rooms are brimming with various plants and wildlife (almost like a scene from *Jumanji*). (Sites, 2014)

It also recently attracted the attention of film director Behn Zeitlin who filmed *Wendy* on the island, a fantasy adventure set in an eco-apocalyptic future where children are in a tribal war precipitated by a pollen that breaks the relationship between ageing and time. Tourists visiting the hotel not only felt they were passing through a film location, now they indeed were walking through a science fiction-scape. 'Plymouth [...] the set of a disaster movie' for travel writer Andrew Eames (2018); the 'hidden ecosystem' and lost relationship with time of *Wendy* is evinced on Montserrat island itself. Elsewhere, he describes visiting this 'eerie, semi-abandoned world' (Eames, 2017) alongside the film crew:

> Entering the exclusion zone is like embarking into the world of one of those Armageddon video games. A warning sign, a prowling police van, a crackling radio and a helicopter buzzing overhead (actually a flight-seeing tour from Antigua). And then the landscape turns monochrome. (Eames, 2017)

Eames senses the buildings sightlessly staring at him 'hollow-eyed', empty and dead though curiously still holding their heads above 'the fossilised tide'. It is a panorama in which he feels like he is a game player.

Tourists and travel readers are attracted to these unsettling juxtapositions (Skinner, 2008c). Tony Seaton (2012: 521) writes of dark tourism as 'that odd conjunction' as tourism is combined with death, disaster and destruction. It is an unexpected coupling that piques the interest. It is not new in the representation of the volcanic or radioactive exclusion zones. Pre- and post-eruption travel writing brought together 'the tropical Irish' (Yogerst, 2007), or 'the Black Irish' ethnic label for many living on 'the Other Emerald Isle' (Messenger, 1994). What Laffey (1995) coined as the humorous incongruities of such unexpected couplings, often betraying Euro-American assumptions, stereotypes and even racism, have taken a more dire turn. They are now paying to visit for all but a brief period of time, the danger zone, the exclusion zone that has been opened up since 2015 for escorted visits with MVO-approved tour guides carrying walkie-talkies, with a police escort and training in rapid evacuation routes (vehicles should be left running if you leave them; they should face the exit route; and sightlines should be maintained with vehicles at all times) (see Figure 17.6). This is 'ash to cash', as the island's premier Donaldson Romeo was quoted (Schuessler, 2016) describing the recent capitalisation upon the volcano. This is science plus fiction as tourists explore the scientists' exclusion zone.

Figure 17.6 Tour guide negotiating access to the exclusion zone (Skinner, 2017)

Sunny Lea guides visitors around the island, focusing upon the volcano (see Figure 17.7). Short tours can visit the MVO. Longer days can combine this with the Montserrat Springs as tourists move from the observatory to the volcano buffer zone. Full day tours can include the observation post on the east of the island and can go into the buried city along carefully pre-set routes. Keri Jones (2016) used similes to describe

Figure 17.7 Guiding the exclusion zone – Plymouth (Skinner, 2017)

Plymouth as the post-nuclear warzone. He writes of driving through 'the wasteland' Daytime Entry Zone with Sunny, to arrive at this 'scene of devastation' where he is shown the remains of half-buried buildings such as Angelo's Supermarket. He was relieved when they returned to the safety of the north. Joining Sunny on several of his tours, I saw and heard first-hand how he introduced and choreographed the visits. Sunny even worked props into his tour guide 'schtick' (cf. Feldman, 2018: 42), formalised routines set up at each of the tour stops: laminated photos of the Montserrat Springs swimming pool before the volcano, binoculars for the view at the veranda assist a set of memories of the different tremors felt through the disaster (seismic vs magmatic), flashlight for exploring empty hotel rooms, examples of curved melted window glass or lightbulbs dripping like icicles tucked away on a ledge behind a supermarket window. The quips and banter are also loosely scripted: they respond to tourist questions but initiate topics and information along the guide's route. Peering through the top of a window into Angelo's we are encouraged to look down at the food labels along the empty aisles, to see the fan moving in the breeze as natural air-conditioning; entering a gloomy room in the Springs Hotel, Sunny jokes, 'why don't you turn on that light?' before discussing the metal corrosion in the room. The main views of Plymouth become an opportunity to contrast the past and the present to show 'the transition', the loss of buildings, shore line, tennis courts and the new outgrowths of land with potential geothermal energy for the future.

Sunny's narrative is biographical as someone who grew up on Montserrat, lived through the ever-ongoing volcanic eruption and even worked at the MVO for a period of time. His information is detailed, personal and scientific; his imagery graphic but cheerful; his stance optimistic and resilient. It is the tourist that fictionalises their proto-narrative of what it would have been like for them if they had gone through the eruptions and were living with the unexpected. At the tour stops, he asks tourists to use their imagination to create a picture of what Montserrat and Plymouth was like. He explains and accounts for the changes, filling in the blanks in our understanding but also around us in the ruins that – if they are not hybrid objects shape-changed by the volcano – are traces of a former existence (a sign, a bathroom, a hotel reception or room, a broken or buried building, a rusted vehicle). This is different to the dusted object at Pripyat. Abandonment is different to ruination, though in both the visitor is able to pick out their own visitway – their touristlines that are unconstructed, non-linear, not carefree. The visits are personal discoveries, some of which are imbued in religious, political or eco-ideology as the future of (that ironic contrastive term) 'mankind' (see Skinner, 2003b). For Sunny the volcanic eruption is an on-going saga that he is living through: 'A lot of people compare it to Pompeii but the difference here is that we had many layers of – and many dozens of – eruptions and so there's much more varied over time. It's greater than even Pompeii because it's not finished yet, it's referred to as a disaster. For the visitors, reactions are varied from the chilled to the chilling, the curious to the sublime: 'It's funny, you don't feel scared, you feel distant from it'; 'a time warp – fixed in time'; 'It's like nature at home. You're dealing with the force of nature'; 'It's also weird that in an area that is full of buildings, it's so quiet. There's no hustle and bustle from the traffic, and no people and no barbecues.' In different ways the dead zone of the exclusion zone is brought to life by the imagination of the guide and their tourist visitors. They project the past onto this empty playground. This temporal retrospection becomes prospective as they look to the future, using a tourism of the apocalyptic space: this 'eerie ghost town' (Reffes, 2017) – a reminder of what once was but now is; this 'lunar landscape' (Reffes, 2017) – an exclusion zone as a forbidden place, foreboding as a warning to avoid this wider calamity in the future.

Concluding Thoughts – Uncovering Exclusion Tourism

The term exclusion tourism that we introduced at the start of the chapter as the most suitable way to categorise the heterogenous character of the visits to such Zones not only conveys the physical inaccessibleness of the sites of tourism interests, but also refers to the particular state of liminality that is characteristic for tourist encounters in such places. In other words: one must exclude themselves, go beyond their regular life, in order to experience the deep novelty of being in the sites such as the Montserrat or the Chernobyl Exclusion Zones. These and other exclusion zones are threatening places. They are limited play zones and vary from other types of exclusion zone such as the demilitarised zone (DMZ), typically a border strip demarcating boundaries. Where the DMZ is a linear buffer to be crossed or passed along, the exclusion zones in our examples are to be viewed and entered and explored under certain restrictions. For exclusion, read exclusiveness for niche tourists typically politically motivated. In both examples the tourist imagination was fed and informed by media representations of the destination as well as apocalyptic allusions and science fiction film references.

We suggest in the above two case studies that there is also a post-apocalyptic aspect to these sites; the devoid-of-human-life nature of these 'ghost towns' has been used as a stimulus for narratives and pictures that strongly affect people's imaginations. Even if some of these narratives mainly help the visitor to focus on the catastrophes themselves – as processes that destroy what we know and what we have in a spectacular manner and to a radical extent – for others a disaster serves only as a pretext to present a personal reflection on the condition of being human: Why do innocents suffer? What is the point of life? How are we able to control the world? When and how will we die? These subjunctive and futuristic reflections are essential regardless of how they are presented in entertaining mass culture. Nevertheless, there is a fundamental difference between experiencing catastrophe in the mediated form of films, books or computer games, and the embodied practices of being in a real place where an apocalyptic disaster has happened and is ongoing. The term 'apocalypse' refers in popular meaning to catastrophes, disasters and tragedies that might be signs of the end of an age. However, it also literally means 'an uncovering', a disclosure of knowledge or revelation. In the case of our examples of exclusion tourism the disclosure of knowledge is possible thanks to the lack of stimulus of ordinary life: they are spectacular in their absence of the everyday mundane – bureaucracy, routine, the qualia of normality. Only fulfilling this condition offers a ground to open all senses to the secrets of post-apocalyptic sites. This is depicted by Alain de Botton (2001: 166, 169) when he writes:

> Sublime places embodied a defiance to man's will [...] sublime places repeat in grand terms a lesson that ordinary life typically introduces viciously: that the universe is mightier than we are, that we are frail and temporary and have no alternative but to accept limitations on our will; that we must bow to necessities greater than ourselves.

We overcome life when we view on these disaster-scapes. The 'adventure tourists', as Emily Brady (2013: 186) refers to them, engage with this sublime as it is more accessible for the contemporary traveller than the Romantic grand tourist of the Georgian era. Where Brady refers to this travelling as an aesthetic that engages with the natural world, in our examples it is more of a reclamation by the natural world, a 'sensual disorder' picking through the ruins of fatality – to approximate Edensor (2005: 169).

Post-apocalyptic sites force their visitors to confront the most fundamental existential questions that life offers us. They represent modern waypoints on a modern Odyssey akin to the Greek tragedy warnings where there is no guilt, only victimhood of fate. In exclusion zones, tourists are more likely to experience this continuity and depth of existence. It comes to them when they explore the sites for themselves. In this way, their visit is a personalisation of place as their find their own path through the place. This is like the reader's imaginative response to a text such as a science fiction novel, or the point of view scope in a STALKER computer game. Whether in Plymouth or Pripyat, the visit to the exclusion zone empowers the visitor and evokes a strong emotional reaction in its extremeness. This negotiation with nature changes the tourist, often adding grit to the green tourist, to rephrase Montserrat's visitor slogan. They return to their homes refreshed but also re-emboldened, vindicated in their concerns for an impending apocalypse now.

Note

(1) This area and its existing legal order are defined in the document: 'On the legal status of the territory which was contaminated by radioactive radiation as a result of the

Chernobyl disaster of 1991' (and subsequent changes), see Про правовий режим території, що зазнала радіоактивного забруднення внаслідок Чорнобильської катастрофи, http://zakon0.rada.gov.ua/laws/show/791%D0%B0-12 [date of access: 3.11.2017].

References

Aspinall, W., Loughlin, S., Michael, F., Miller, A., Rowley, K., Sparks, R. and Young, S. (2002) The Montserrat volcano observatory; its evolution, organization, role and activities. In T. Druitt and B. Kokelaar (eds) *The Eruption of Soufriere Hills Volcano, Montserrat from 1995 to 1999* (pp. 71–91). London: The Geological Society of London, Memoir 21.

Banaszkiewicz, M., Kruczek, Z. and Duda, A. (2017) The Chernobyl exclusion zone as a tourist attraction. Reflections on the touristification of the zone. *Folia Turistica* 44, 145–169.

Banaszkiewicz. M. and Duda A. (forthcoming) To be a S.T.A.L.K.E.R. On architecture, computer games and tourist experience in the Chernobyl Exclusion Zone, draft manuscript under review.

Beeton, S. (2016) *Film-Induced Tourism* (2nd edn). Bristol: Channel View Publications.

Brady, E. (2013) *The Sublime in Modern Philosophy: Aesthetics, Ethics, and Nature.* Cambridge: Cambridge University Press.

Brown, S., Bell, J. and Carson, D. (1996) Apocaholics anonymous: Looking back on the end of marketing. In S. Brown, J. Bell and D. Carson (eds) *Marketing Apocalypse: Eschatology, Escapology and the Illusion of the End* (pp. 1–20). London: Routledge.

De Botton, A. (2001) *The Art of Travel*. London: Penguin Books.

Dobraszczyk, P. (2010) Petrified ruin: Chernobyl, Pripyat and the death of the city. *City* 14 (4), 370–389.

Donovan, A. and Oppenheimer, C. (2014) Science, policy and place in volcanic disasters: Insights from Montserrat. *Environmental Science & Policy* 39, 150–161.

Donovan, A., Oppenheimer, C. and Bravo. M. (2012) Contested boundaries: Delineating the 'safe zone' on Montserrat. *Applied Geography* 35 (1), 508–514.

Eames, A. (2018) The Caribbean as it used to be: How Montserrat is coming alive again, 14 March. See https://adventure.com/montserrat-caribbean-opening-tourists/ (accessed 28 November 2018).

Eames, A. (2017) Montserrat rises out of Soufrière Hills' ashes, 17 March. See https://www.ft.com/content/2654e834-0804-11e7-ac5a-903b21361b43 (accessed 8 March 2019).

Edensor, T. (2005) *Industrial Ruins: Space, Aesthetics and Materiality*, Oxford: Berg Publishers.

FCO (2019) Foreign travel advice: Montserrat – natural disasters (volcanos). See https://www.gov.uk/foreign-travel-advice/montserrat/natural-disasters (accessed 26 February 2019).

Feldman, J. (2018) Mediation as a practice of identity: Jewish-Israeli immigrant guides in the Christian Holy Land. In J. Feldman and J. Skinner (eds) *Tour Guides as Cultural Mediators – Special Issue of Ethnologia Europaea /Journal of European Ethnology* (Museum Tusculanum Press) 48 (2), 41–54.

Fukuyama, F. (1992) *The End of History and the Last Man*. New York: The Free Press.

Goatcher, J. and Brunsden, V. (2011) Chernobyl and the sublime tourist. *Tourist Studies* 11 (2), 115–137.

Hicks, A. and Few, R. (2015) Trajectories of social vulnerability during the Soufrière Hills volcanic crisis. *Journal of Applied Volcanology* 4 (10), 1–15, open access: https://appliedvolc.biomedcentral.com/articles/10.1186/s13617-015-0029-7.

Johnston, J. (1995) *Jumanji*. Culver City, CA: Tristar Pictures.

Jones, K. (2016) *I Visit Plymouth In Montserrat – The Caribbean's Modern-Day Pompeii, Great Destinations Radio Show.* See http://greatdestinationsradioshow.com/2016/10/26/i-visit-plymouth-the-abandoned-capital-of-montserrat/ (accessed 20 June 2017).

Kasdan, J. (2017) *Jumanji: Welcome to the Jungle.* Culver City, CA: Columbia Pictures

Klein, N. (2007) *The Shock Doctrine: The Rise of Disaster Capitalism.* New York: Metropolitan Books.

Kuśmierczyk, S. (2012) *Księga filmów Andrieja Tarkowskiego.* Warszawa: Wydawnictwo Skorpion.

Laffey, S. (1995) Representing Paradise: Euro-American Desires and Cultural Understandings in Touristic Images of Montserrat, West Indies. Unpublished MA Anthropology Thesis, Texas University, Austin.

Lyng, S. (2005a) Introduction: Edgework and the risk-taking experience. In S. Lyng (ed.) *Edgework: The Sociology of Risk-taking* (pp. 3–16). London: Routledge.

Lyng, S. (2005b) Sociology at the edge: Social theory and voluntary risk-taking. In S. Lyng (ed.) *Edgework: The Sociology of Risk-taking* (pp. 17–50). London: Routledge.

MacCannell, D. (1989) *The Tourist – A New Theory of the Leisure Class.* New York: Schocken Books.

Manjikian, M. (2012) *Apocalypse and Post-politics: The Romance of the End.* Plymouth: Lexington Books.

Messenger, J. (1994) St. Patrick's Day in 'The Other Emerald Isle'. *Eire – Ireland,* Earrach Spring 1994, 12–23.

Michigan Tech (1997) Montserrat volcano risk map: September 1997. See http://www.geo.mtu.edu/volcanoes/west.indies/soufriere/govt/miscdocs/images/vrm_sept1997_3c.gif (accessed 26 February 2019).

MVO (2019) Montserrat hazard level system zones. See http://www.mvo.ms/pub/Hazard_Level_System/HLS-20140801.pdf (accessed 26 February 2019).

MVO (2018) MVO Weekly report for the period 21 to 28 December 2018. See (http://www.mvo.ms/), http://www.mvo.ms/pub/Activity_Reports/2018/20181228-MVO_Weekly_Report.pdf (5 January 2019).

Mycio, M. (2005) *Wormwood Forest: A Natural History of Chernobyl.* Washington, DC: Joseph Henry Press.

Petryna A. (2002) *Life Exposed: Biological Citizens after Chernobyl.* Princeton: Princeton University Press.

Pezzullo, P. (2007) *Toxic Tourism: Rhetorics of Travel, Pollution, and Environmental Justice.* Tuscaloosa: University of Alabama Press.

Pora, P. and Olsen, B. (2014) Imaging modern decay: The aesthetics of ruin photography. *Journal of Contemporary Architecture* 1 (1), 7–56.

Reffes, M. (2017) Touring Plymouth, Montserrat, the Pompeii of the Caribbean, *TripSavvy,* 10 December. See https://www.tripsavvy.com/plymouth-montserrat-pompeii-of-the-caribbean-4050597 (accessed 10 March 2019).

Rogers, M. (2016), Mysterious Montserrat: Volcano-buried city, Beatles legacy, USA Today, 5 July. See http://www.usatoday.com/story/travel/experience/caribbean/2016/06/28/montserrat/86437812/ (accessed 1 September 2016).

Sather-Wagstaff, J. (2011) *Heritage that Hurts: Tourists in the Memoryscapes of September 11.* Walnut Creek, CA: Left Coast Press.

Schuessler, R. (2016) 'Ash to cash': Montserrat gambles future on the volcano that nearly destroyed it. *The Guardian,* 28 January. See http://www.theguardian.com/world/2016/jan/28/montserrat-volcano-british-territory-geothermal-energy-tourism-sand-mining (accessed 30 January 2016).

Seaton, A. (2012) Thanatourism and its discontents: An appraisal of a decade's work with some future issues and directions. In T. Jamal and M. Robinson (eds) *The SAGE Handbook of Tourism Studies* (pp. 521–542). London: SAGE Publications Ltd.

Sekuła, P. (2014) *Czarnobyl: Społeczno-gospodarcze, polityczne i kulturowe konsekwencje katastrofy jądrowej dla Ukrainy*. Kraków: Wydawnictwo Szwajpolt Fiol.

Sites, P. (2014) How not to be a typical tourist: Visit Montserrat, *The Weekend Roady*, 2 April. See https://weekendroady.com/2014/04/02/how-not-to-be-a-typical-tourist-visit-montserrat/ (accessed 24 July 2017).

Skinner, J. (2018) Plymouth, Montserrat: Apocalyptic dark tourism at the Pompeii of the Caribbean. *International Journal of Tourism Cities* 4 (1), 123–139. https://doi.org/10.1108/IJTC-08-2017-0040.

Skinner, J. (2015) Let's get gritty on the Black and Green, *MNIalive.com – Global Caribbean Media*, 21 May 2015. See http://www.mnialive.com/articles/let-s-get-gritty-on-the-black-and-green (accessed 4 March 2019).

Skinner, J. (2008a) Ghosts in the head and ghost towns in the field: Ethnography and the experience of presence and absence. *Journeys: International Journal of Travel and Travel Writing* 9 (2), 10–31.

Skinner, J. (2008b) The text and the tale: The difference between scientific reports and scientists' reportings on the eruption of Mount Chance, Montserrat. *Journal of Risk Research* 11 (1–2), 255–267.

Skinner, J. (2008c) Glimpses into the unmentionable: Montserrat, tourism and anthropological readings of 'subordinate exotic' and 'comic exotic' travel writing'. *Studies in Travel Writing* 12 (3), 167–191.

Skinner, J. (2003a) Anti-social 'social development'? The DFID approach and the 'indigenous' of Montserrat. In J. Pottier, A. Bicker and P. Sillitoe (eds) *Negotiated Local Knowledge: Power and Identity in Development* (pp. 98–120). London: Pluto Press.

Skinner, J. (2003b) Voyeurs, voyagers and disaster tourism from Mount Chance, Montserrat. In D. Macleod (ed.) *Niche Tourism and Anthropology* (pp. 129–144). Glasgow: University of Glasgow Press.

Stone, P. (2013) Dark tourism, heterotopias and post-apocalyptic places: The case of Chernobyl. In L. White and E. Frew (eds) *Dark Tourism and Place Identity: Managing and Interpreting Dark Places* (pp. 79–93). London: Routledge.

Strugatskyi, A. and Strugatskyi, B. (1977) *Roadside Picnic*. MacMillan, New York.

Tarkovsky, A. (1979) *Stalker*, USSR: Mosfilm.

Thoms, S. (2016) Montserrat – 5 days in a modern Pompeii: Plymouth – town of God, *ShaneThoms*. See www.shanethoms.com/#!shane-thoms-visual-journal/c11sn (accessed 16 May 2016).

Tweedie, N. (2006) Sun, sea … and sulphur. *The Telegraph*, 25 February. See http://www.telegraph.co.uk/travel/sunandsea/734651/Sun-sea…and-sulphur.html (accessed 24 July 2017).

Yablokov, A., Nesterenko, V.B. and Nesterenko, A.V. (eds) (2010) Chernobyl: Consequences of the catastrophe for people and the environment. *Annals of the New York Academy of Sciences* 1181. New York: John Wiley & Sons.

Yankovska, G. and K. Hannam (2014) Dark and toxic tourism in the Chernobyl exclusion zone. *Current Issues in Tourism* 17 (10), 929–939.

Yaroshinskaya, A. (2017) (ed.) *Chernobyl: Crime without Punishment*. London: Routledge.

Yogerst, J. (2007) The other Emerald Isle: A Caribbean St. Paddy's celebration. *Island Magazine* reproduced on MSNBC. See http://www.msnbc.msn.com/id/17248491/ (accessed 28 February 2009).

Young, R. (1995) *Colonial Desire: Hybridity in Theory, Culture and Race*. London: Routledge.

18 Hotel Anthropocene

Martin Gren and Emily Höckert

Chapter Highlights

- Presents a day at the Hotel Anthropocene, in the midst of an emerging ecological catastrophe.
- Uses fiction to investigate the future in terms of paradoxes and contradictions.
- Illustrates how the future is given figure and is being enacted in everyday practice.
- Invites critical reflections on tourism in the context of planetary ethics.
- Provides a leaflet companion for further reflection and indicative reading.

Preamble

Hilary: 'Charlie, are you sure that this is really the right place? Do you see the sign over there? It says Hotel Anthropocene, but I booked us for Hotel Holocene!'
Charlie: 'Well, the GPS says we are at the right location, the address is still the same, Modern Tourism Drive 238, right?'
Hilary: 'Yes it sure is. I guess they must have changed the name, I wonder why… I mean, what was wrong with "Holocene"?'
Charlie: 'Beats me, but then I don't even know what Holocene means. Anyway, as long as the hotel has not been downgraded, who cares? I mean, we booked it because it was advertised as a luxury retreat with all-inclusive. Rare to find these days.'

SCENE 1 – 14:00 Check-in at the Hotel Lobby

Charlie and Hilary are trying to find the reception desk in the hot and crowded hotel lobby. They stop by a big aquarium where a young guide speaks to a group of charter tourists. The aquarium serves as 'a conservatory for some of the fishes who used to have their habitat in the coral reefs outside Australia'. Hilary is eager to hear more, but Charlie pulls her towards a hotel staff looking person who is busy answering questions from a gathering of families; 'Yes, the gym and the souvenir shop are both open all around the clock', 'Of course our breakfast includes a chocolate fountain and exotic fruits, as promised in our brochures', and 'You can find timetables for all the meals and for activities like yoga, meditation, Spa and kids-club in the information centre just over there. Unfortunately our Infinity Pool is temporarily closed'. The group joins a collective sigh of disappointment.

Charlie: 'This cannot be true! They advertise the hotel by this very special pool, it is supposed to give you an "infinite experience". And I really, really need to jump into a pool and cool down.'
Hilary: 'I know, me too! I envied the fishes in that aquarium... Well, at least until I heard about their sad demise.'

Hilary and Charlie sneak through the moblike crowd of hotel guests and line up for the queue to the reception desk.

Receptionist: 'Welcome to Hotel Anthropocene, we wish you a nice and relaxing stay! Could you please fill in your passport number, address, and your carbon-footprint on this form?'
Middle-aged man: 'What do you mean? My "carbon footprint"?'
Receptionist: 'It's the amount of carbon your own travel to the hotel has generated. It's new environmental rules and regulations, and we are hoping to become certified as an eco-friendly hotel. I apologise for the inconvenience.'
Middle-aged man: 'This is really ridiculous! I drive an electric car and I have solar panels on the roof of my house. Now I'm just a tourist you know, and I fly only once a year when I'm on vacation... You should take the bloody footprints of those other real polluters instead!'
Receptionist: 'Please take your time; you can calculate your carbon footprint on the tablet over here. No need to hurry, it will still take some time before you can get access to your room.'

Hilary and Charlie also begin to calculate the carbon-footprints of their travel. They list; 'taxi to the train station', 'train trip' and 'rental car'. Since guests with a low carbon-footprint are granted certain privileges, the receptionist gives them a voucher for massage. He also offers VIP bracelets that allow them special prices at the hotel bar – as a compensation for the closed Infinity Pool. He casually, in passing, informs them that a part of the hotel is temporarily being used as a climate refugee-camp, before proceeding to the final information.

Receptionist: 'I must encourage you to look at our information screen over there. In addition to the hotel program and the restaurant menu, it provides you with updated data and diagrams about the internal climate of the hotel, everything from temperature to the weather and statistics about the use of the hotel's resources. As you can see, we are in the middle of high season, and most of the diagrams are therefore in red. This means that we are using much more resources than we should. Therefore, please remember to take short showers, and enjoy your stay!'

The receptionist then turns, now with a worried look on his face, to one of the secretaries nearby. Hilary overhears the secretary saying; 'We don't have enough rooms', just before they both walk away to one of the remotely located office rooms where the following conversation takes place.

Receptionist: 'What do you mean, that we have run out of rooms? We have the two annexes two annexes with plenty of space and beds.

Secretary: 'Haven't you heard? About one hundred rooms or so are out of order. They're too hot, and part of the water pipe just broke, I think the engine has been running dry due to the draught. We also need to lodge new guests in other rooms in order to find space for the climate refugees.'

Receptionist: 'What? I mean, I have heard that there might be new groups of refugees coming, but we can't accommodate all of them here. I mean, we don't have enough space. And how should we be able to sustain them when we don't even have enough for ourselves?'

Secretary: 'I know, but it seems that there is no end to it. Things seem to be escalating. But remember that you cannot say anything at all to the guests at this stage. Under no conditions, whatsoever. The general manager will sack you immediately. Think of your kids, and I have family too. It's troubled times for all of us, you know.'

SCENE 2 – 14:40 Corridors and Hotel Rooms

Hotel Anthropocene has recently been redesigned by its new owners. There is now a new 'Tourism Holocene Room' with representations of tourists travelling on return tickets between homes and destinations on the Earth's surface. There are also hints about a rapidly emerging 'Anthropocene tourism', and in a leaflet it is stated that the Earth of the Anthropocene is 'not the Globe or the Earth's surface, but an Earth system in which the human species as a collective geoforce itself is a part'. A sketchy visual drawing tries to give figure to a subspecies that by its own fossil-driven mobility for luxury purposes propels its own demise by adding greenhouse gases that warms up its own internal living space. The name of the subspecies? *Homo Touristicus*.

Most of the guests pass the 'Tourism Hotel Holocene Room' without entering, but many notice the artistic pictures in the hotel corridors. Pinned on the walls, in tribute of famous guests who had once stayed at the former hotel, are drawings and photos of beings and species from different epochs who all have left their footprints; from *Mr. Dilophosaurus* to *Elvis Presley*, from *Sophora Toromiro* plants to *Kleopatra*, from *Olof Palme* to *J. F. Kennedy*, the list goes on.

Big sister:	'Mom, do you think that they will be able to open the Infinity Pool soon again?'
Mom:	'Honey, I sure hope so. But what if you take a nice, long, bath when we get into the room?'
Big sister:	'But they told us at the reception not to waste water.'
Mom:	'I know, but I am sure that everybody else is doing the same. It doesn't make much of a difference if we alone try to save some water.'
Little brother:	'416; look, here is our room!'
Big sister:	'Wow, come and look; we are staying in a "tiger room"! I can read what it says here; t-h-e l-a-s-t J-a-v-a-n T-i-g-e-r. The last Javan Tiger! Mom, why it say "the last"?'
Mom:	'Well, it's a long and complicated story. Start unpacking your bags instead. I wonder why they have hanged up these unsettling pictures even in the hotel rooms. I hope we won't have nightmares of extinct tigers roaming all over.'
Dad:	'I kind of like this other one, that with a lonely polar bear. It's originally painted by… Marina Zurkow.'
Little brother:	'Come and see; we have a very big bathtub… and lots of ants!'
Dad:	'Ants in the bathroom? We need to get some strong spray and kill them. And the air-condition isn't working either. What kind of hotel is this?'
Little brother:	'Daddy, don't step on them! They live in here!'
Mom:	'I mean, seriously, there's no way we can share a room with a tiger, a depressing polar bear and ants! I think that there was a housekeeping room next to the elevator, I bet they have some ant poisoning spray there. I will go and have a look. There's no point in calling the reception, they seemed so extremely busy.'

SCENE 3 – 16:00 Pool Bar

Among the activities the hotel offers their guests is the quite recently added 'Last chance tourism', a trip to a nearby glacier, which gives 'a unique glimpse of a disappearing world'. At site a well-educated guide will 'give a small talk on how the vanishing glaciers are visible signs in real time of global warming and climate change'. Very few sign up, presumably not eager to find out more about their own dire future. In addition, there is actually not much left of the glacier, so there is little to tickle the visual tourist gaze.

A local destination attracts more of the guests. The pool bar is getting lively during the afternoon happy hour. A drink during daytime signals that one is traveling on leisure island, far away from the inertia of everyday territory and mundane routines. The bartender announces that there will be a Karaoke contest in an hour, and gets ready to take orders.

Bartender: 'Mojito, sure! Unfortunately, we have no ice, but I will do my best…'

Hilary: 'No ice? Is this part of being eco-friendly? We were told that the hotel is trying to get an ecological certification.'

Bartender: 'Not really, the ice-machine is out of order, there seems to be something wrong with the hotel system at the moment. But that's a good idea, we could sell these as ecological drinks!'

Charlie: 'And remember to leave out the straws, they're plastic you know!'

James (guest):   'Sorry to intrude, but we might have other issues to worry about than lukewarm, strawless Mojitos. Have you guys seen that red message on the screen over there?'

Charlie: 'Hmm, let me see. Unfortunately, the hotel is temporarily not working as it should. In case you happen to be an expert in fixing damaged hotel systems, please contact any member of the staff.'

Hilary: 'What do they mean by experts Charlie? Don't they have their own expertise? And hotel systems? They should be more specific about what needs to be fixed.'

Nicole (friend of James):   'Yes, and this is not something that we guests should get involved with, is it James? Excuse me, bartender, when does the Karaoke start?'

Two other guests are sitting at a table next to the bar, and they have also heard about the announced problems somehow related to the hotel system. They are especially concerned about what to tell their daughter.

Jennifer
(leaning over to her husband):   'What should we tell Sandra? The Hotel is one thing, but we can't really tell her about the other stuff that is going on. I heard on TV that it's a severe drought in the whole region, worse than last year, record-breaking again.'

Franco (Sandra's father):	'No, we can't tell her about that now. I mean, that would be like destroying her future, she's only eleven. It terrifies me, I simply don't know what to say to her in these troubled times.'
Jennifer:	'But she understands anyway, you know, the way we behave. She asked me this morning why I look so sad all the time.'

Other guests in the pool-bar area are also concerned, although for different reasons.

Older lady:	'I am staying here with my children and grandchildren, and I must say that we're all very disappointed. This hotel is certainly not as good as we thought.'
Less older lady:	'I know! I have heard so many good things about Hotel Holocene, but there are new owners now. They do not manage the hotel very well, I can tell you that for sure!'
Older lady:	'This is so different from the tourism brochures where the pools and beaches are empty, but people are literally fighting over too few sun-chairs and trash is flying around. I would never have imagined that the Infinity Pool would be empty, there's only small pebbles on the bottom of it. And this bar serves cocktails with no ice, you can't even get a proper dry Martini!'
Tor Nolonger:	'Excuse me, I happened to overhear your conversation, and I couldn't agree more. If this is the tourist experience we pay for, then I want my money back! The Hotel, frankly, it sucks! But I'm afraid it's a telling sign of a business in decline.'

SCENE 4 – 18:00 Dinner

Dinner is a prime time for every tourist. However, there is not as big a culinary variety as the guests had expected. In fact, there is not even enough food for everyone, and the dishes run out before all the guests get a chance to try them. Nevertheless, money can still sometimes pave the way, and those who are able to pay enough extra often do get food.

Mother:	'Waiter, excuse me, would it be possible to get some vegetables for my children?'
Waiter:	'No, I am sorry madam. We have to be very strict, it's all rationed you know. Until recently, we have managed to be self-sufficient, but no longer.'
Mother:	'But there is only some crumbs of bread and bits and pieces of pasta left. And no desserts.'
Waiter:	'We can only use what we get from our own farmland, which is not much. The hotel is suffering from the severe drought, all crop yields are record low. I am sorry, really sorry.'

Several parents are beginning to raise their voices in desperation. Children are crying and the atmosphere is becoming hostile.

Upset father:	'I want to talk to the hotel manager! My children are hungry! You must fix food for them and us. At this point we can soon eat whatever you have.'
Head waiter:	'I am really sorry, but….'
Another dad:	'Then order food from someplace else! Whatever! And the batteries will soon die. Without their iPads, and no food, God knows what will happen to the children!'
Head waiter:	'Sir, there is no home delivery to this hotel. In fact, no one is able to deliver any food to us under these circumstances.'

As the heated exchange continues, one of the hotel guests carries food out from the kitchen in a couple of take-out-boxes. He tries to walk out slowly in secret, but when receiving attention from the upset crowd he runs away.

A grandmother:	'Who was that? And what was in those boxes? So there is still food after all? Who was that food for?'
Waitress:	'There are some who have paid an extra-fee for additional food. You are also welcome to do that, but it will cost you quite a lot.'

Charlie and Hilary are following the happenings at a distance from their table. Hilary pushes away her dessert, and stares on the plastic flower in the waterless vase. Charlie makes an effort to cheer her up by singing along with the troubadour, who has just begun to play in an attempt to mitigate the chaos; 'Money, money, money, must be funny, in the rich man's world'.

Hilary:	'Of all possible songs, why did he choose this one…'
Charlie:	'We have paid with our own money Hilary, and we deserve a great experience. I want us two to make the most of it, even if we have to look the other way and pretend. Isn't that what all tourists do anyway? I mean, you go somewhere else, and you know that is not your real place and life.'

Hilary:	'Probably, and I know that it's the two of us that are the most important of all. No matter where we are, but I just have this too real uncanny feeling that there is some very disturbing things going on in this place. Seriously, listen… the parrot has stopped talking. And there are no butterflies or anything. Look at the palm trees dropping their leaves, and there is that strange smell, don't know what it is. Despite the piña coladas, the yoga and massage, this place fills me with a strange melancholy.'
Charlie:	'I must admit that I feel bad for eating so much from the buffet, while many of the families only got bread and whatever else. The staff should have told us beforehand that there would not be enough food for all.'
Hilary:	'But you noticed immediately that there was very little food at the buffet, and you went there as quickly as you could in order to secure the shrimps and avocado that we paid lots of extra for.'
Charlie:	'Please do not try to make me feel worse. I am sure they are about to get food for everyone. There must be food – this is a hotel, for heaven's sake!'
Hilary:	'I don't mean to say that it's your fault Charlie… I guess it's all our fault together. I mean, all that food shortage, crop failure, and climate change….'
Charlie (interrupting):	'Ha, so it's all our fault now? And why should I believe in all that nonsense? Next year will be cold and rainy as usual, and then everybody will… Aah, it doesn't matter, and I don't want to have that discussion while we're on vacation. Let's just go play table-tennis, or whatever, and try to cheer up! We deserve that, we have both been working so hard over the last months.'

SCENE 5 – 20:00 Common Room

At 20:00 the hotel manager makes an announcement that is broadcasted over the internal speaker system. Guests are being informed that 'the fire-alarm is likely to go off during night-time because of small fires', and that 'smoking is now prohibited everywhere on the hotel premises'. The hotel manager also urges all guests 'to try to stay calm', and they are kindly asked to go to the Common Room where further information will be provided.

Hotel manager:	'Thank you all for gathering here. As you might have noticed, the hotel has encountered some problems. There is no real cause for alarm, but we nevertheless need to inform our guests.'
Chief technical adviser:	'Our main concern is a sudden release of a methane, you have probably not been able to smell it. It's related to a leakage somewhere, and it's bound to increase the amount of greenhouse gases in the internal

atmosphere of the hotel. We have noticed that some of our plants are not doing very well, and we are a bit concerned over how all this might affect people's well-being.'

Assistant technical adviser (interrupts): 'By all respect sir, but I think it is now time to speak the truth. In reality we have ample evidence of a severe hotel system damage – and with foreseeable dire consequences. The truth is that the hotel is currently approaching a dangerous tipping point. It can all flip rapidly, which means that we will not have a hotel anytime soon!'

Technician: 'Yes that is all true, and many of us have tried to communicate this to the hotel management for a long time. We have told them that it's a problem they need to deal with now, not in a future time. I mean, if you mess up the energy balance, from the inside of the hotel, then you run into problems. It only takes high-school physics to understand and calculate that. But the admin guys don't understand, they all have degrees in business and economics.'

Assistant technical adviser: 'Many of us have begged the owners and the management to take action. But now, I am afraid, it might already be too late.'

Charlie: 'I would like to ask the manager, if all this has been foreseen, if you have been aware of this for such a long time, then why on Earth have you not done anything to stop it!?'

Hotel manager: 'Don't say that I have not done anything, because that is absolutely not true! I have written down all data received from our technical staff, and delivered them to the hotel general manager. In fact, you must have seen parts of the data yourself, if you have followed our continuous updates on the screen in the lobby. But it's all above me you know, and we're not the only hotel in the chain. I have also told our general hotel manager, repeatedly, that we cannot continue with luxury services and products, that's simply not sustainable for our hotel. Moreover, a group of external hotel consultants have over decades written extensive annual reports, taking into account extensive technical data and statistics. Their future scenarios have been well known, all the way up the CEO and the company board!'

Assistant the hotel manager: 'I must strongly emphasise that a huge part of our problems was also inherited from Hotel Holocene. I mean, they had in many ways already designed the future for Hotel Anthropocene, there was only so little that we could change in a short period of time.'

Part-time electrician: 'I have seen some reports, at least those summaries that have been distributed to our coffee room. They have always included warnings about future consequences for the hotel if we do not do this and that. However, I could not imagine that all this would be happening so soon and so rapidly, and on this magnitude.. I don't know what to believe, is it that bad with the hotel?'

Chief technical adviser: 'To be really honest, the hotel is so damaged that I do not even know where to begin. Those in charge of the modelling have always said that the future is uncertain, but my experience is that uncertainty is surely not on our side. The future of the hotel they have been talking about, I mean the bad scenario, is already here and now. It seems that all along, at least as long as I have been working here, the hotel has not been moving towards the future. Instead, the future has actually been moving towards us.'

Father (whispers): 'Mary, what are we supposed to tell the children? That they will have no future? That we're all bound to become extinct? I mean, it could be all over for all of us very soon Mary.'

Mary: 'Dear God, dear God…'

Head chief (whispering): 'I don't know what we could do.'

Charlie: 'You hire engineers, technical guys, they know science, and they can come up with technological solutions. If you cannot repair the hotel system, you can build a new one! Nothing is impossible!'

Intern: 'Please notice that we're also trying to become certified as an eco-hotel, although this will take some time, there's quite a lot of bureaucracy you know. It will be very good for the hotel, there is a huge market for sustainable tourism.'

Very angry guest: 'Are you totally insane? How can you talk about marketing this as an 'eco-hotel'

Architect (responsible for interior design): when this entire hotel is in the midst of an ecological catastrophe! It's about to implode from the inside! What is that you do not get, you maroon?!'

'I think that this hotel should have been renovated a long time ago, before the name was changed to Hotel Anthropocene. My guess is that it will also be difficult for you to get that eco-certificate. I mean, the hotel in itself is one thing, it may be in a bad state, but I have also heard about lots of other complaints. You cannot sustain a business if you as treat your guests like you seem to do!'

SCENE 6 – 01:00 Night and Early Morning

The emergency meeting in the Common Room is eventually dissolved. The bar is once again getting crowded, and the bartenders are busy pouring tequila shots. Why wouldn't it be possible to forget all about the common problems, and just try to enjoy and think positive? The future is another time, and the present is here and now. And why worry? If the apocalyptic prophesies about the future destiny of the hotel really were to be true, then people would not be in the bar drinking, dancing and singing Karaoke, would they?

During night-time, many guests are forced to leave their rooms because of the broken air-conditioning and water-damages due to a sudden burst of torrential rainfall. Some families sleep in sunbathing chairs, others on couches in the lobby. Many have been awake, some talking to themselves, some with others. One of the guests has been running up and down in one of the corridors while screaming; 'We have no more future', 'We will all die soon!', until he was eventually knocked out by a forced act of medication.

In the morning, the fishes in the aquarium are floating upside down, but no one bothers much about them. On the internal solar-powered monitor, dire facts and diagrams about the current situation with the hotel system are on display. All in bold red. An atmosphere of melancholic trauma haunts the lobby, which is filled by people wanting to check out.

Benny: 'This is awful! My drunkenness is turning into a hangover. I feel like I'm paralysed. I cannot have a panic attack now...'

Stephanie: 'Try to calm down. We just need to check out. There has to be another place that we can go to. You need a breath of fresh air. That will make you feel better.'

Nature tourist:	'I wish that I was not wearing my party outfit from yesterday, should have geared up at least with some hiking boots and a Fjällräven backpack. It feels like I have stayed at this hotel forever. So where are we heading, is it back to nature now?'
Cultural tourist:	'Yes, sounds good! We can make use of you, since you seem to know everything about nature. A perfect and safe guide that can take care of us outside of this horrendous hotel enclave.'
Nature tourist:	'At your service! Although I can vividly imagine myself running 'into the wild', eating poisonous berries and die. Instead of becoming a quoted nature-geek, I will be awarded a Darwin-prize. Ha!'
Another tourist:	'We can look for Arctic blueberries – the ones that they used put here into the breakfast smoothies. I know what those look liked during the happy days of the Hotel Holocene! Oh dear, how I miss those good old times!'
Post-humanities scholar:	'Have you guys heard about Timothy Morton, this British philosopher who has been described as the 'philosophical prophet of the Anthropocene'. Anyway, he argues that the problem with the present ecological planetary mutation is that we are only trying to be ecological instead of living our ecological knowledge.'
Master student (social science):	'I have read some of his works, and I agree. We know enough, and the most important thing is that we must start to do something now. We cannot continue to live as if the common future of the hotel was something located in another time over there. If we keep on thinking of the future like that, it is bound to come and get us in the present before we know it.'
Tourist (talking quietly to a travel companion):	'I don't really believe in all that about an ecological catastrophe facing the hotel system. I mean, take tourism, if that was all really true, then of course everybody would be up their arms combatting. Tourists would take action against tourism, instead of contributing to the so-called planetary ecological crisis by travelling to hotels over the world.'
Tourist (the companion):	'And the whole tourism business would not be in full swing if we all were on the future trajectory towards 'Hothouse Hotel', especially if it still can be avoided. Of course,

>   the business would have already taken
>   responsibility for its share by pulling the
>   emergency brake on all its carbon emissions,
>   and whatever else they say we need to mitigate'

Meanwhile, an angry man is screaming at the receptionist, demanding his money to be returned. He also requests a booking at another hotel as soon as possible.

Receptionist: 'I understand, but there is simply no other hotel available. We have many people currently engaged in repairing damages and searching for solutions. I remember that you told me Sir, that you were an engineer. Could you not consider staying and help us out with the hotel system? Of course, you will get a deduction on your room.'

Charlie: 'Hilary, we gotta get out of this good damn place!'

Hilary: 'But Charlie, where can we really go to? Do you remember that old song by the Eagles, I forgot the name of it, but it has a line, something like this; "You can check out any time, but you can never leave..."'

Charlie: 'It's "Hotel California".'

Hilary: 'Yeah, that's the one. And this is "Hotel Anthropocene".'

Charlie: 'Then I do no longer want to check out, I want to leave.'

Hilary: 'Don't you understand Charlie, the whole point is that you cannot leave.'

Concluding Arguments

This chapter has been an attempt to explore fiction as a method for investigating conceptualisations and practices of the future in the domain of tourism and climate change. In the beginning of our fictional exercise, we felt uneasy about the characters whose thoughts contradicted themselves and others around them. Eventually, however, we began to realise how contradictions and paradoxes in fact often characterise the conceptualisations of the future and the practices that surrounds 'it'.

For that reason, we also feel that fiction as a method of writing is valuable. Instead of erasing contradictions, tensions, paradoxes and the like, fictional writing makes them visible. This becomes particularly evident when contrasted with the narrative that characterises traditional research papers in social sciences, including the realm of tourism. According to a dominant linear logic the reasoning goes from A to B, from premises via reasoning to the final destination of conclusions. If there were to be contradictions between knowledge claims, then it signals problems in the analysis. When one is not supposed to change premises underway, paradoxes and messiness become enemies. The same holds for many understandings and attempt to map the future.

The chapter brakes away from a linear uniform narrative of the future, as it is populated by voices that express different understandings and conceptualisations of the future. Furthermore, we conclude that this is also a fundamental feature of how the future is being practiced. In everyday life, including at a hotel, the enactment of the future is complex and includes a myriad of factors: information, too much information, where one is located, real or virtual access to other places, the agencies of non-humans, feelings, emotions, existential threats, education, social status, to name only a few of them. The way humans handle the future is equally complex. In the face of planetary climate change in real time some of the hotel guests try to keep up the spirits, others withdraw, and many seem to walk on by as if little or nothing has changed. Hotel Anthropocene, we believe, signifies a major paradox that cuts into the heart of climate change and our common planetary future. If what climate science claims is true, then how come that everyday life and practices continue as business-as-usual?

A Leaflet Companion
Preamble

The Anthropocene is the suggested name of a new geological epoch where the collective of humans is distinguished as a geological force and as a part of the functioning of the Earth System. Contrary to the geological epoch of the Holocene, which corresponds to a relatively stable state of the Earth System favourable for humans, the Anthropocene ushers in dire predictions of an Earth System state that includes the potential of abrupt planetary climate change with disastrous consequences for human civilisation. In a similar vein, one could think of a 'Holocene tourism' denoting tourism as a social phenomenon, which spatially evolves on the Earth's surface, and has nature/the environment as an outside. 'Anthropocene tourism' refers to tourism as also being a geo-physical phenomenon, and therefore implicated in the functioning of the Earth System. Finally, what is both stated and imagined in relation to the Anthropocene can also be figured by fiction.

Bonneuil, C. and Fressoz, J-B. (2015) *The Shock of the Anthropocene: The Earth, History and Us*. Verso: London and New York.
Bristow, T. (2015) *The Anthropocene Lyric: An Affective Geography of Poetry, Person, Place*. Hampshire: Palgrave Macmillan.
Gren, M. and Huijbens, E. (2015) *Tourism and the Anthropocene*. London: Routledge.
Trexler, A. (2015) *Anthropocene Fiction: The Novel in a Time of Climate Change*. Charlottesville and London: University of Virginia Press.

Scene 1 – 14:00 Check-in at the Hotel Lobby

A social ontology of tourism has hitherto dominated tourism research and tourism studies, one example being the conceptualisation of tourists as seeking social experiences outside of the parameters of their ordinary social life. Contrary to this, the physical ontology of tourism is here emphasised and understood as a geo-force. Consequently, tourists also become drivers of contemporary planetary climate change, illustrated by them having to account for their carbon footprints as part of the check-in procedure. This also raises general questions about how to monitor and control future consequences of tourist behaviour and actions in the new planetary regime.

Gössling, S., Hall, C.M., Peeters, P. and Scott, D. (2010) The future of tourism: Can tourism growth and climate policy be reconciled? A mitigation perspective. *Tourism Recreation Research* 35 (2), 119–130.

Mann, M.E. and Kump, L.R. (2015) *Dire Predictions: Understanding Climate Change*. Hong Kong: DK Publishing.

Scene 2 – 14:40 Corridors and Hotel Rooms

In the context of our planetary future, there is an understandable concern for what will happen with the next human generations. Yet, that is also problematic in its human-centeredness, since it provides an anthropocentric narrative that overlooks the history and future of the more-than-human guests who co-habit the Earth. In other words, it creates an image of Earth as a home hosted merely by humans. In the times where humans are causing the eradication of other living things, there exists an urgent need for alternative environmental and planetary ethics that decentre the humans and enhance kinship among human and non-human agencies. This means joining the ongoing efforts to envision more welcoming and caring relations among human and more-than-human agents within tourist practices. Including, we may add, ants.

Bar-On, Y.M., Phillips, R. and Milo, R. (2018) The Biomass distribution on Earth. *Proceedings of the National Academy of Sciences* 115 (25), 6506–6511.

Haraway, D. (2016) *Staying with the Trouble: Making Kin in the Chthulucene*. Durham and New York: Duke University Press.

Hamilton, C. (2010). *Requiem for a Species: Why We Resist the Truth about Climate Change*. London: Earthscan.

Hamilton, C., Bonneuit, C. and Gemenne, F. (eds) (2015) *The Anthropocene and the Global Environmental Crises: Rethinking Modernity in a New Epoch*. London and New York: Routledge.

Kolbert, E. (2014) *The Sixths Extinction – An Unnatural History*. New York: Picador.

Valtonen, A., Salmela, T., Höckert, E. and Rantala, O. (2019) Envisioning proximity tourism through new materialism. *EGOS 2019 short paper proposal*.

Scene 3 – 16:00 Pool Bar

Tourism, not least in the perspective of our common planetary future, is not only threatened by the prospect of an emerging environmental catastrophe but is also one of its driving forces. This leaves tourism in a sort of paradoxical state. For example, while tourism emissions are making the glaciers melt, there are tourists who travel to witness the remaining cryosphere (i.e. frozen water) before it is too late. While some seem to live in denial about the ongoing ecological mutation, others put their trust in technofixes, such as biofuels and solar power. Furthermore, a growing number of people seek more ecological options guided by environmental certifications, choose slower travel alternatives, proximity tourism or staycations. Like other people, tourists are also inevitably experiencing 'solastalgia', 'climate anxiety', or lurking melancholic feelings of living in planetary troubled times.

Jeuring, J. and Haartsen, T. (2016) The challenge of proximity: The (un)attractiveness of near-home tourism destinations. *Tourism Geographies* 19 (1), 118–141.

Lemelin, H., Dawson, J. and Stewart, E.J. (eds) (2011) *Last Chance Tourism Adapting Tourism Opportunities in a Changing World*. London: Routledge.

Mann, M.E. and Toles, T. (2016) *The Madhouse Effect: How Climate Change Denial is Threatening our Planet, Destroying our Politics, and Driving Us Crazy*. New York: Columbia University Press.

Maher, P.T., Gelter, H., Hillmer-Pegram, K., Hovgaard, G., Hull, J., Jóhannesson, G.Þ., Karlsdóttir, A. and Pashkevich, A., Rantala, A. (2014) Arctic tourism: realities and possibilities. See https://arcticyearbook.com/images/yearbook/2014/Scholarly_Papers/15.Maher.pdf (accessed 10 June 2020).

Scene 4 – 18:00 Dinner

The scarcity and fragility of common resources requires us to rethink questions of ethics, responsibility and justice in tourism settings. Whose concern is it when tourism businesses use much of the fresh water on an island; that is, when infinity pools become dry? What happens when birds' and butterflies' run out of food? How do hosts and guests negotiate hospitality and use of natural resources? Reflecting on contemporary neoliberal values and Eurocentrism in tourism practices, as well as in curriculums and scholarly discussions, the most common approach to good life seems to be an individualistic cosmology in which everything begins from the free self, as if we were all tourists entitled to tourism mobility and consumption. However, the ongoing ecological catastrophe cannot be tackled by merely celebrating individual protagonists living in a touristified world apart from earthly attachments. Instead, it challenges us to turn towards relational modes of living and being well among ourselves, and together with non-humans whose existences we depend on for our own survival on the planet.

Höckert, E. (2018) *Negotiating Hospitality. Ethics of Tourism Development in the Nicaraguan Highlands*. London: Routledge.

Grimwood, B., Caton, K. and Cooke, L. (eds) (2018) *New Moral Natures in Tourism*. London: Routledge.

Kingsolver, B. (2012) *Flight Behaviour*. New York: HarperCollins.

Smith, M. (2009) Development and its discontents: Ego–tripping without ethics or idea(l)s? In J. Tribe (ed.) *Philosophical Issues in Tourism* (pp. 261–277). Bristol: Channel View Publications.

Veijola, S., Germann Molz, J., Pyyhtinen, O., Höckert, E. and Grit, A. (2014) *Disruptive Tourism and its Untidy Guests. Alternative Ontologies for Future Hospitalities.* New York: Palgrave MacMillan.

Scene 5 – 20:00 Common Room

In the Anthropocene, humanity, or the human species, is facing a wicked universality. There is no other planet to escape to, and the ecological mutation is now an interdependent planetary common for all humans. We often refer to a 'common future', but 'common' is to be read with caution since also the future will play out in geographically and socially differentiated ways. It has also become increasingly evident that our notion of the future, not least when issues like 'climate change' and 'global warming' are addressed, are problematic. The future has often been imagined as something 'over there', that is, as if located in another time and space. However, the present ecological mutation at planetary scale can no longer be allocated to another time. Climate change is happening in the historical and geological here and now. Moreover, if we do not act now things will get much worse. In other words, whereas we used to believe that we were moving towards a better future, we are now instead facing a dire future that is moving closer and closer towards us. Also, our actions are intimately related to our understanding of the future. Should we act now, or will the problem be solved in the future? Although many may still believe that the problem can still be fixed, and that various technological innovations and solutions will help us, technology alone will not save us.

Biermann, F. (2014) *Earth System Governance: World Politics in the Anthropocene.* Cambridge: The MIT Press.
Hamilton, C. (2013) *Earthmasters: The Dawn of the Age of Climate Engineering.* New Haven and London: Yale University Press.
IPCC (2018) *Special Report: Global Warming of 1.5 °C.* Geneva: World Meteorological Organization.
Latour, B. (2017) *Facing Gaia: Eight Lectures on the New Climatic Regime.* Cambridge: Polity Press.
Lovelock, J. (2014) *A Rough Guide to the Future.* London: Allen Lane.
Moore, J.W. (ed.) (2016) *Anthropocene or Capitaloscene? Nature, History, and the Crisis of Capitalism.* Oakland: Kairos.
Wilson, E.O. (2016) *Half-Earth: Our Planet's Fight for Life.* New York and London: Liveright Publishing Corporation.

Scene 6 – 01:00 Night and Early morning

It can now be explicitly revealed that 'Hotel Holocene' is our metaphor for all kinds of tourist destinations and local hostels that offer dwellings for guests on the Earth's surface. It resonates with the obsolete idea of modern tourism as a particular form of socially driven spatial mobility, and where the so called nature/the environment is an outside of tourism.

Considerations of its future thus belongs to tourism alone, or in relation to external environmental affairs. 'Hotel Anthropocene' denotes instead a conceptualisation of tourism without such an outside. Hereinafter there will be no more 'tourist bubbles', since also tourism itself is part of the functioning of the Earth System, the overall planetary host that provides all earthly guests with the means and possibilities for common terrestrial hospitality.

Hamilton, C. (2017) *Defiant Earth: The Fate of Humans in the Anthropocene*. Cambridge: Polity.
Latour, B. (2018) *Down to Earth: Politics in the New Climatic Regime*. Cambridge: Polity Press.
Morton, T. (2010) *The Ecological Thought*. Cambridge: Harvard University Press.
Morton, T. (2013) *Hyperobjects: Philosophy and Ecology after the End of the World*. Minneapolis and London: University of Minnesota Press.
Tsing, A. (2017) *The Mushroom at the End of the World: On the Possibility of Life in Capitalist Ruins*. Princeton and Oxford: Princeton University Press.

Acknowledgements

We wish to express our gratitude to Anu Valtonen, Tarja Salmela and Outi Rantala from the research group 'Intra-living in the Anthropocene', whose careful thoughts helped us to envision the story of the hotel. Many thanks to Michael C. Hall for providing us with a particular hotel key.

Part 5
Concluding Thoughts

19 Developing a Theoretical Framework of Science Fiction and the Future of Tourism: A Cognitive Mapping Perspective

Ian Yeoman, Una McMahon-Beattie and Marianna Sigala

Chapter Highlights

- Use of cognitive mapping for developing an ontological and epistemological framework of science fiction and the future of tourism.
- Science fiction is the basis of explanatory claims from an ontological perspective.
- The core epistemological concepts of *Plurality of the Future(s)*, *Disruption and Transformation, Hyperreality of Authenticity, DysTopia, Liminality, Scepticism* and *Narrative*
- The core manifestations of science fiction portrayed in tourism futures include *Technological Singularity has Arrived: Westworld; Sustainability;* and *COVID-19 and Pandemics*

Introduction

COVID-19 has fuelled and intensified global discussions about the 'next normal' in the economy but also in tourism specifically. At the core of such discussions is the future of tourism through sustainability (Duedahl, 2020) and the word 're-imagination' is now right at the heart of scenario planning, science fiction and futures studies. Indeed, with a changing world, it is important to reinvent methods and approaches when conditions demand new ways of thinking and re-imagining (Fergnani & Song, 2020). In this COVID-19 context intensifying the need and stress for tourism re-formations, science fiction offers plenty of opportunities to activate and visualise the richness of our imaginations. Scenario planning also has its limits (Curry & Hodgson, 2020; Yeoman & McMahon-Beattie, 2018, 2020): the politics of prediction, limited rationality, the challenges around envisaging collapse and the similarity of scenarios. However, fictional images of the future are powerful and influential in shaping our image of tourism. For example, from a gaze theory perspective (Cohen & Cohen, 2012; Curry & Hodgson, 2020; Mars *et al.*, 2017; Urry, 2011) we can use image to imagine the future of tourism, whether it is sex tourism in Amsterdam, destination images or the impact of climate change. The use of science fiction allows us to think the unthinkable or what Dator (2009) describes as transformation or collapse, which are the dimensions of the future often captured through science fiction.

This concluding chapter aims to capture the essence of the book by providing a reflective overview of all the previous discussions around the theoretical contribution of science fiction to the visualisation and shaping of the future of tourism. To achieve this, first, a series of cognitive maps are developed to visually represent the discussion topics and their inter-relations as presented within each chapter. Then, by using a Computer Assisted Qualitative Data Analysis (CAQDAS) called DECISION EXPLORER (DE), we developed a conceptualisation of science fiction and tourism by taking an ontological and epistemological perspective.

Cognitive Mapping Perspective

Cognitive maps (also known as mental maps, mind maps, cognitive models or mental models) are a type of mental processing composed of a series of psychological transformations by which an individual can acquire, code, store, recall and decode information about the relative locations and attributes of phenomena in their everyday or in a metaphorical spatial environment (Eden & Ackerman, 1998). Applied as a research methodology, they are used to represent cognition of the researched thoughts through a series of links visualised in a map or conceptual framework. In other words, cognitive maps can be used to unravel the structure of the knowledge base of a research field. Jones (1993: 11) states that a cognitive map:

> ...is a collection of ideas (concepts) and relationships in the form of a map. Ideas are expressed by short phrases which encapsulate a single notion and, where appropriate, its opposite. The relationships between ideas are described by linking them together in either a causal or connotative manner.

In this chapter, we use the method of Eden and Ackermann (1998), who adopted the Personal Construct Theory or Personal Construct

Psychology (PCP) in order to apply cognitive mapping for mapping concepts and their interrelations from the strategic management and management science fields (Kelly, 1977). PCP was developed by the psychologist George Kelly (1955) in the 1950s and helped patients to uncover their own personality 'constructs' with minimal intervention or interpretation by the therapist. The repertory grid was later adapted for various uses within organisations, including decision-making and interpretation of other people's worldviews. Eden and Ackermann's approach to cognitive mapping centres on the idea of concepts. These are short phrases or words which represent a verb in which ideas are linked through as cause/effect, means/end or how/why. Consequently, a cognitive map is a representation of a person's perceptions about a situation in terms of bipolar constructs, where the terms are a contrast with each other. For example, 'the invention of the steam train' may lead to 'holidays by the seaside....congestion and pollution'. The result is not unlike an influence diagram or casual loop diagram, although it is explicitly subjective and uses constructs rather than variables (Mingers, 2014). Eden and Ackermann (1998) suggest that cognitive mapping can also be used to record transcripts of interviews in a way that promotes analysis, questioning and understanding. However, the literature on the application of cognitive mapping (Yeoman, 2004) is bastardised as researchers adapt the theory based upon their own skills and research philosophies.

Decision Explorer

A CAQDAS approach, according to Silver and Lewins (2014), assists in the automation of processing data, speeding and capturing of concepts. It helps the modeller to view relationships of phenomena and data through the ability to trace and track data. It also provides a formal structure for notes and memos to develop an analysis platform, which is consistent with grounded theory (Corbin, 2015). DE is an interactive tool for assisting and clarifying problems (Huff & Jenkins, 2002), using the principles of cognitive mapping (Ackermann, 2011; Eden & Ackerman, 1998) within the realm of CAQDAS. DE allows a visual display and analysis of cognitive maps in such a manner that it permits 'multiple viewpoints', 'holding of concepts', 'tracing of concepts' and 'causal relationship management'. It is a rich interactive tool that allows for the movement of concepts and connections for the modeller to be able to identify turning points. This allows the modeller to draw conclusions and construct a meaningful future piecing together to research and produce a close set of practices and interpretations that can present a series of findings which 'make sense' (Weick, 1989; Yeoman, 2004). The most important feature of DE is the ability to categorise concepts, values and emergent themes (Eden & Ackerman, 1998). It allows the modeller to elicit data and code concepts, for example, using 'set management' commands.

Overall, DE a process empowering the modeller to emerge or stand back from the data. In other words, it is a data analysis tool enabling its users to think openly and flexible in order to identify and visualise multiple 'scenarios' and inter-relations between concepts so that they can create a story that makes sense. This approach to modelling and map building is well documented by authors in tourism research. For example, it has been used to explore emergent themes in: family tourism (Schänzel & Yeoman, 2014); the future of events (Yeoman et al., 2014); a conceptualisation of food tourism and futures (Ellis et al., 2018; Yeoman et al., 2015); demography trends (Yeoman & Watson, 2011); and the future past of tourism (Yeoman & McMahon-Beattie, 2019).

The Contribution of Each Chapter

In this section, we identify the contribution that each chapter makes by developing a cognitive map that identifies its key concepts and inter-relations, and which in turn reflect their summative meanings and 'stories'.

Chapter 2: Science Fiction and the Future of Tourism

Science fiction, known as sci-fi or SF, is a broad genre which often contains speculations based on current science and technology. It contains elements of fantasy, utopia, dystopia, structured with narratives, plots, stories in which its imaginary elements – sometimes conceivable other times not (Yeoman, 2012a). The chapter sets the scene for the book by examining what is science fiction and exploring the role of science fiction in academic research and tourism futures. The core concepts associated with this chapter (Figure 19.1) include 'science fiction', 'explanation', 'future(s)…future, 'multiple' and 'truth… explanation'.

From an ontological perspective, the chapter discusses the role of truth and explanation in science fiction writing thus drawing on Bhaskar's (1978) writings of realism. Weak signals are often past or current developments/issues with ambiguous interpretations of their origin, meaning and/or implications. They are unclear observables, warning us about the probability of future events. Ontology is concerned with the study of being and assumptions are concerned with what constitutes reality, in other words, what is (Scotland, 2012). Thus, researchers need to take a position regarding their perceptions of how things really are and how things really work. One way to look at the future from an ontological perspective is Bergman et al.'s (2010) ontological classification of the future which is based upon two dimensions of the future, namely truth claim and explanatory claim with each having two dimensions. Bergman et al.'s classification sets out to create an ontologically grounded typology of future states (or what Bergman et al. call forecasts). Using a classification of truth and

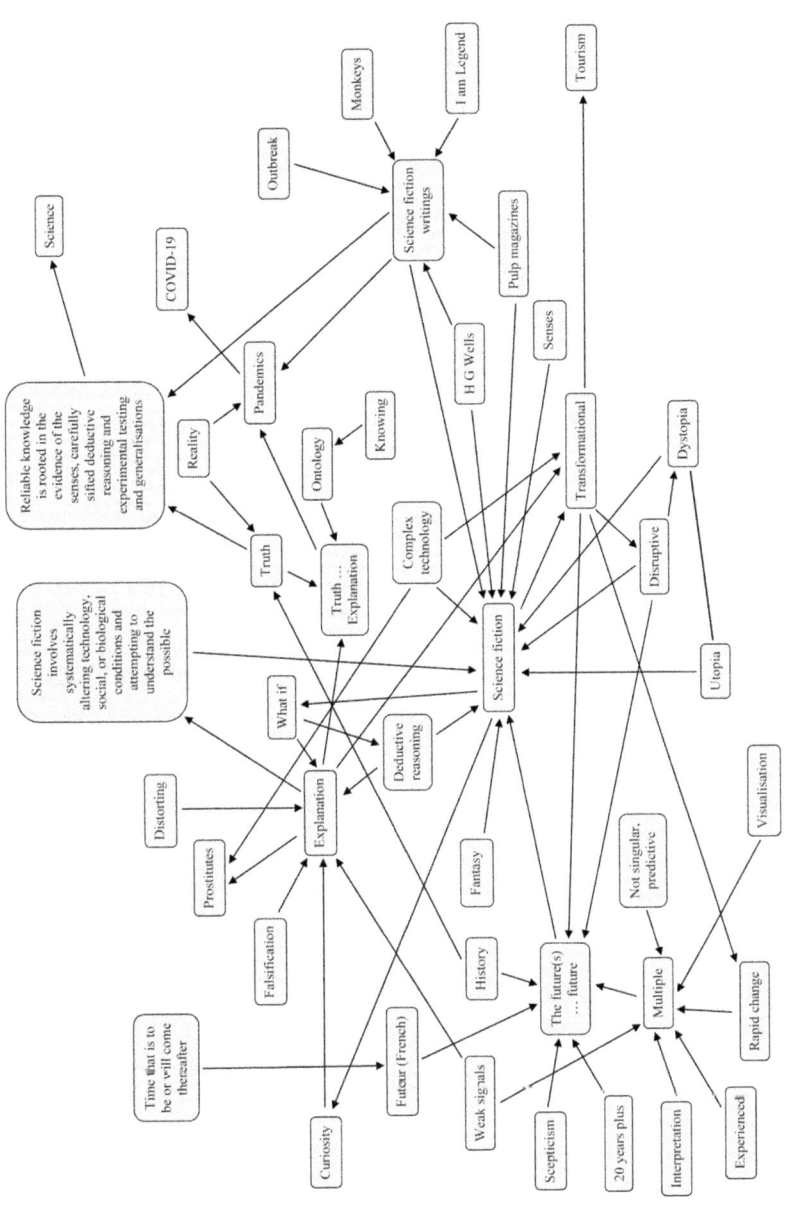

Figure 19.1 Science fiction and futures theories

explanatory claims, the chapter presents two cases studies of science fiction related to COVID-19 and sex tourism.

The value of this chapter is that, firstly it explores a theoretically framed science fiction in futures studies and tourism futures from an ontological (truth and explanation) perspective. Secondly, it discusses the use of science fiction in scenario planning as weak signals and the emancipatory power of the 'what if' question. Thirdly, it examines the role and purpose of utopian and dystopian futures in scenario planning and futures research as creating visionary futures and effective approaches to crisis management.

Chapter 3: The Future, the Devil and the Deep Blue Sea

Figure 19.2 identifies the core concepts in this chapter as 'ocean tourism', 'futures orientated research', 'future(s)…future', 'transformation' and 'tourist playground'. The chapter opens a conversation about what is unique and potentially transformative in the development of new undersea encounters through tourism and leisure. It addresses, in broad terms, the potential arising from oceans as cultural territories and the value of understanding the role and nature of tourism as a significant industry in delivering this potential. While oceanic depths have maintained a long-held fascination, inspiring imaginary visits, futures and possibilities, the 21st century describes a time of unprecedented access to these worlds and tourism as a key facilitator of this access. Submerged people, transport, accommodation and attractions occupy more oceanic space and a significant amount of this is activity is fuelled by the development of tourism infrastructure and experiences (pods, hotels and attractions). While these are expanding the world-wide 'blue economy', the idea of the ocean as an enduring resource that is open for exploitation increasingly meets the resistance of vital oceanic ecologies and processes expressed as a growing sentiment of caution and nurture that is often shared by tourists who encounter oceans and 'discover' their inherent value. Hence the tourism industry dives into these shared oceanic places occupied by existing and emerging paradigms and industries and bathed in the ubiquitous, shifting presence of science and technology.

The chapter also emphasises the mundane yet profound observation that as the tourism industry and oceanic environments become more closely entwined, they produce experiences that do not often conform to land-based ways of doing tourism. In contrast to land-based tourism, undersea tourism negotiates an unfamiliar and somewhat unnatural environment that often speaks to the limitations of experience and knowledge. It evokes ideas of the cyborg tourist, post-human realities and necessarily new or altered methods for the delivery of tourism experiences. Vastly altered environments demand vastly altered behaviour and reasoning and this is where the devil lies in the deep blue sea for the tourism industry and

Developing a Theoretical Framework of Science Fiction and the Future of Tourism 261

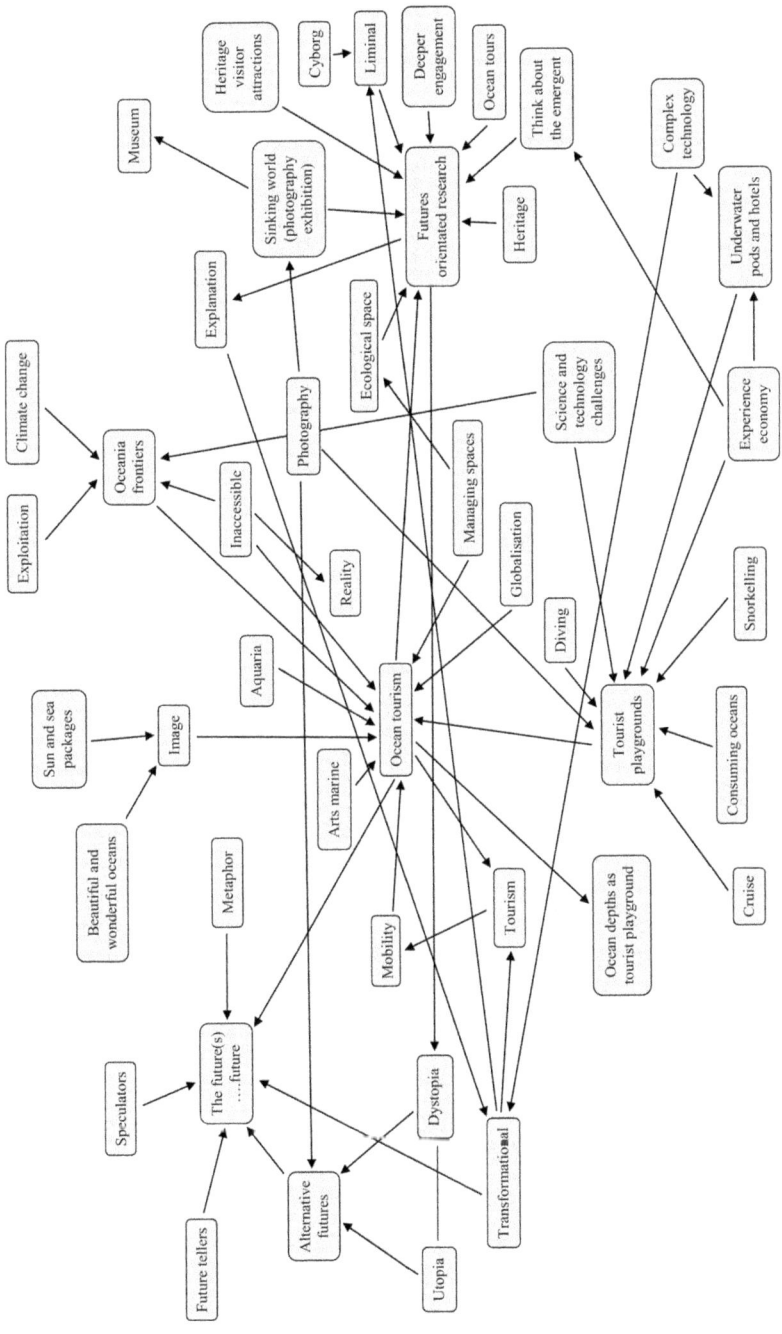

Figure 19.2 The future, the devil and the deep blue sea

scholars alike. This is also where the transformative capacities of tourism meet the equally transformative capacities of oceanic space to produce experiments that are refined into experiences that ferment enduring relationships and collective futures under the sea.

Chapter 4: Prosuming Existential Authenticity in Dystopian Spaces

This chapter develops a narrative which acknowledges firstly that utopia and dystopia are not opposites, then explores the tourist experience in the 21st century as a rather banal activity with little reward and much opportunity to disappoint. The core concepts that appear in this chapter as highlighted in Figure 19.3 are 'presumptions & co-creating', 'places of heritage...history' and 'disappointment of authenticity'.

The chapter considers that disappointment is an issue which can be addressed through consulting other people for ideas using websites such as Tripadvisor. This is one example of many where Web 2.0 has delivered a model of prosumption where consumers are employed free of charge by DMOs to carry out their marketing. The fact that some of this disappointment arises from a failure to find the authenticity that MacCannell (1976) talked about in 1976, the chapter considers the role of self within the experience and posits that self-existential authenticity is the end goal of the tourist. Post-tourists are considered in this context as they embrace the fakery and frippery of the tourism industry, and this is exemplified through a discussion about the nature of the heritage tourism sector.

The chapter suggests that some of our dissatisfaction with heritage experiences is an analogy for a wider issue – that most of what is there to be explored has been seen and shared widely. As a result, those who are keen to explore do so in new ways which are very similar to tourism, but also very different. As the chapter uses urban environments to contextualise discussion, the idea of dystopian destinations is discussed, giving rise to a discussion around urban subcultures and the evolution and growth of urban exploration as an alternative tourism activity.

Through this discussion the chapter raises questions about the future of urban tourism products, the challenges of commercialising dystopia and the opportunities provided by urban exploration to explore new meanings of embodiments within tourism.

Chapter 5: Space Tourism – Science Fiction Becoming a Reality

For decades, space tourism has featured in science fiction stories and Hollywood movies, enhancing audiences' imagination to visit space. However, only a handful of space tourists have been able to witness the curvature of the Earth, as the industry takes its first steps to becoming a member of the larger tourism industry. Rapid innovations in technology

Developing a Theoretical Framework of Science Fiction and the Future of Tourism 263

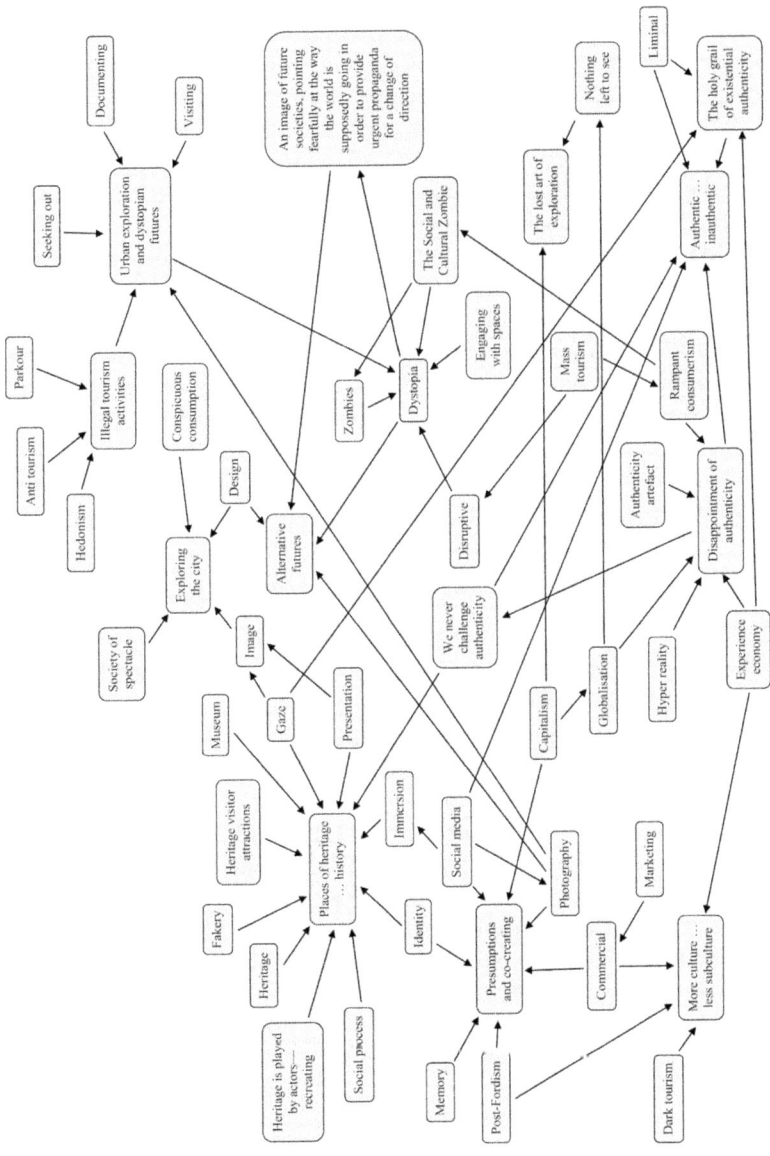

Figure 19.3 Prosuming existential authenticity in dystopian spaces

have facilitated new solutions for the development of the space tourism industry. Suborbital space travel, whereby passengers are transported above the Earth to experience weightlessness and star sighting, will be made available in the near future. The goal is to develop space tourism that is accessible to the masses; however, the high cost of a journey has so far limited the pioneering stage to the wealthy elite.

Space tourism represents a new sector of adventure tourism that has been subjected to very little future perspectives research within academia. This chapter is thus of great utility, both for the nascent space tourism industry as well as for researchers of tourism futures. From Figure 19.4 we can see the core concepts of 'space tourism', 'wealthy tourist', 'planning for the future' and 'transformational'.

The developing space tourism industry needs to act according to predicted consumer megatrends, such as environmental consciousness trends in consumption and travelling. This can be achieved with the assistance of tourism futures research methods and modelling. The chapter explores the development of space tourism, before investigating futures approach methods. Based on available data, a new postmodern model, the Future Space Tourism Roadmap, is introduced. The futuristic space tourism visions that emerge from the data are subsequently explored, and finally the concluding arguments are presented.

Chapter 6: A Life Without Limits: Design, Technology and Tourism Futures in *Westworld*

This chapter considers the multifaceted relationship between design, technology and tourist futures through the lens of Jonathan Nolan's *Westworld* TV series. From the emergence of artificial intelligence (AI) and additive manufacturing technologies (3D printing) to the relationship between historical computational technologies (the player piano), *Westworld* considers the relationship between design, technology and the future of tourism. The chapter is focused on a number of core concepts (Figure 19.5) as 'Westworld', 'human interaction' and 'transformational'. In a post-capitalist theme park in which money, life and human interaction are simulated and rationalised under the algorithmic and database logic of the computer game, Westworld invites the viewer to contemplate a tourist future governed by automation and source code. Specifically, Westworld proposes a post-humanist tourist experience. From the replacement of Da Vinci's Vitruvian Man with a 3D printed biological artificial intelligence to an environment in which control of source code provides godlike power (Poser, 1990), Westworld lays claim to a tourist future governed by an unreality that exists beyond industrial capitalism and the tourist gaze (Urry, 2011). Ultimately, Westworld invites us to think of a tourist future in which design is centre stage as automation disrupts the labour of service industries. Westworld imagines an industry fulfilling one

Developing a Theoretical Framework of Science Fiction and the Future of Tourism 265

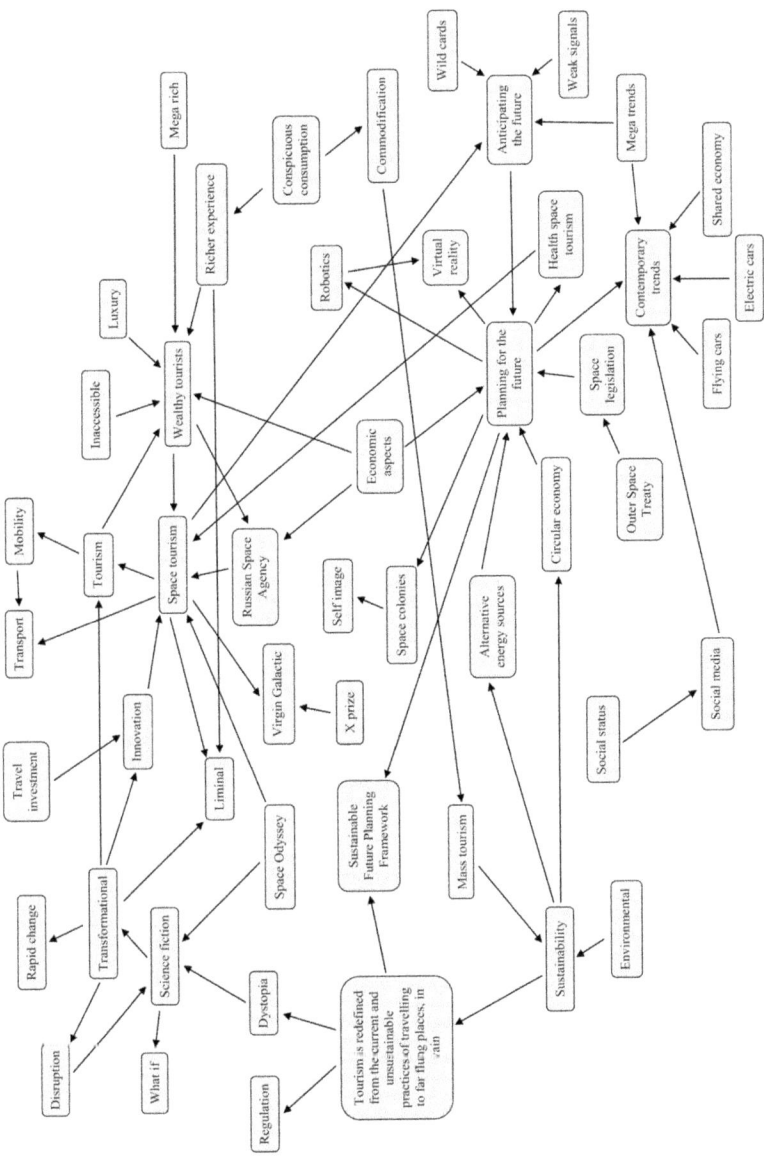

Figure 19.4 Space tourism – science fiction becoming a reality

266 Part 5: Concluding Thoughts

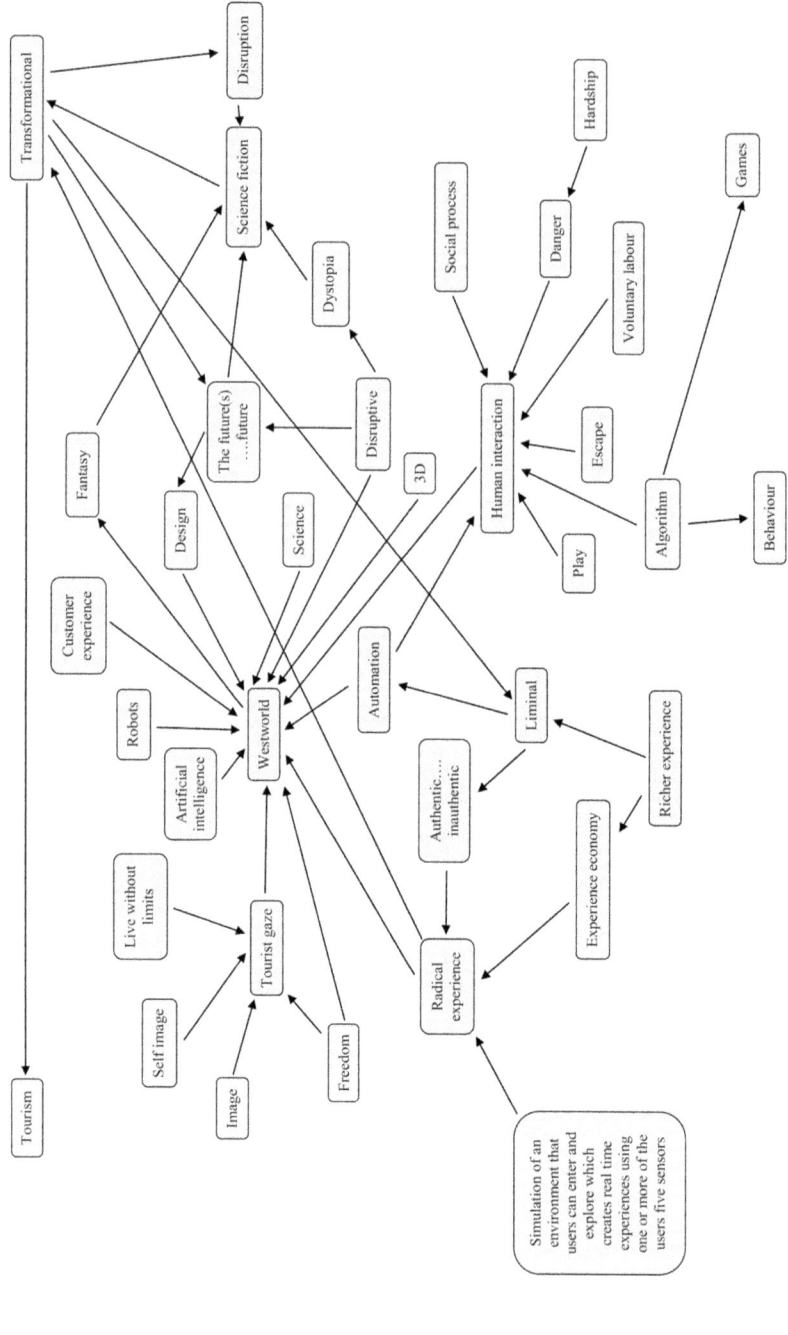

Figure 19.5 Design, technology and tourist futures in *Westworld*

of the few services traditionally assumed to be impervious to automation: human interaction. In so doing it invites us to consider what this disruption in labour will do for the balance of the industry and suggests that its likely implication is an environment in which user experiences, while apparently freer ('Live without Limits' is the sentiment stated by the promotional video and repeated in some form by the other characters throughout), while in fact being far more curated and designed. Throughout Westworld runs a tension between the design of experiences, the disruptive potential of technology to deliver such experiences and the desire of tourists to feel free from the moral, social, economic and political constraints of their daily lives. This chapter further considers the implications of this for tourism.

Chapter 7: Harry Potter and the Future of Tourism

The Harry Potter phenomenon has captured audiences worldwide. The dominant themes of the fight between good and evil, facing personal, educational and moral challenges while coming of age, and the complex context of power, authority, courage, love and friendship resonate across countries, continents, cultures and age groups. Harry Potter is considered one of the most significant pop cultural phenomenon of all time and provides fans across all age groups and cultures with a relatable tool to translate its values of love, courage, friendship and tolerance to real life. Popular culture goes far beyond simple entertainment but establishes norms, contributes to innovations and induces social change (Kidd, 2007). This implies that Harry Potter, as the most relevant pop culture phenomenon in the globalised world, has the potential to facilitate much more than the escape into fictional realms, indeed can encourage and contribute to social change. This chapter uses the Harry Potter phenomenon to highlight the potential impacts of popular culture for the future of tourism.

The Harry Potter series discusses themes of prejudice, discrimination, power, authority and morality, providing readers with a framework to assess what is acceptable and good in the world. These concepts are highlighted in Figure 19.6 as 'Harry Potter', 'society' and 'future of tourism'. It has been shown that readers and fans of the series show more positive attitudes towards stigmatised groups, possess more evolved perceptions of morality and are argued to be equipped with a stronger sense of agency and perceived control to change the world around them. Through fan activism in the form of the global Harry Potter Alliance, this has already been harnessed for a variety of philanthropic and charitable purposes linked to fictional storylines. As many of the current challenges in and negative impacts of tourism can be linked to the themes dominant in the book series, it provides an opportunity to draw parallels between fiction

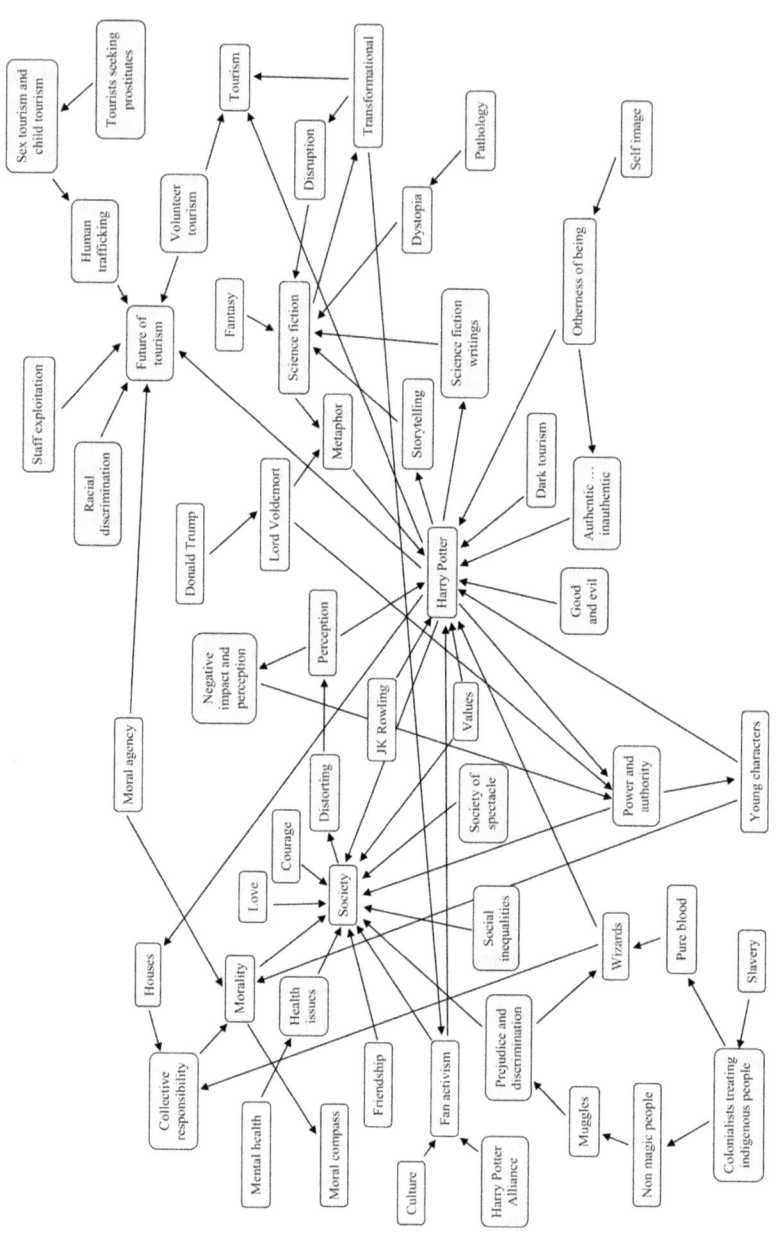

Figure 19.6 Harry Potter and the future of tourism

and reality to reduce tourists' moral disengagement in an effort to change the future for the better. This showcases the power of popular culture and the relevance of fiction for the world we live in, a concept predominantly neglected yet potentially powerful.

Chapter 8: Wildlife Tourism in 2150: Uplifted Animals, Virtual and Augmented Reality and Everything In-between

The value of this chapter is in its discussion of current research in both tourism and other disciplines in order to make an educated guess about how the future will be. The chapter highlights the following core concepts (Figure 19.7) as 'virtual and augmented reality', 'staged wildlife experiences' and 'wealthy tourists'. Environmental bubbles are a topic of research in tourism, but not in regard to the choice of these enclaves because the tourists feel afraid of potential harm, while other less impacting risks are researched more often (such as strange food and language). For all tourist categories, it is very likely that in the future the tourists will avoid risks, in different forms. Wealthy people will be able to isolate themselves from risks completely or to have the thrills, but without the risks. In this chapter, this is not argued only based on the science fiction scenarios, but on research outlining the different aspects of perceived risk, fear of crime and concerns related to terrorism and disease. The presented studies are not in agreement with each other, because of different factors: different cultural perceptions, subjective aspects and the kind of information tourists are exposed to. However, perceived risk is indeed capable of purporting changes in behaviour. Considering the growing presence of news outlets, social media and crime TV programs, the author predicts that people will perceive a heightened risk as the future progresses and demand for no-risk bubbles will rise. At the same time, even people who ignore or seek risks when travelling will also be able to enjoy a suitable vacation. The implications of these choices could be major: tourism would become an even more isolated phenomenon from the places in which the resorts are placed. For instance, the positive impact on local economies would be reduced to close to zero and the demand for certain destinations could shift greatly because people would travel to places they would have never travelled to before. This research is important for tourism studies in general, since it considers many different research results that could have enormous impacts on the tourism industry. However, through the conceptualisation of science fiction examples, this chapter is also relevant for future studies because it presents a novel way to see a phenomenon that is already present in tourism (the environmental bubbles), but that could develop in unexpected ways deeply influencing the tourism industry as we know it today.

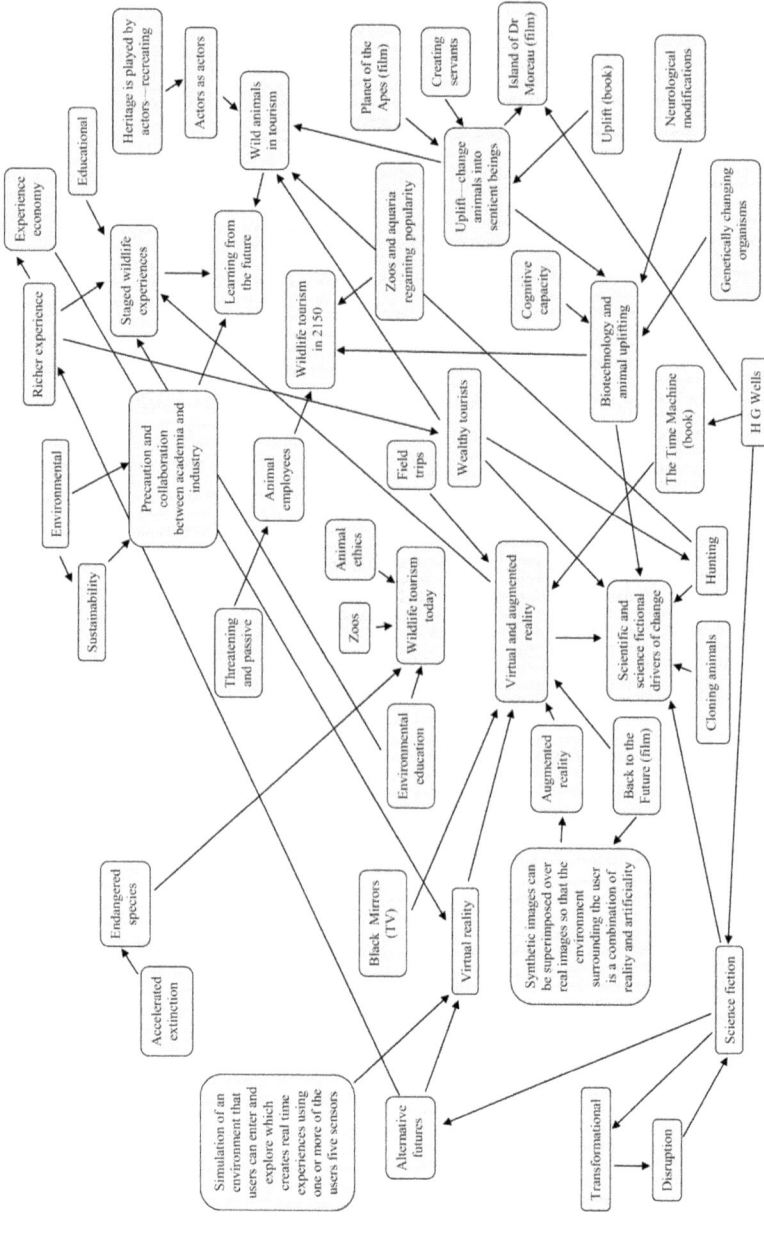

Figure 19.7 Wildlife tourism in 2150: uplifted animals, virtual and augmented reality and everything in-between

Chapter 9: Tears in the Rain: Tourism in the World of *Blade Runner* and *Total Recall*

'I have seen things you can't possibly imagine' utters the character Roy Batty in the film *Blade Runner* (1982). This chapter addresses the worlds and their technology as depicted in the science fiction works *Blade Runner* and *Total Recall* (both films based on the literary publications of sci-fi author Philip K. Dick (1968, 2002)) and their impact and influence on tourism. Realms populated by replicants (artificial humans), flying cars, holograms and memory implants of experiences and places that were never actually visited. These films are portrayed as the core concepts in Figure 19.8. Parallels are drawn with our own world and how advancing technological developments bring us ever closer to the visions depicted in these once science fiction worlds. Our airlines and hotels have been experimenting with robots for check-in, online chatbots to deal with customer queries, keyless check-in through smartphones in hotels, the development of self-driving cars, the increasing use of VR (virtual reality) and AR (augmented reality), developments in AI (artificial intelligence) and the increasing connectivity among myriad devices (dubbed the Internet of Things) mean we are changing the face of tourism, enabling people to see and do things that not very long ago, they couldn't possibly imagine. This brings enormous benefits and the possibility for great positive change. It also carries with it a dark side, with many inherent dangers. Some, such as the threat to people's jobs and livelihoods from AI programs, robots and machines of various kinds, are already apparent. Yet further dangers lurk in the background. As such this chapter addresses where the rapid increases in technology innovation and development currently being experienced are leading us within the realm of tourism. Using science fiction works such as *Blade Runner* and *Total Recall* and the worlds they portray, the chapter takes their depiction of technology and examines from a futures perspective how this will shape the tourism industry and the tourist experience in our own world.

Chapter 10: Destination of the Dead: The Future for Tourism?

This chapter discusses how zombism may be used as a metaphor for tourism focusing on the 'economic zombie', 'environmental zombie' and 'the social and cultural zombie' (Figure 19.9). As noted by the chapter both tourism and zombism result in an irrevocable change in the participant, 'the otherness of being'. This change is not only confined to the participant but also occurs in the location where tourism and zombism happen.

As overtourism spreads across the globe, the similarities between a zombie plague and tourism increase too. The basic premise of tourism having an impact upon the environment, society, culture and economy of a

272　Part 5: Concluding Thoughts

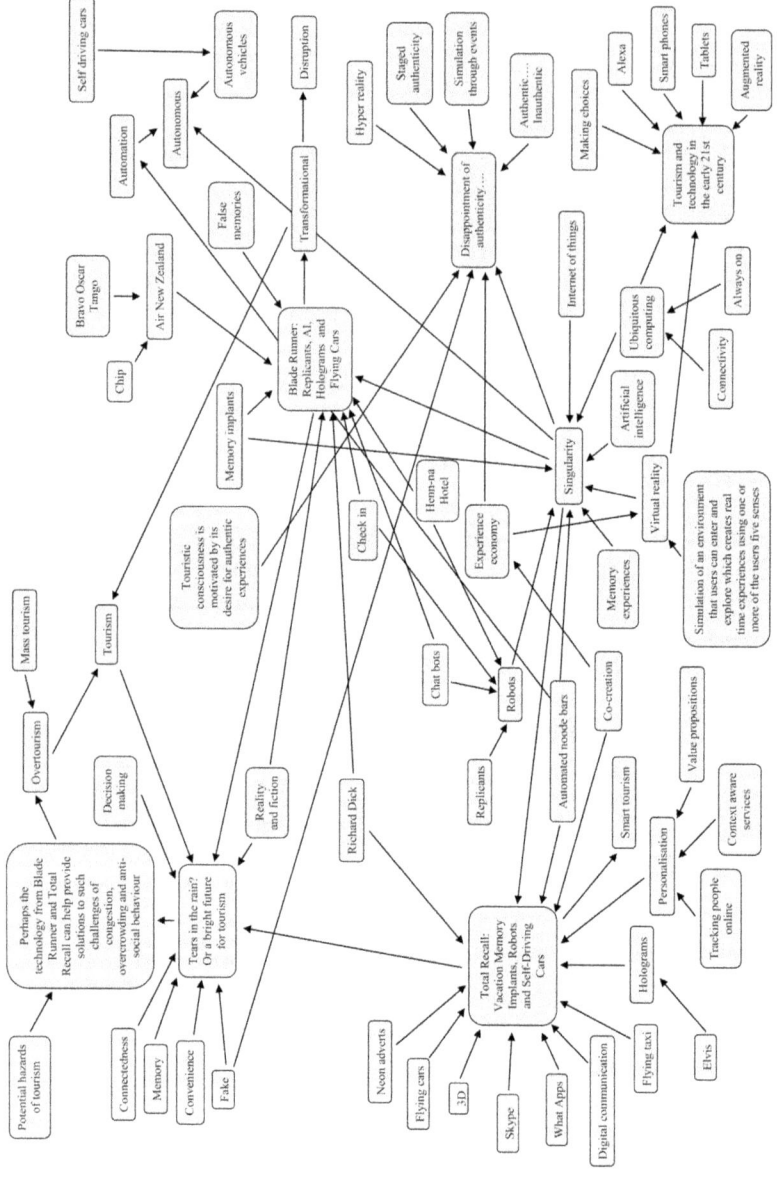

Figure 19.8 Tourism in the world of *Blade Runner* and *Total Recall*

Developing a Theoretical Framework of Science Fiction and the Future of Tourism 273

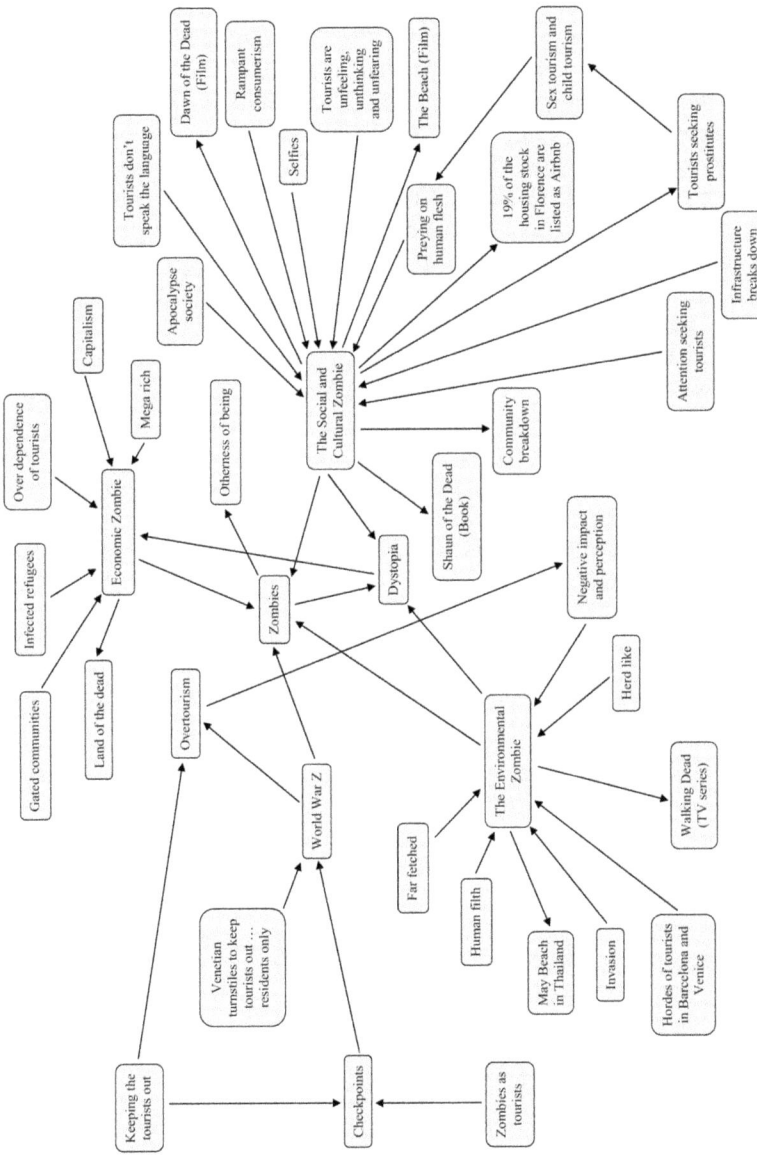

Figure 19.9 Destination of the dead: the future for tourism?

region is juxtaposed to examples from the zombie genre with specific real-life case studies used to illustrate this point. These examples range from the response of the Venetian government to the 'invasion' of tourists expected over a long weekend in May 2018 to the increase in deaths by ultimate selfie-seeking tourists. The similarity is also exemplified by the following notable quotes, one from the famous zombie film *Dawn of the Dead* (Romero, 2004) and one from the well-cited academic, Mishan (1969). The character Peter in the film states, 'When there's no more room in hell, the dead will walk the earth' (Scott, 1982), while Mishan (1969: 140), notes in relation to the scale of tourism, 'what a few may enjoy in freedom the crowd necessarily destroys for itself'. Ultimately, as the world-famous zombie film director, George A. Romero, points out, zombie stories are a way of describing how the world responds to a global issue – the continued unprecedented growth in international tourism and how governments and populations react is therefore, perhaps, the ultimate zombie story.

This is not just a whimsical study. This chapter proposes that potential resolution to overtourism can be derived from the solutions to the zombie threat identified in films, literature and television programmes. These solutions could be adopted (and indeed, in some instances, are being adopted) by Destination Management Organisations. At the very least, the theory of zombism as a metaphor for tourism can be used to stimulate discussion regarding the threat of overtourism and how it may be tackled and managed.

Chapter 11: Holidays with Inspector Maigret: Mixed Reality Adventures as Value Drivers in Future Tourism

Scenario technique is a well-known approach used to deriving potential scenarios of the future. Many studies in this field analyse the different potential futures along selected decision criteria. Such studies give great insights into the variety of potential scenarios and help us to understand which key decision variables might trigger a concrete scenario.

In this chapter, Bingemer follows another path with the goal being not to analyse what the potential futures might be. Instead, this chapter focuses on one concrete future scenario (i.e. the future implementation of mixed reality (MR) adventures in tourism) and follows a narrative approach to guide the reader into a new format of MR that might influence future tour operating packages. By using storytelling, the reader is able to empathise with this new world much quicker than in a purely academic writing style. This helps tourism managers to better understand the opportunities and threats of a MR adventure. In the case selection, we focus on a scenario that has face validity as it is based on technologies that exist but potentially need to be further improved for the implementation depicted in this chapter. Thus, by using a scenario that centres on the need for technological improvement instead of a complete futuristic scenario, we extend the managerial view on how the future might look like. The chapter is unique in that it describes one concrete potential future. For this concrete future, implications for the improvement of customer and user experiences as well as for marketing and sales of such an offer are discussed and challenges and solutions are proposed to the reader. The chapter discusses how tourism partners (e.g. gastronomy) could leverage such offers to derive own new value propositions around a MR adventure. Like Pokémon Go, where gastronomy has jumped on the train by attracting Pokémon players into their bars and restaurants, a MR adventure could be the foundation of a wider value-creation scheme. In Figure 19.10 the core concepts are identified as 'quantum leap MR adventure based on Maigret and the dead girl' and 'mixed reality has the potential to disrupt the way we do city travels and we combine fictional stories'.

Chapter 12: Digital Destinations and Avatar Tourists: A Futuristic Look at Virtual Reality Tourism and Its Real-World Impacts

This chapter offers a futuristic look at virtual reality (VR) tourism over the coming centuries. It draws on inspiration from science fiction stories, referencing them at relevant times throughout the chapter. The core concepts that are represented in this chapter (Figure 19.11) are 'future of VR', 'liminal' and 'untethered'. Nevertheless, the chapter remains rooted in a realistic imagining of what VR technology and VR tourism might look like, as it is based on existing VR technology and cutting-edge research. The chapter begins with an overview of present-day VR technology, which is followed by an examination of the futuristic directions this technology is heading. This look at VR technology covers all five human senses and extends to futuristic possibilities including neural implants. In sum, it is posited that VR experiences eventually will become nearly indistinguishable from real life. Next, the chapter highlights the distinct ability VR experiences have to elicit powerful emotional responses. With such powerful experiences in mind, the chapter then discusses the phenomenon of VR tourism, with a focus on why it is reasonable to anticipate VR tourism experiences eventually will compete with real tourism experiences and alter tourism purchase patterns more generally. This argument is partly made by highlighting how the current limitations of VR tourism will be mitigated as the technology advances, yet the potential benefits will remain. Also, it is pointed out that VR tourism will not have to be perceived as 100% authentic in order to impact tourism decision making. Finally, the chapter examines various key implications of the eventual emergence of VR tourism, as relating to a variety of different issues. For example, VR tourism will prove to be an economic benefit for some destinations/attractions and a negative development for others, it may impact users' mental health and physical well-being, it will give rise to new gatekeeping distribution platforms, and it will raise important questions regarding privacy and data tracking.

276 Part 5: Concluding Thoughts

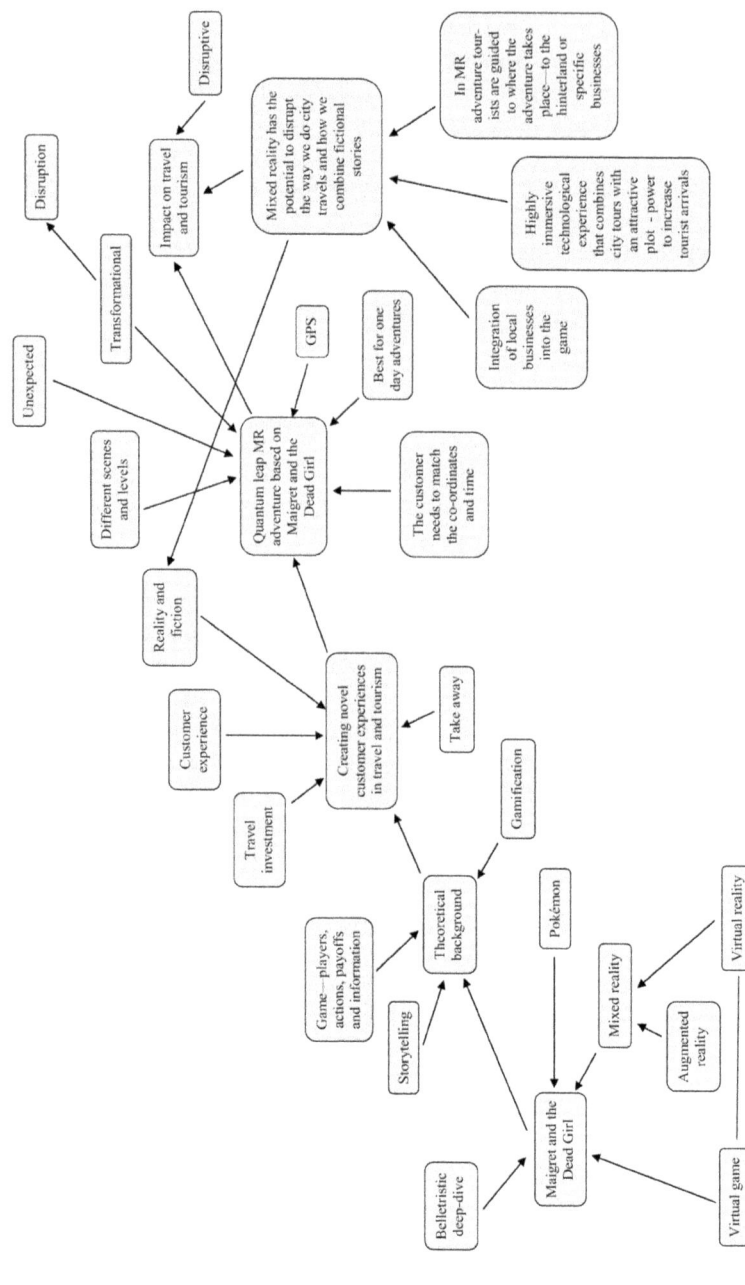

Figure 19.10 Holidays with Inspector Maigret: mixed reality adventures as value drivers in future tourism

Developing a Theoretical Framework of Science Fiction and the Future of Tourism 277

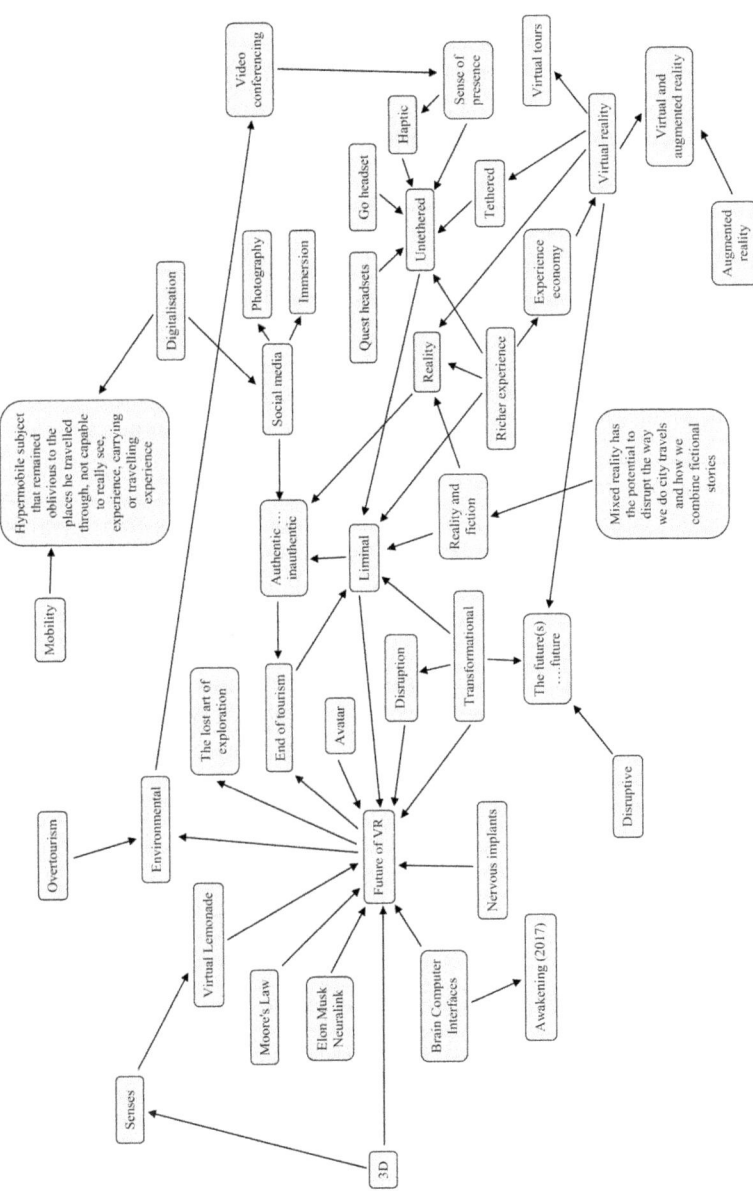

Figure 19.11 Digital destinations and avatar tourists: a futuristic look at virtual reality tourism and its real-world impacts

Chapter 13: The 'Safety Bubble' and the Future of Enclave Tourism

In Figure 19.12 the key concepts identified include 'environmental bubble', '*Elysium* (2013)', 'risk', 'science fiction' and '*Rick and Morty* (2017)'. Environmental bubbles are a focus of research in tourism, but not regarding the choice of these enclaves because the tourists feel afraid of potential harm; other impacting risks such as strange food and language are researched more often. For all of Cohen's (1979) tourist categories, it is very likely that in the future tourists will avoid risks, in different forms. People who are willing to pay will be able to isolate themselves from risks completely or to have the thrills, but without the risks. In this chapter, this is not argued only based on the science fiction scenarios, but on research outlining the different aspects of perceived risk related to destination choice and information gathering. The presented studies are not in agreement with each other, because of different factors: different cultural perceptions, subjective aspects and the kind of information tourists are exposed to. However, perceived risk is indeed capable of purporting changes in behaviour. Considering the growing presence of news outlets, social media and crime TV programs, the author predicts that people will be more willing to pay for risk-free vacations. The influence of all the information sources can be used to steer the tourists to risk-free enclaves. This can be done by underlining the risks. The implications of this can be major for the future of enclave tourism: people would pay to have a risk-free vacation, isolating the enclaves totally from the local populations, eliminating all the potential benefits deriving from their presence. Through the conceptualisation of science fiction examples, this chapter is also relevant for future studies because it presents a novel way to see a phenomenon that is already present in tourism (the environmental bubbles) with science fiction lenses. The way in which enclaves may develop in the future due to concerns related to perceived risk and the information the tourists receive through different channels calls for a reflection on the potential positive and negative implications of its realisation. This is not only a concern for future research but also for destination developers.

Chapter 14: The Coming of the Fugue and the Blind Tourist?

In Chapter 14, Reid and colleagues ask if mass tourism is morphing into overtourism, and can this be conceptualised as an emerging plague of zombie tourists, and what kinds of tourism futures might come of it? The chapter (Figure 19.13) is focused on a number of core concepts including 'zombies', 'overtourism' and 'hypermobile subject that remained oblivious to the places he travelled through, not capable to really see, experience, carrying or travelling experience'.

Overtourism is not only unsustainable; it is the logical outcome of capitalism and thus, a token signifying that the capitalist system is well and

Developing a Theoretical Framework of Science Fiction and the Future of Tourism 279

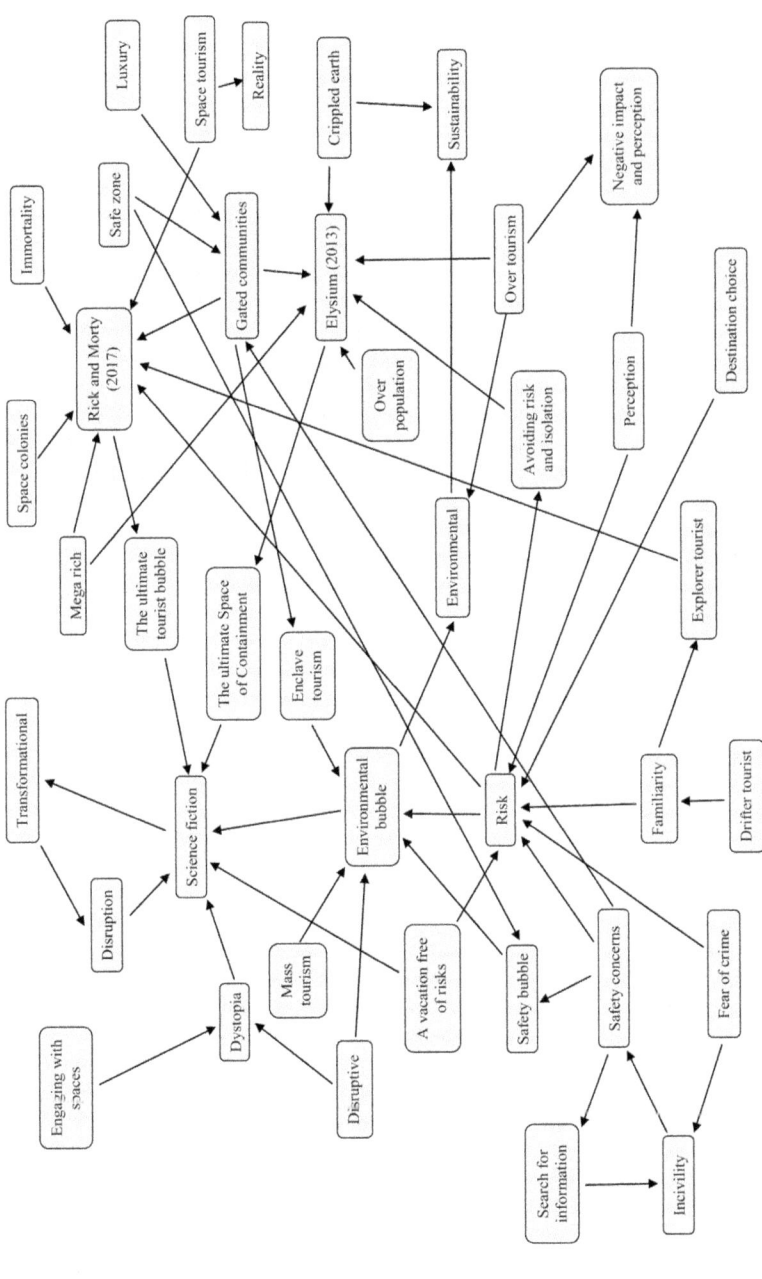

Figure 19.12 Safety bubble and the future of enclave tourism

280 Part 5: Concluding Thoughts

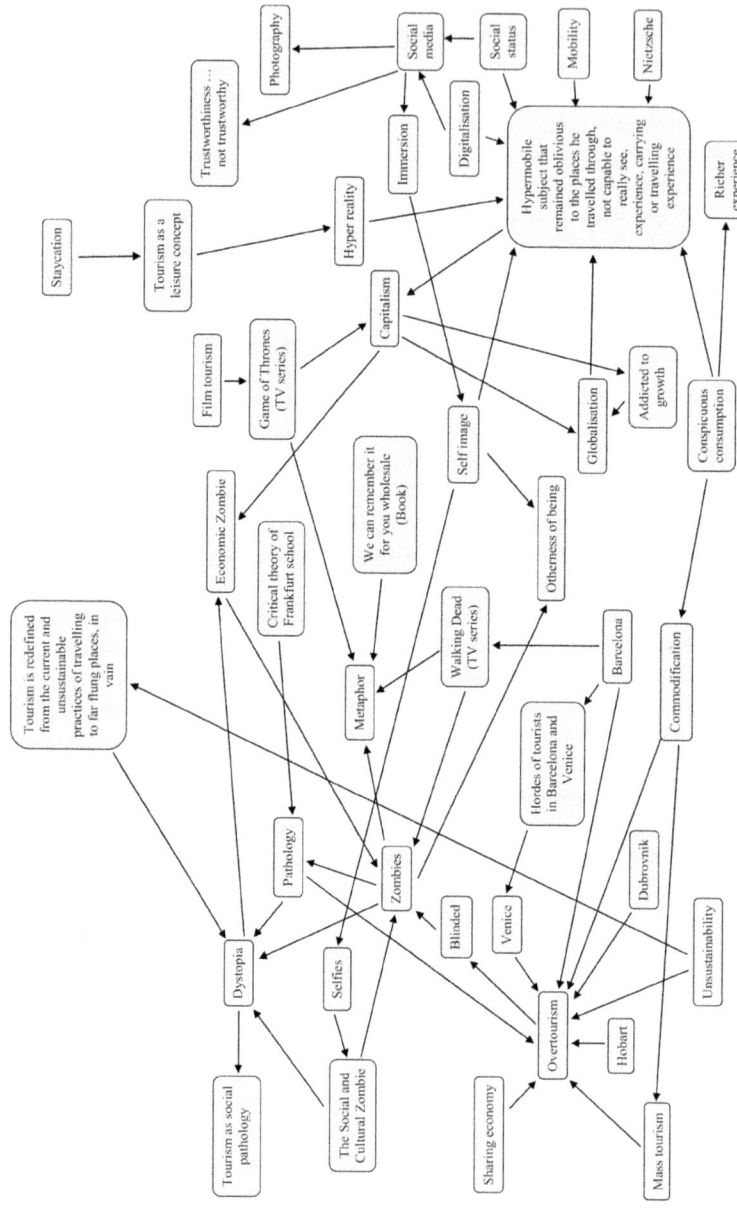

Figure 19.13 The coming plague of the fugue and the blind tourist?

alive even though it is threatening everything else on Earth. We outline a drastic narrative of this unsustainable phenomenon, characterizing it as a pathological condition that chisels-out a zombie tourist, well-travelled but at the same time oblivious to the tourist destinations he or she passes through. It is further argued that the entrance of social media and digital portable technological devices have increased this pathological state as it has added a self-centric, narcissistic dimension into the set of touristic practices to a degree that the zombie tourist even runs an increased risk of ending up dead. Secondly, the chapter presents two bifurcated scenarios presenting possible trajectories for future tourist practice, both less reliant on physical long-distance travelling: the implantation of digital memories in the individual tourist's mind and consciousness and 'staycation' tourism, i.e. short trip close to the home of the tourist. These two more optimistic scenarios bring some relief to the environmental situation generally and the social situation of overtourism already emerging at many destinations. But, at the same time, these two scenarios are nevertheless embedded in a pathological capitalism and are perhaps bound to create new societal and environmental problems, possibly bringing new kinds of unsustainability.

Chapter 15: Technological Frontiers: From the Wild West Myth to the Dystopia of the Westworld's Post-human Theme Park

This chapter explores what happens when the tropes of the 19th century frontier meets the science fiction trope of the sentient android on the 'technological frontier.' The concept of the 'Wild West' of America has long attracted tourists with forms of staged authenticity (MacCannell, 1976). The 19th-century frontier is frequently associated with a legendary narrative of pioneering individualism and adventure, leading to romantic tourism, re-enactments and cowboy theme parks. The television series *Westworld* is highly relevant to future tourism studies because it is set in a theme park which is hosted by androids and visited by wealthy guests requiring bespoke tourism.

From Figure 19.14 the key concepts are clustered around 'Fast authenticity' 'West World' and 'science fiction'. Issues of authenticity are central to this discussion, nowhere is the difference between copies and fakes more clearly shown than in discussions of the post-human (Braidotti, 2013) and hyperreality (Baudrillard, 2001; Eco, 1986). The chapter deconstructs aspects of the television series narratives, plots and interviews with key designers, introducing the idea that authenticity of the future may be hyper-staged (adapted from MacCannell, 1976). Hyper-staged authenticity is typical of western science-fiction, marrying the two central strands of nostalgia with futuristic technologies. Authentic period details which are intended to enhance the immersive experiences of tourists use new technologies to duplicate the past in *Jurassic Park*-style scenarios, moving ever-closer to resurrecting original sources.

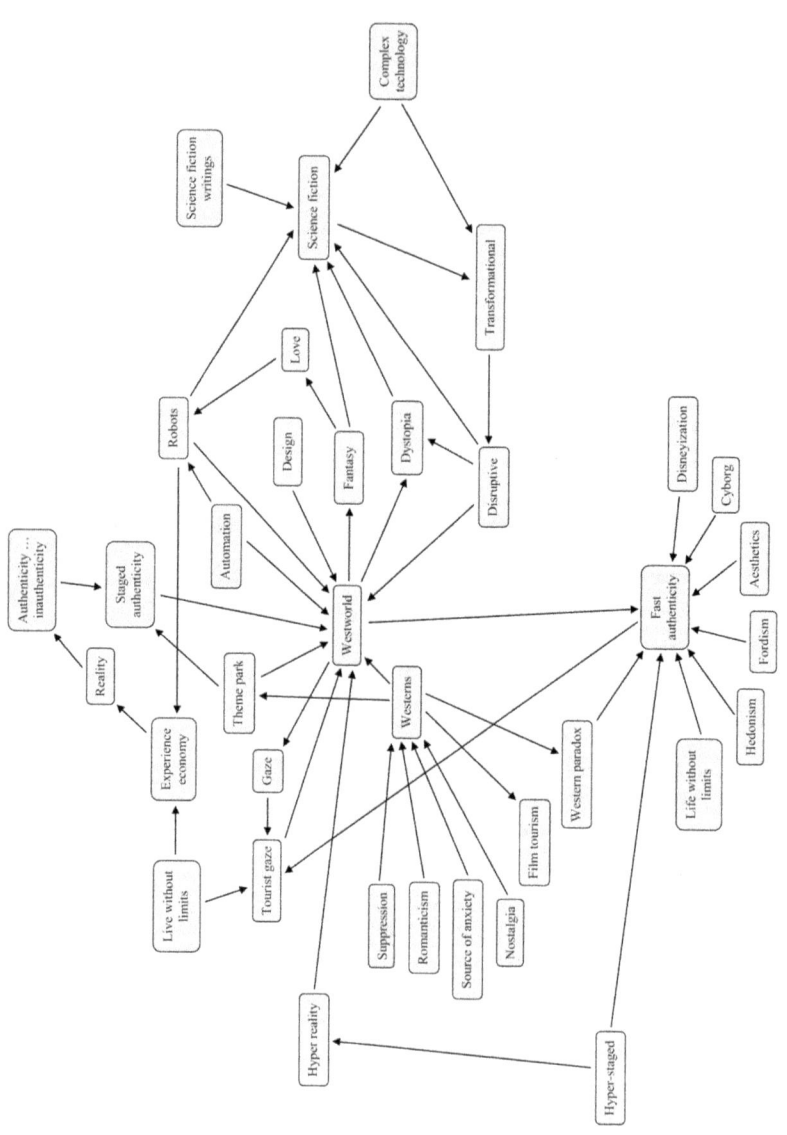

Figure 19.14 Technological frontiers: from the Wild West myth to the dystopia of the Westworld's post-human theme park

Westworld appears to offer a utopic future consumer dream of 'fast authenticity,' comprising transformative and immersive experiences, but this masks a system of surveillance and enhanced control. The technology forges a new dystopia; the out-of-control theme park, where the visitors seeking safe risks are locked into a more authentically brutal wild west nightmare. Unethical decisions repeat the past mistakes, the aggressive suppression of Native American peoples on the frontier is recreated in the abuse and oppression of the android hosts and the resultant chaos as, inevitably, the post-humans take over the tourist attraction. The theme parks in Westworld act as metaphors for change, demonstrating how, as technologies become more sophisticated, future tourism trends will reflect the not simply the advantages, but the fear of a post-human and post-Anthropocentric universe.

Chapter 16: The Future of Music Concerts and Tourism in Dystopian Times

Chapter 16 by Wright offers original and novel research on the future of music and tourism, focusing on benefit concerts as a means of supporting people living in dystopian realities. It considers how tourism and benefit concerts can ensure global exposure and support to suffering communities. The core concepts highlighted in this chapter (Figure 19.15) include 'dystopia', 'benefits of concerts and tourism…future considerations' and 'an imaginal piece or state of affairs in which conditions are extremely bad, especially in which results from the contamination of the trend to the extreme'. To explore this, the chapter considers the role and power of music in society, on individual and collective levels. It recognises the influence of music and musicians and its ability to engage wide audiences around the globe. Significantly, it highlights the power of music and its capability in driving cultural change, and that culture can reflect the music industry of any given time throughout history. This is increasingly powerful in society as musicians can reach out and influence audiences of billions of people through social platforms and via the media. It also addresses the meaning and characteristics of what a dystopian reality consists off and the potential for looming dystopian realities in the future. Subsequently, he suggests how future benefit concerts, driven by the power of music and tourism could have the ability to bring about relief to dystopian communities. The idea proposed is, can benefit concerts become a more common scene in the tourism market as a tool for supporting communities? If so, going forward key aspects need to be addressed, as there is great responsibility on the part of organisers, musicians, media and tourists in the way they promote, engage and ultimately support communities living in dystopia. While benefit concerts have often been a reactive approach to supporting victims in need, this chapter suggests that rather than operating as a contingency plan, the industry could establish more

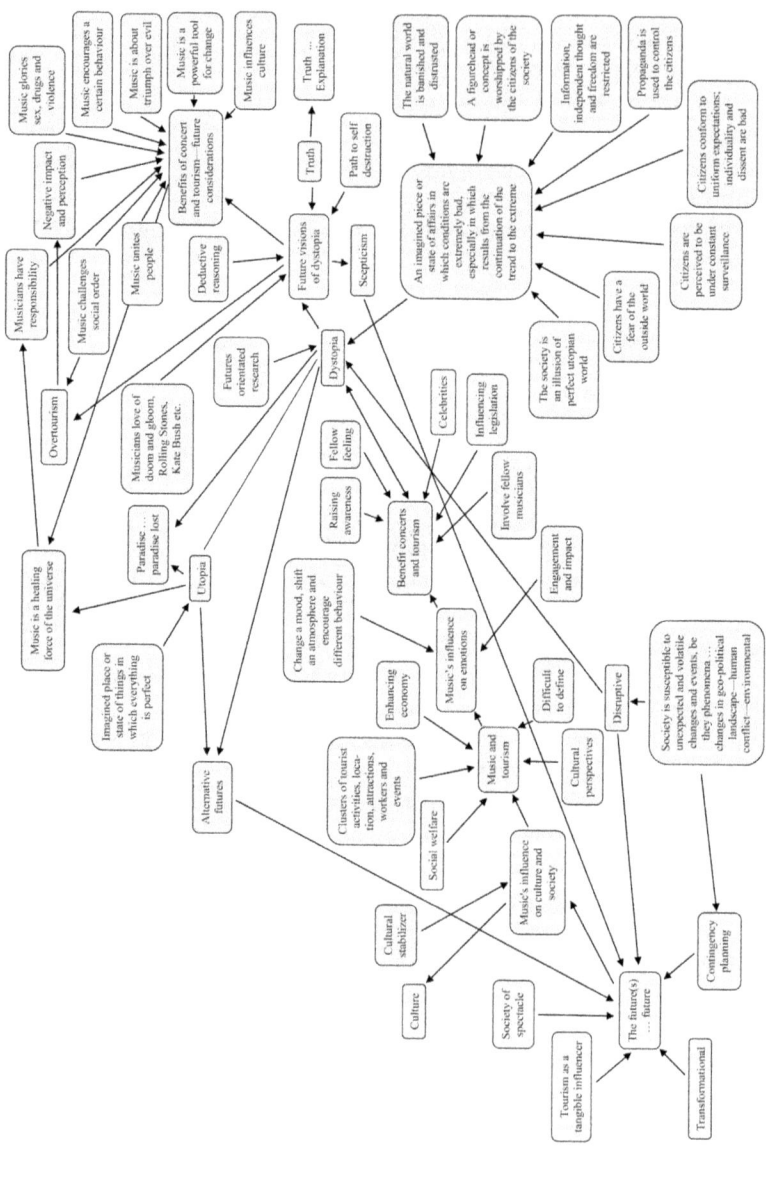

Figure 19.15 The future of music concerts and tourism in dystopian times

frequently organised benefit concerts to not only support people living dystopian realities, but as a means to drive positive social and cultural change.

Chapter 17: Exclusion Tourism: Sci-Fi Stalkers and Subjunctive Plays in Apocalyptic Destinations from Chernobyl to Plymouth, Montserrat

Chapter 17 by Banaszkiewicz and Skinner examines venues that had been destroyed or abandoned but in their apocalyptic ruin have emerged as conflicted tourist attractions associated with science fiction. Figure 19.16 illustrations the core concepts of 'exclusion tourism', 'uncovering exclusion tourism', 'Chernobyl Exclusion Zone tours' and 'apocalypse is a revelation, a sense of ending when only a few struggle to survive'. There is no doubt that post-apocalyptic landscapes are particularly attractive for science fiction texts (books, films and video games). As tourist destinations, they elicit reactions of emptiness, unevenness, loss and desolation that are counterbalanced by awe, fascination and a sublime enjoyment that (Manjikian, 2012) refers to as 'the romance of the end'. The apocalyptic – whether 'man-made' or 'natural' – are dystopic locations where the future subjunctive can be played out by temporary visitors who court the uniqueness of their exclusion.

Specifically, the chapter engages anthropologically through participant observation with two apocalyptic destinations: the Chernobyl Exclusion Zone (Ukraine) and the Plymouth, Montserrat Exclusion Zone (Eastern Caribbean), both destinations for tourists seeking to engage with their paradoxical aesthetics. The Chernobyl Exclusion Zone, a case of exclusion through radioactive leak, of modernity's ills, is frequented by 'stalkers' – a community of illegal urban explorers whose archetype was presented by Andrei Tarkovsky in *Stalker* (1979), a film loosely based on the novel Roadside Picnic by Boris and Arkady Strugatsky (Strugatsky & Strugatsky, 1971). This contrasts with visitors to the island of Montserrat that is riven down the middle by an on-going volcanic eruption. Tourists to Montserrat contrast 'the green' of the north with 'the gritty' of the south and the ruin of Plymouth, 'a modern-day Pompeii', exploring the former capital and imagining by reference to films such as Jumanji (Johnston, 1995) a new politics from the hybridity of its nature-wrought debris. Together, these case studies of the apocalyptic show how science fiction texts have established a new form of tourist practice: 'exclusion tourism'.

Chapter 18: Hotel Anthropocene

Hotel Anthropocene is a new hotel advertised as a luxury all-inclusive resort. Until recently it was operated under the name of 'Hotel Holocene', but due to the widespread attention towards 'the Anthropocene' the new

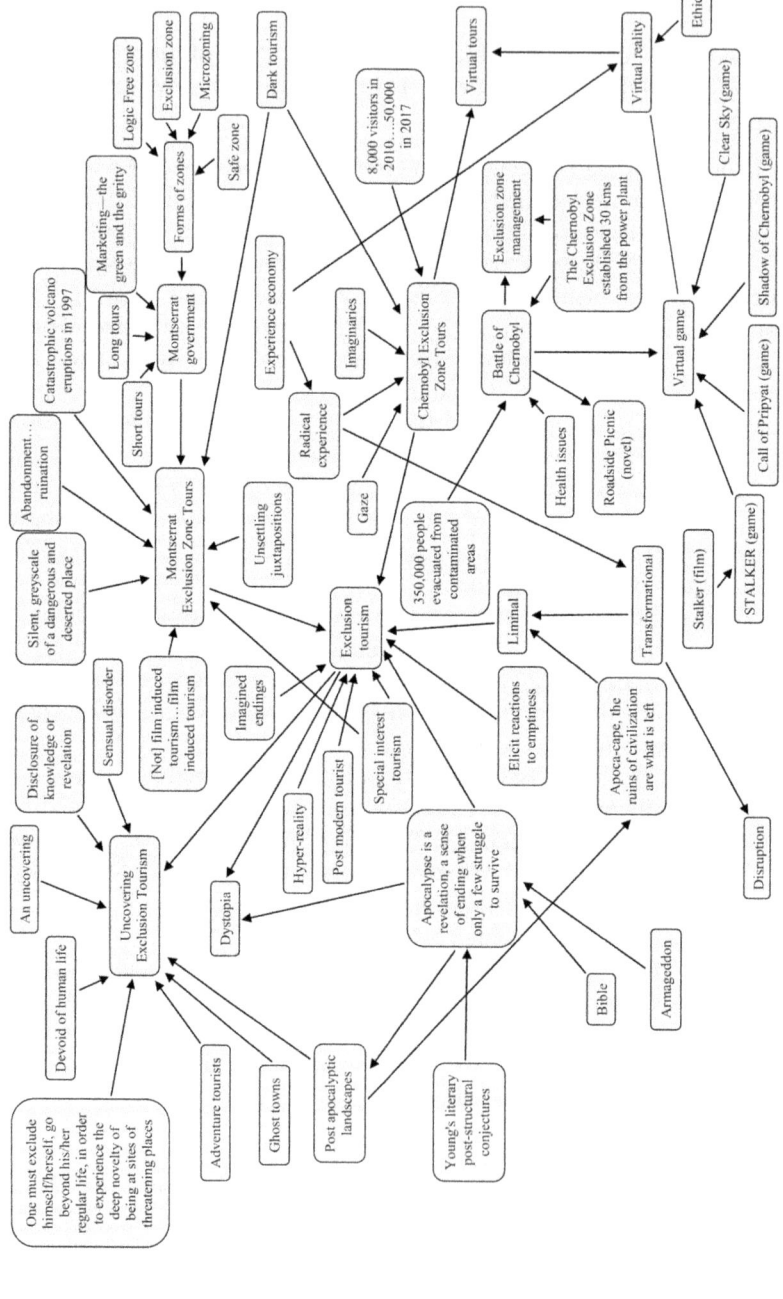

Figure 19.16 Exclusion tourism: sci-fi and subjunctive plays in apocalyptic destinations from Chernobyl to Plymouth, Montserrat

owners decided to change its name. The hotel guests are experiencing deep dissatisfactions as they slowly begin to realise that this is a badly managed hotel on the verge of collapsing. The dialogues and heated discussions in the hotel lobby, corridors, rooms, pool bar, restaurant and common room disclose the cognitive dissonance in tourism settings. While many tourists are aware of the evolving ecological crisis, they have little or no idea what to do in practice.

The purpose of the chapter is to question the idea of environmental crises as something that we assume we will have to face in the future with the key concepts identified in Figure 19.17 as linear range of 'scenes....' According to IPCC reports, ecological researchers, environmental philosophers, among others, we are already in a world of climate change (Change, 2014; Chaturvedi & Doyle, 2010). During the fictional day at the Hotel Anthropocene the guests exhibit different ways of knowing, feeling and sensing ecological crises in real time. The readers are invited to critically reflect on tourism and planetary ethics. The chapter draws attention to the fundamental paradox of tourism, where search for well-being and hedonistic joy also contributes to an acceleration of mass-extinction. The story of the hotel is used to problematise the conceptualisation and practical use of 'the future' as a time and place located somewhere else, not in the here and now of the present. It also disrupts the idea of touristic bubbles, as without entanglements and responsibility for our common planetary state.

Towards the end the guests are beginning to realise that they are able to check-out from Hotel Anthropocene, but they cannot leave. The story of the hotel also raises questions about our responsibilities as researchers and teachers when using fiction as method for producing and sharing knowledge about our current and future planetary situation in the Anthropocene. How do different kinds of stories tune us in, or out? What kinds of storytelling and listening should we engage in order to contribute to more caring, planetary sustainable and peaceful forms of co-existence at our one and only common 'hotel'?

Developing an Aggregated Cognitive Map of Science Fiction and Tourism

The purpose of this section is to demonstrate how a construction of the aggregated cognitive map of the whole book took shape, which in turn was used for developing a theoretical framework of science fiction and tourism based upon the virtues of ontology and epistemology. Because of the complexity and subjectivity of the construction, the section is only an illustration of the process to guide the readers' understanding of how the process happened. As can be seen in the preceding section, all the chapters have an individual cognitive map. The merger of the individual cognitive

288 Part 5: Concluding Thoughts

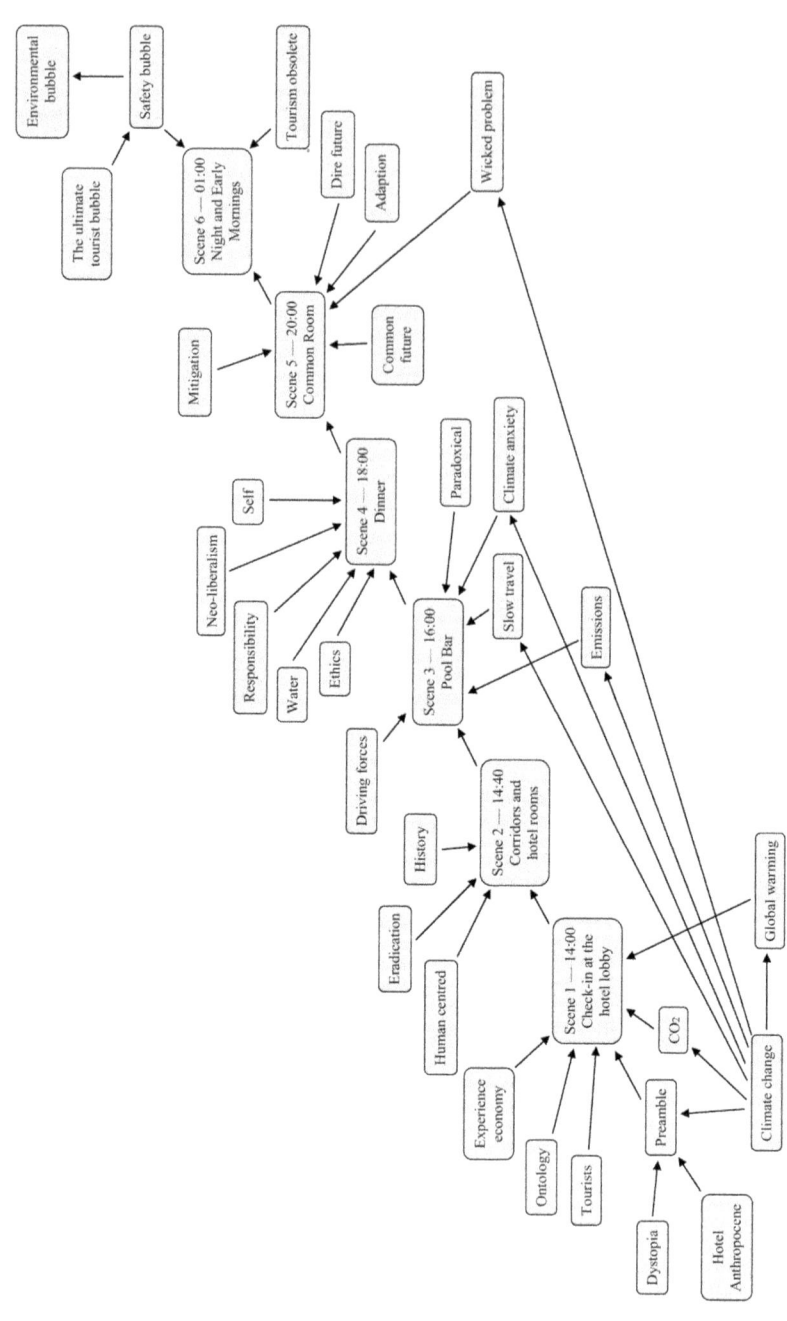

Figure 19.17 Hotel Anthropocene

maps into an aggregation reflects the process in which the researchers immersed themselves into the individual maps and search for concept connections driven by semantic similarity. This allows the drawing out of key concepts from each individual map and remapping the concept in DE. Once this is complete and after several iterations, an aggregated cognitive map is formed.

As this aggregated map is complex, with numerous concepts and (inter-) connections, various features of the DE were used for breaking up the aggregated map into viewpoints. By doing this, the researcher can build, explore and reflect on these maps as component parts of the total aggregated map.

First the 'central' command looks at specified band levels which are connected to the concepts. This allows the researchers to view the importance of the length of linkage between concepts. Each concept is weighted according to how many concepts are traversed in each band level. Fundamentally, the central command shows how many concepts are dependent upon one concept. Exhibit 19.1 demonstrates this view.

The 'domain' (Exhibit 19.2) command performs a hierarchical domain analysis which lists each concept in descending order of the linked density around that concept. Those concepts with the higher link density are listed first. The importance of the 'domain' command highlights the importance of the closeness of the local links between concepts. The researchers used both the 'central' and 'domain' commands to identify the most important concepts in order to explore and construct maps. Further, both the 'central' and 'domain' commands identify a number of concepts to map in which the modeller makes a judgement to construct and explore these concepts while holding them as a central view.

The cotail (standing for 'composite tails') is an analysis which searches through the model to find those 'potential' options which have more than one outcome, i.e. more than one consequence leading from them (see Exhibit 19.3). Starting from the bottom of the model – the tails – the analysis traces up each chain of argumentation until it meets one of these branch points. It flags this as a composite tail and starts with the next chain of argumentation (Jones, 1993).

Using these three commands – 'central', 'cotails' and 'domain' – the researchers explored the concepts using DE commands such as 'show unseen links' with which the modeller is able the find connections between concepts and thus, start to build a cognitive map, explore links and reflect upon them. Other commands within DE can be used to recall multiple concepts that surround other concepts. This process is repeated several times until several views make sense to the researchers. Therefore, a series of concepts were identified as significant that are common to both the domain and the central analysis. These concepts where 'explored' and 'mapped' using further commands in DE, which ultimately resulted in

Cent Scores Calculated...

13 dystopia
150 from 340 concepts.

128 Science fiction
144 from 337 concepts.

0 (deleted)
132 from 326 concepts.

309 Disruptive
131 from 327 concepts.

134 The future(s) ... future
126 from 297 concepts.

339 Westworld
122 from 291 concepts.

167 Transformational
120 from 287 concepts.

112 Liminal
118 from 288 concepts.

0 (deleted)
111 from 264 concepts.

149 Tourism
108 from 267 concepts.

191 Futures orientated research
103 from 261 concepts.

319 Exclusion tourism
99 from 246 concepts.

99 Experience economy
99 from 240 concepts.

173 Ocean tourism
96 from 235 concepts.

490 Harry Potter
95 from 225 concepts.

Exhibit 19.1 Central Command

nine viewpoints which determine the concepts associated with tourism and science fiction: *The Essence of Science Fiction, Authenticity, Dystopia, Plurality of the Future(s), Disruption & Transformation, Liminality, Tourism Portrayed as Science Fiction, Technological Singularity has arrived: Westworld, and Sustainability.*

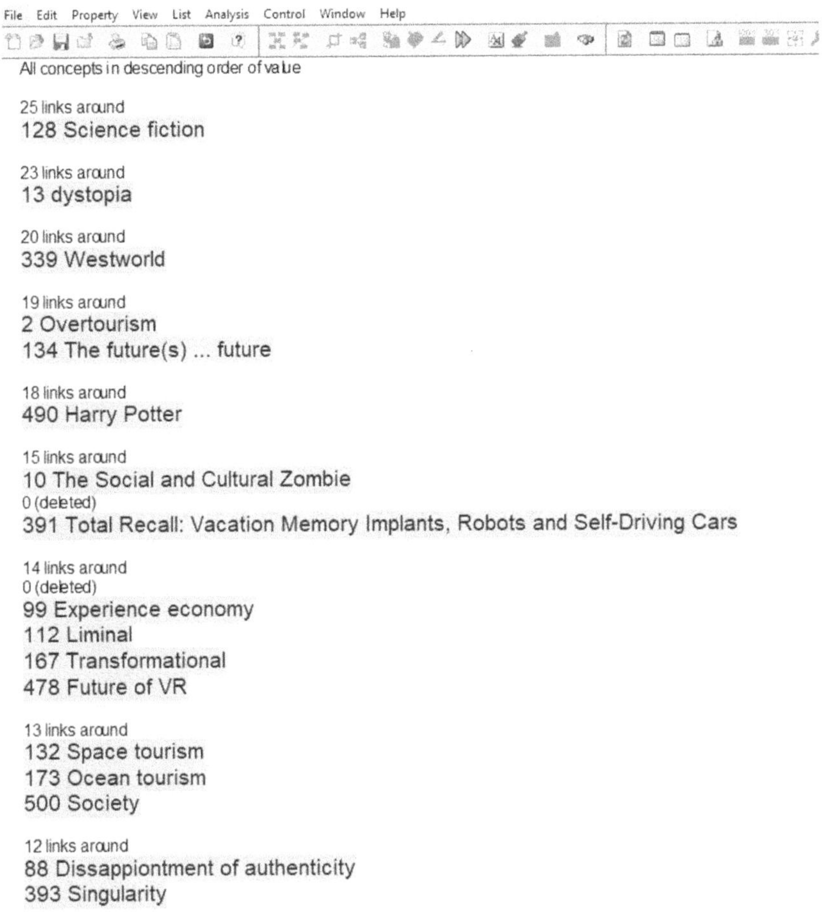

File Edit Property View List Analysis Control Window Help

All concepts in descending order of value

25 links around
128 Science fiction

23 links around
13 dystopia

20 links around
339 Westworld

19 links around
2 Overtourism
134 The future(s) ... future

18 links around
490 Harry Potter

15 links around
10 The Social and Cultural Zombie
0 (deleted)
391 Total Recall: Vacation Memory Implants, Robots and Self-Driving Cars

14 links around
0 (deleted)
99 Experience economy
112 Liminal
167 Transformational
478 Future of VR

13 links around
132 Space tourism
173 Ocean tourism
500 Society

12 links around
88 Dissappiontment of authenticity
393 Singularity

Exhibit 19.2 Domain Command

Science Fiction and Tourism

In this section, we discuss the viewpoints identified through the process of cognitive maps (described in the previous section) that determine a conceptualisation of tourism and science fiction. These viewpoints are (Figure 19.18): *The Essence of Science Fiction, Authenticity, Dystopia, Plurality of the Future(s), Disruption & Transformation, Liminality, Tourism Portrayed as Science Fiction, Technological Singularity has arrived: Westworld* and *Sustainability.*

The following sections focus on each viewpoint by providing their individual cognitive maps to analyse and discuss the concepts and their inter-relations that characterise the knowledge base of each viewpoint as described within the chapters of this book.

Analysing model, and calculating Cotail results - Please wait a moment...

Branch points of style standard
1 World War Z
2 Overtourism
7 Keeping the tourists out
9 The Environmental Zombie
10 The Social and Cultural Zombie
11 Economic Zoombie
12 risk
13 dystopia
38 Capitalism
43 Mega rich
46 Conspicuous consumption
48 Social status
51 Digitalisation
62 Barcelona
65 Metaphor
86 heritage
88 Dissappiontment of authenticity
100 post-Fordist
101 Commerical
102 Identity
103 social process
105 Heritage visitor attractions
107 memory
108 Presentation
111 Heritage is played by actors - recreating
112 Liminal
117 Society of spectacle
118 Image
125 Hedonism
126 Dark tourism
128 Science fiction
130 Explanation

Exhibit 19.3 Cotail

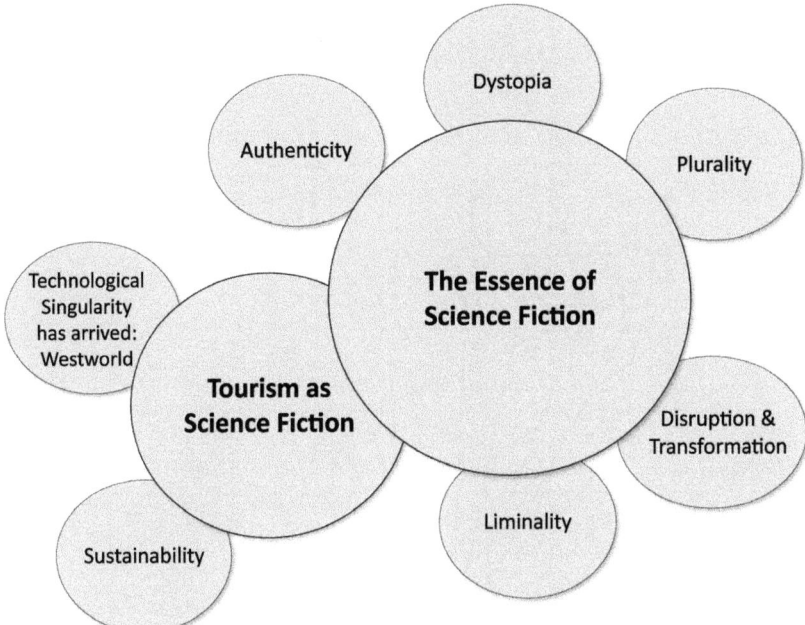

Figure 19.18 Conceptualisation of science fiction and tourism

Viewpoint 1: The essence of science fiction

COVID-19 has highlighted the frequency of disruption and uncertainty in a modern world, which also characterises the reality of the next normal. On the other hand, prediction and rationality are often outcomes from the scenario planning process. This needs to be challenged. To operate and thrive under disruption and uncertainty, we need to readdress the balance of certainty towards a stronger emphasis on uncertainty and the unthinkable, hence the importance of science fiction.

Why is scenario planning often perceived as too rational? Maybe it is the lack of scepticism in management research, but which Dator (2009; Dator & Yeoman, 2015) insists is the essence of any good futures studies. Furthermore, fictional images of the future are powerful and influential in shaping the public's images of the future (Fergnani & Song, 2020), as they help us think the unthinkable.

Figure 19.19 is clustered around the concepts of 'explanation' and 'scientific and science fictional drivers of change'. Explanation is what Bergman *et al.* (2010) call 'explanatory claims' where the author of a statement explicitly indicates mechanisms as causes behind the events or states that they are forecasting. For many decades, the concept of explanation with the philosophy of science was presumed to revolve around the idea of universal laws of science. Events were to be 'explained' in terms of being instances deduced from general covering laws, which were

294 Part 5: Concluding Thoughts

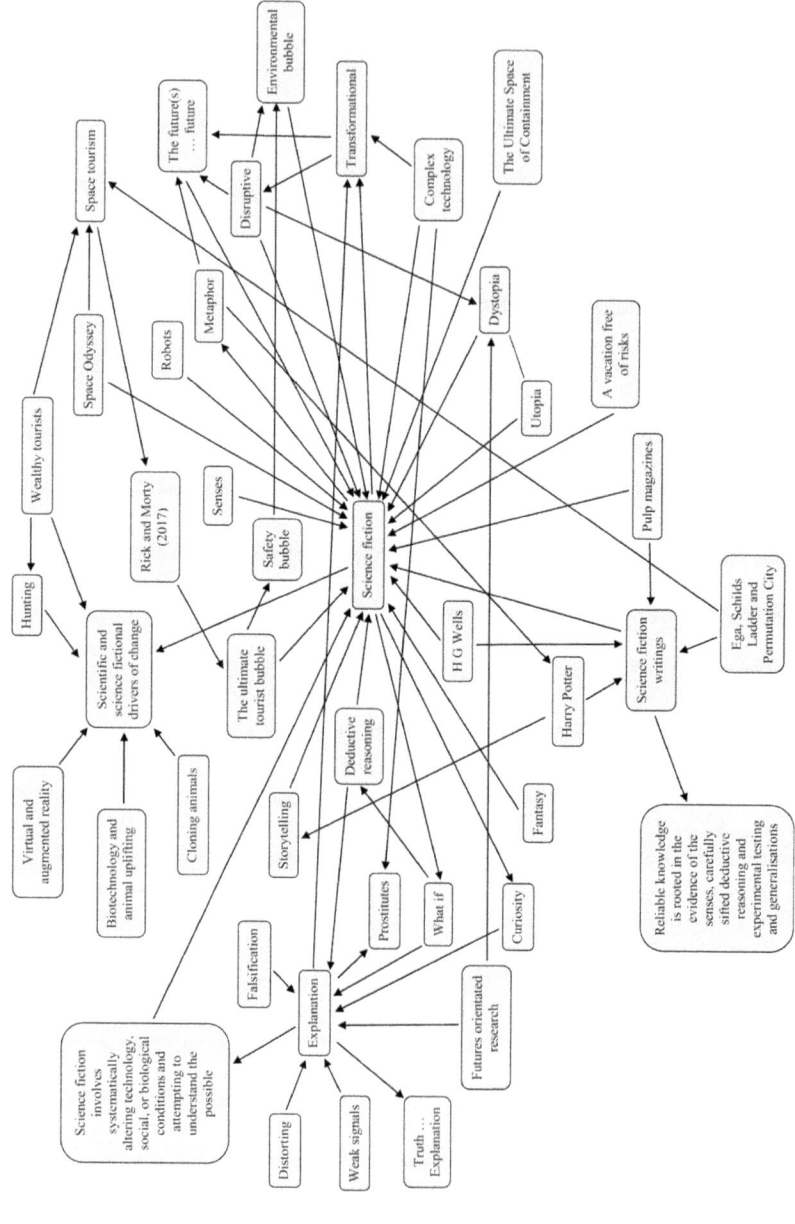

Figure 19.19 The essence of science fiction

themselves developed through some form of induction from observable empirical examples. Bergman *et al.* (2010) advocate the writings of Bhaskar (1978), taking a critical realism perspective which describes the interface between natural and social worlds. According to Mingers (2014), Bhaskar does not explicitly define what he means by 'mechanism' and sometimes refers to them as structures. Fundamentally, mechanisms exist in a real ontological sense independently of how they may be known or described by others. They can be higher order, hence having a sense of being with properties that can be physical, social or conceptual. However, they are defined by causality rather than a perceptual criterion. Thus, science fiction is about explanation of how a future could occur. This casualty is relatively weak and often portrayed through narrative and stories. The use of narrative to portray a future scenario links to Rand's (1967) theory of objectivism, which argues that reality exists independently of consciousness through a sense of perception. A sense of perception and hence reality is created through the story.

As many of the chapters in this book explain, a sense of consciousness, imagination and perception is created through the narrative (e.g. Picken's chapter (3) talking about the deep blue sea or the Bolan's discourse in Chapter 9 about *Blade Runner* and *Total Recall*). Thus, perception creates explanation and plausibility. This narrative, as seen in Figure 19.19, draws out a number of concepts which appear across a range of chapters, including the concept of 'storytelling'. In Chapter 4, Robinson reminds that tourism has always been about storytelling with storytellers finding meaning (explanation) through events and images often related to history and culture. Robinson goes on to link narrative with the sense of falsification and fabrication of science fiction worlds, as these worlds create over-aestheticised images of decay which link to dystopian worlds, darkness and technology. As such, over-aestheticised images also link to concepts of 'distortion' and 'falsification'.

The focus on technology is evident throughout the book, with Chapters 6 and 15 covering the adult fantasy of Westworld; here, as is often the case, the focus is on transformational technology. Indeed transformation is a core concept of Dator's (2014) alternative futures and it is seen as radical and associated with new creation. In Chapter 5, space tourism, often highlighted by science fiction writers, represents a world that is very different from the present. Toivonen uses it as an example of transformation; transformation is equal to innovation, which is inherent in the tourism, given the nature of competition and SMEs. In Chapter 9, Bolan demonstrates how technology trends such as Alexa, Siri and Cortana have become transformational technologies blurring fiction and reality, which are now reshaping tourist behaviours. Bolan argues that science fiction writers such as Philip K. Dick raise the issue of *what is reality* – hence linking to the concepts of authenticity and liminality. Robinson in Chapter 4 argues that the tourism gaze is fundamentally about fantasy and illusion (Cohen & Cohen, 2012). In science fiction, this fantasy is often dark and unnatural, with characters using unfamiliar senses (see Yeoman & McMahon-Beattie in Chapter 2 in relation to robot prostitutes). Fantasy is about escapism and technologies like AR and VR allow us to experience immersive fantasy, which is a theme discussed in many chapters of this book.

'Weak signals' are also an important feature of science fiction. These can be viewed as embryonic seeds of the future as science fiction possesses threads of truthfulness, but the key is explaining how this future could occur. Many authors in this book attempt to do this, for example in Chapter 5, Toivonen provides various scenarios of future space tourism experiences or McEntee and colleagues (Chapter 10) explain how the destination of the dead can become a possible tourism future. From the editors' perspective, the value of science fiction relates to the question 'what if' which is exemplified in Chapter 14, whereby Reid and Ek positions science fiction as a series of stories, images, events and circumstances in a dystopian world of zombie horror. But what if this dystopian scenario was the future... What if, just like COVID-19, it became a reality?

Viewpoint 2: Authenticity

Futures research, in particular scenario planning, creates context and a sense of reality (Keough & Shanahan, 2008; Schnaars, 1987; Yeoman, 2012b). In their narratives, scenario planners build upon weak signals (Robertson & Yeoman, 2014; Robertson *et al.*, 2015) in order to bring a degree of authenticity. They account for specific situations through stories which create a structure for understanding (Moriarty, 2012). It is through these stories that truth resonates with readers, as to a certain degree, writers of science fiction draw parallels with reality (Stableford, 2003). In a sense, science fiction writers and those involved in scenario planning are creating a personal story of the world and its future. They are creating a narrative of 'this is how I see the future' (Winter, 2002).

This leads us to the notion, that truth is in fact 'truth to self' or authenticity. Thus, science fiction has a degree of (subjective) truthfulness as it is argued that it is based upon science but with a degree of inventions, lies, exaggerations and delusions or what Dator (2009) calls scepticism. So, what is authenticity and how does it link to science fiction and tourism? Maccannell (1973) suggests that tourists are in search of 'authentic experiences', a search that defines tourism and sets it apart from everyday contemporary existence. In Figure 19.20, we see several clustered concepts including 'authenticity', 'reality', 'disappointment of authenticity', 'reality' and 'authenticity...inauthenticity'.

Reality features prominently in science fiction. According to Abbott (2009) one of the key elements of art cinema is realism (real locations, problems, sexuality and characterisation). While the genre is by its very

Developing a Theoretical Framework of Science Fiction and the Future of Tourism 297

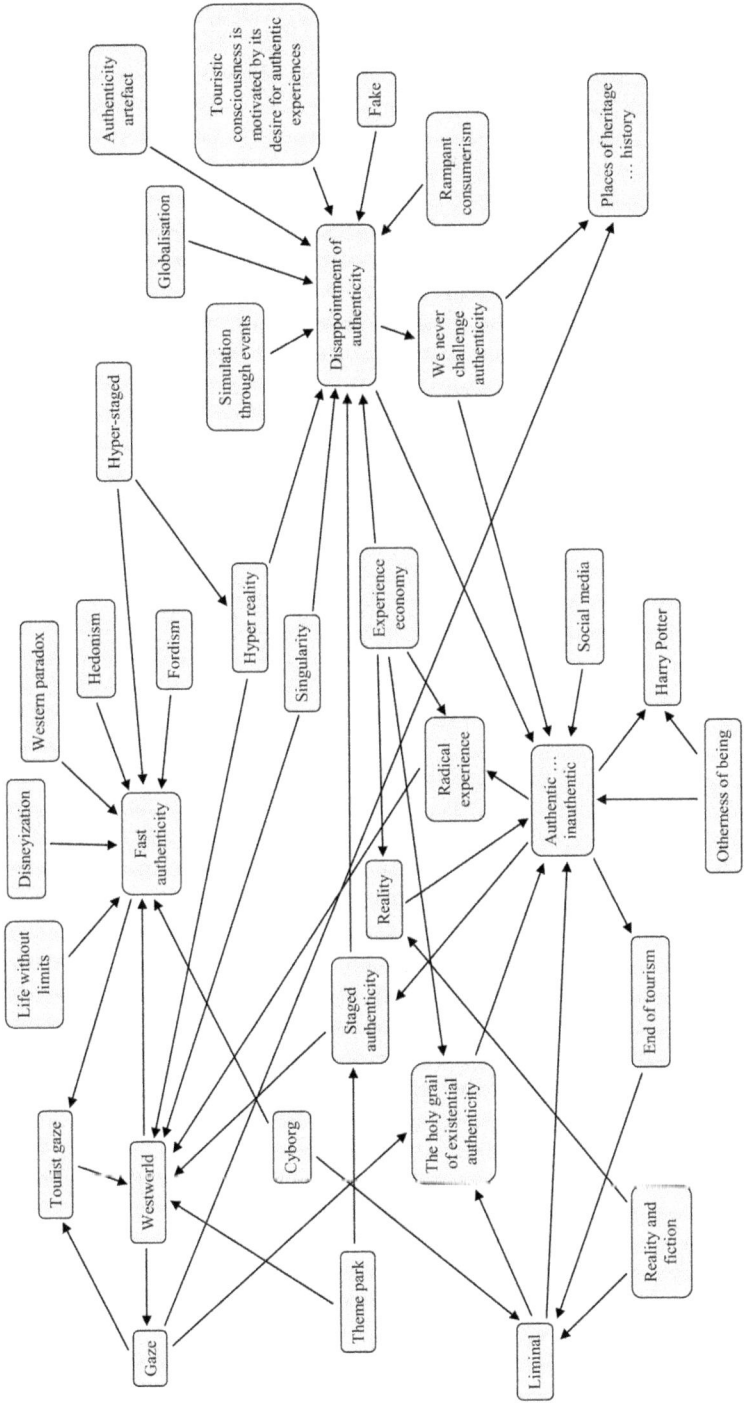

Figure 19.20 Authenticity

nature speculative and often explores futuristic and/or fantasy narratives such as time travel, space exploration, or alien visitation/invasion, art-house science fiction association with concepts of realism distinguishes it from more mainstream examples of the genre. The use of real locations is often mixed with realistically presented futuristic sets, creating a believable and recognizable environment in which to set the narrative. For instance, while imagining what space travel might be like in the future, Kubrick's (1968) *2001: A Space Odyssey* grounds its presentation within real experience. The journey to the orbiting lunar station is presented as a spaceflight, replete with attendants wearing Pan Am uniforms, serving refreshments and defying the weightlessness of their environment through their gravity shoes. Presenting the future as a 'realistic' or believable extension of the present provides the film with a sense of plausibility.

So, is science fiction reflective of elements of authenticity? Robinson argues in Chapter 4, that the authenticity is the Holy Grail of tourism, as travel is predominately about the search for authenticity by 'consuming' and/or interacting with the place and culture. However, authenticity is something that is highly contested given modern hyper consumerism. What we are seeing today is tourists embracing fast authenticity, manufactured authenticity, or false authenticity. Fundamentally, it can be argue that tourism is theatre in which falsification of an authentic experience is presented as a gaze (Gillespie, 2006) and which contributes to contrived scenes and pseudo-events. This is highlighted in Chapter 6 by Gurevitch who argues that 'Westworld' is both a world of fantasy and authenticity, with a focus on the design of experiences. Likewise, Bolan in Chapter 9 highlights the staged authenticity of *Blade Runner* and *Total Recall* as an example of touristic fantasy. In fact, he argues that authenticity should be dismissed as people's views and knowledge of the world are based upon imaginary coming from various forms of media such as, fiction and films which in turn are inherently false. Guttentag in Chapter 12 acknowledges that virtual reality experiences through films like *Inception*, *The Matrix* and *Avatar* push the concepts of reality and authenticity to their boundaries, because of the conscious experiences of immersion that are created through VR. Maybe the answer lies in the concept of validity and its overlap with authenticity whereby validity is based upon what a person is entitled to believe and feels is genuine (Winter, 2002). Fundamentally, it is what the reader can afford and wants to believe, as for some, the authenticity of tourism is science fiction (Guttentag, 2010; Mars *et al*., 2017).

Viewpoint 3: Dystopia

Dystopia is the opposite of utopia (Shklar, 1965). Utopias paint a picture of perfect worlds (Yeoman *et al*., 2015) and so, they are often seen as a form of pure escapism in an ideal world in which people feel happy and comfortable. Indeed, in terms of the tourist gaze, tourism is often portrayed as a paradise, beautiful beaches and stunning landscapes (Urry, 2011). On the other hand, dystopias portray a negative (and so, unwanted) image of future societies. They are worlds of contingency, conflict, and uncertainty (Page et al., 2006; Tribe & Liburd, 2016). We associate them with misery and cruelty and they are depicted as troublesome worlds where nobody wants to live, unless of those gaining pleasure by viewing and/or experiencing pain and misery. Dystopian fiction often portrays a glimpse of potential futures in the form of a warning of imminent dangers and a proactive call for action and behaviour change. Indeed, whether dystopias satisfy a demand for pleasure or 'education', the production and consumption of many dystopian science fiction pieces or art (e.g. films, fables and novels) has been long been mushrooming. Consequently, when a disaster or crisis happens, it is easy to make connotations and connections between the 'predictive' power of such art works with the catastrophic reality. For example, as the COVID-19 outbreak emerged, it was like reading Koontz's (1981) science fiction novel *The Eye of Darkness* which tells the story of a virus called Wuhan-400 which accidentally leaks from a research laboratory and sweeps the world. Mayurika (2020) has noted that COVID-19 has held the entire world hostage, producing a resemblance to the post-apocalyptic world depicted in many science fiction texts Similarly, Canadian author Margaret Atwood's (2004) classic novel, *Oryx and Crake*, set in a future where genetic engineering rules the world refers to a time when, 'there was a lot of dismay out there, and not enough ambulances' (Mayurika, 2020) – a prediction which is reflective of our current predicament.

Tourism writings of Yeoman and colleagues about pandemics (Page et al., 2006), war (Yeoman et al., 2005) and disease (Yeoman et al., 2005), all portray potential dystopian futures where signals are explored and include a call to action. More broadly, the consumption of dystopian futures is often associated with dark tourism, which is a concept that is right at the centre of tourism practice and theories. For example, by becoming a metaphor of catastrophic and deadly disasters, Farkic (2020) explains how dark places are described as a dystopic place which attracts tourists with various motivations (e.g. nostalgia, learnings, cultural links, personal experiences etc.). Lennon and Foley (2000) brought dark tourism in the mainstream and today, dark tourism provides an opportunity for members of the society to reflect upon death and juxtapose reflections of an inevitable mortality with those of conscious and even hyperreal experiences (Podoshen et al., 2015; Preece & Price, 2005). Indeed, several authors (Sigala & Steriopoulos, 2021; Zheng et al., 2020) and various theories (e.g. the mortality mediation model, Light, 2017) discuss the opportunities afforded by dystopian dark tourism places and experiences to elicit emotional experiences and trigger cognitive mechanisms that engage the visitors into a reflective, meaning-making and ultimately, life-changing process. These transformational effects of dystopian dark tourism experience are good evidence of the role of dystopia in enabling people to visualise and imagine tourism futures by reflecting and using the dark past as a metaphor of a future 'dystopia' reality. In other words, through metaphors, tourism uses dark places as a dystopia to imagine and visualise 'the future of the past'.

In Figure 19.21 we see the concepts of 'future visions or dystopia', 'disruption' and 'urban exploration and dystopian futures' prominent. Robinson, in Chapter 4, draws us into the past through forgotten worlds where attractions portray death, with urban exploration of derelict hospitals and asylums which have become symbols of dystopia through the modernity of dark tourism. In Chapter 17, Banaszkiewicz and Skinner go even further with dystopia being connected to an apocalypse which is characterised by 'death', 'destruction', 'chaos' and 'carnage'. What resonates in this chapter is the importance not of heritage, but the emptiness associated with Chernobyl and a post-apocalyptic landscape. In Chapter 14, Reid and Ek present a dystopian tale depicting the onset of the pathological fugue condition of zombie tourists. They link this to Nietzsche (1986) characterisation of the blind tourist as a hypermobile subject that remained oblivious to the places he or she travelled through, not capable of really seeing, experiencing, and living a life based on his or her travelling experiences. In Chapter 15, Lovell and Hitchmough extent this apocalyptic world further based upon HBO's *Westworld* series where an utopic future consumer dreams of 'fast authenticity', which actually forges a new dystopia; the out of control theme park, where the visitors are locked into a more authentically brutal, truly 'wild' west nightmare bridging hyper-real versions of authenticity embodied with death, immortality, sex and unethical formations. It is an immersion in a life without limit and violent delights. In Chapter 18, Gren and Höckert's thought-provoking account of Hotel Anthropocene explores a series of contradictions and paradoxes between the natural world and tourism from an ecological and climate change perspective. Paradox research has been proposed by Sigala (2020) as a way to re-imagine and reset 'better' tourism futures in the post-pandemic period, as paradoxes enable one to re-conceptualise concepts by deconstructing and re-constructing them to make better sense (i.e. paradox research enables analysis by paralysis). Dystopia situations like COVID-19 have mushroomed paradoxes and/or enabled us to see the various of paradoxes that exist, e.g. the co-existence of the belief in human rights but people's acceptance of mobility restrictions and tracing apps.

Viewpoint 4: Plurality of the future(s)

Pluralism is a term used in philosophy, meaning 'doctrine of multiplicity', which is often used in opposition to monism and dualism. In ontology, pluralism refers to different ways, kinds or modes of being (Bergman *et al.*, 2010). In epistemology, pluralism is the position that there is not one

Developing a Theoretical Framework of Science Fiction and the Future of Tourism 301

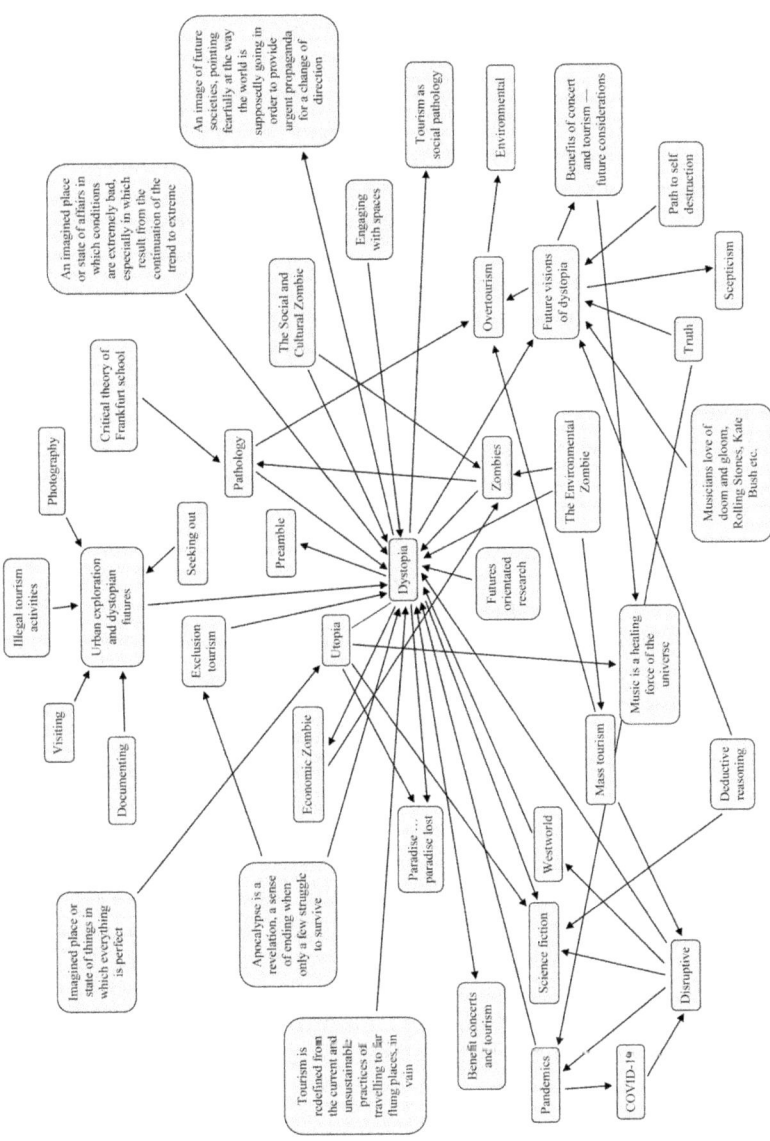

Figure 19.21 Dystopia

consistent means of approaching truths about the world, but rather many (Searle, 1995). Simplistically, pluralism implies that there are many beings and many truths of their modes of being, and in turn each one needs to be seen from different lenses and perspectives. The multiplication of the ontological and epistemological view of the beings and their truths creates the complexity and multiplicity of tourism futures.

In science fiction, plurality is often associated with cosmic pluralism, the plurality of worlds or simply pluralism, and it describes the philosophical belief in numerous 'worlds' (possibly an infinite number), in addition to Earth, which may harbour extra-terrestrial life. Indeed the 20th century the emerging debates about the existence of extra-terrestrial life and intelligence, ensured that science and science fiction enjoyed an increasingly symbiotic relationship. Not only did pluralism find a voice in fiction through narratives about aliens, but fiction also inspired science to broach questions in the real world. The myth and the associated expectation of the third encounter, developed through the fictional accounts by H.G. Wells (1902), Stapledon (1930) and Clarke (1972), influenced the development of the UFO phenomena, which in turn further transformed and strengthened the belief in extra-terrestrials (Brake, 2006).

In future studies, pluralism refers to more than one future derived from multiple sources of knowledge (Bell, 1993). Today, plurality is accepted as the norm in future studies. This is evident from Figure 19.22 in which the concepts of 'future(s)....future' and 'futures orientated research' are dominant. Amara's (1981) typology of probable, possible, and preferable futures is goal orientated and conceptualises pathways of multiple futures, which is replicated in many theories of the future. Similarly Dator's (2009) first law of futures states that, 'the future cannot be predicted but alternative futures can, and should, be forecasted'. Van der Heijden's (2002) scenario planning method is also based upon four scenarios or futures. Thus, throughout this book the contributors do not write about *the future* but *futures* and all these futures fundamentally proposition a series of alternatives with which to view the future of tourism. Hence, contributions from chapters in this edited collection represent a series of alternatives to the present. These are grounded in cosmic pluralism which attributes knowledge that is diversified and a speculation of extra-terrestrial activity (thus linking to scepticism). This could be interpreted as authors presenting a particular future that is singular as science fiction writers are writing a particular story rather than a range of scenarios as Dator and Yeoman (2015) positions future studies. Thus, the continuum of 'futures (s)...future' as seen Figure 19.22.

Viewpoint 5: Disruption & transformation

Whoever bears the responsibility for strategic decision-making in an organisation knows the paralysing effect of uncertainty, particularly in

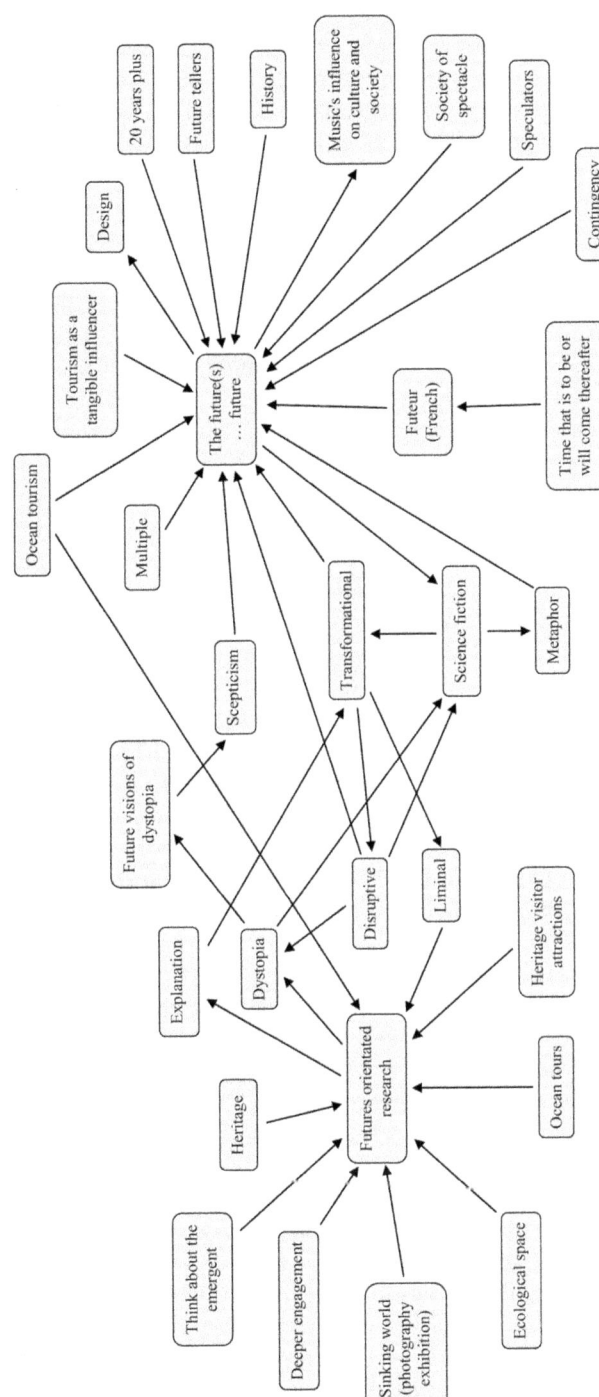

Figure 19.22 Plurality of the future(s)

relation to the impact of the everchanging external environment. Complexity and uncertainty combined has been the essence of COVID-19, being disruptive and transformational. This crisis of COVID-19 has disrupted every mode of work, personal and social life causing prolonged 'temporarily' changes in the economic, sociocultural, environmental, political, ethical–moral systems of the society that have unknown long-lasting effects and so, uncertain, futures. To many, disruptions are change agents challenging and shaking established equilibria as well as transformational stressors accelerating the need and urgency to find new equilibria and re-set the next futures. The case of COVID-19 is a good evidence of a disruption accelerating and driving change by restructuring and recalibrating dynamics, priorities, values in economic, sociocultural, political, ethical and moral world (Sigala, 2020).

Literary scholars love catastrophe, disaster, and disruption. From Maurice Blanchot's seminal *Writing of the Disaster* (Blanchot, 1995) to work about ecocriticism (Rigby, 2015), catastrophe and disaster have been discussed extensively in literary scholarship. However, the formal challenges raised by catastrophe have not been investigated as thoroughly or systematically from a narratological perspective. Often cited examples include *Oryx and Crake* (Atwood, 2004) in which a disillusioned bioengineer, unleashes a 'hot bioform' that kills most humans. Similarly, *The Plague* by Camus (1948), published first in 1947, astonishes the reader with the vibrancy of the description of the disease outbreak. Indeed, one could not mention disruption and catastrophe without reference to COVID-19, as it is a dystopian science fiction narrative of the most horrific kind, with no immediately discernible way out of the unfolding human tragedy. The anxiety, fear and disruption portrayed in Camus' book mimics that experienced by society in the current COVID-19 crisis. It is indeed, this very scenario has been depicted in many science fiction novels and films but we have failed to use this foresight to adequately prepare ourselves for today's struggle (Grech, 2020).

In the 21st century we have entered a nascent paradigm shift where science fiction has become science fact, with technological disruption the main driver to transformation (Ocident et al., 2020). This argument is supported by the concepts of 'disruption' and 'transformation' in Figure 19.23. Novel technologies are one of the principal means of surprising enemies or competitors and of disrupting established ways of doing things (National Research Council. Committee on Forecasting Future Disruptive, 2010). Military examples of surprise include the English longbow, the Japanese long lance torpedo, the American atomic bomb, stealth technologies, and the global positioning system (GPS). Business examples include the telephone (Bell), business computers (UNIVAC and IBM), mobile phones (Motorola) and recombinant DNA technologies (Genentech). Until the 1970s, technological innovation tended to come from a limited number of well-established 'techno clusters' and national

Developing a Theoretical Framework of Science Fiction and the Future of Tourism 305

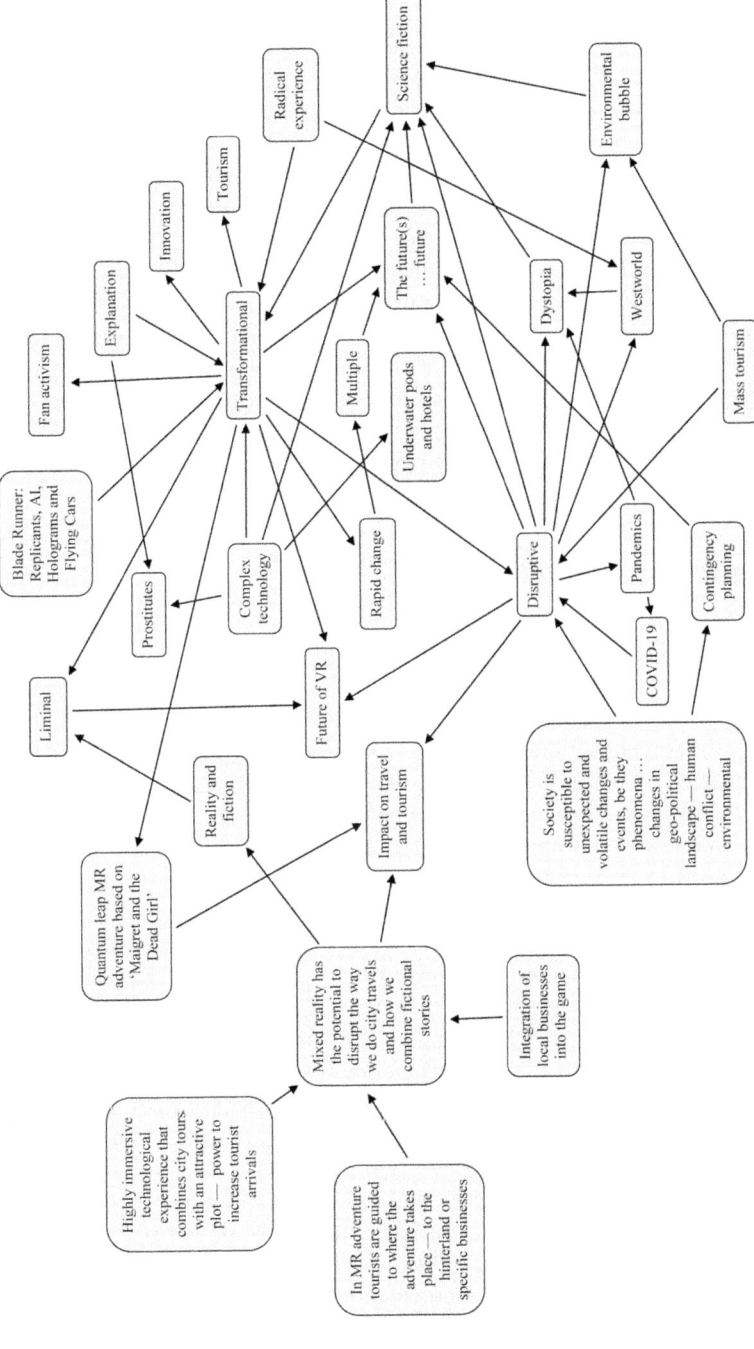

Figure 19.23 Disruption & transformation

and corporate laboratories. Westworld is a futuristic adult fantasy theme park based upon Michael Crichton's science fiction novel and the 1973 film of the same name portraying a frontier themed 19th-century theme park based upon recreating authenticity. It is based upon the principles of technology singularity (Kurzweil, 2005). This type of disruption and transformation is clear in several chapters. Gurevitch in Chapter 6 uses Westworld as an example of automation as a disruptor for the service industries whereas Lovell and Hitchmough in Chapter 16 focus on the transformational change to the tourist experience brought about by technology, questioning what the essence of reality is. These are similar issues raised by Bolan in Chapter 9 through the science fiction films *Blade Runner* and *Total Recall*. Bingemer in Chapter 11 explores this further by examining MR as a driver of future tourism from an experience perspective. Using Inspector Maigret as a case study, he demonstrates that gamification can provide tourism business with the opportunity to overlay stories with locations. Guttentag demonstrates this disruption further in Chapter 12, highlighting the distinct ability that VR experiences have to elicit powerful emotional responses.

Viewpoint 6: Liminality

One of the roles of science fiction in research is to push out knowledge boundaries (Yeoman & Beeton, 2014) as theories may hinder us. The use of stories which are the basis of scenario planning and fiction assists us to look beyond the boundaries, reflecting on new knowledge and creating lightbulb moments. Thus, science fiction takes us beyond what knowledge is to an area of liminality and the unknown (Dredge & Jenkins, 2011).

According to Turner (1998), liminality refers to the indescribable time when the individual has left the past but has not yet arrived at his/her destination. This passage of time also marks a spatial change, for completing a rite of passage typically allows the participant to claim a new place in the cultural terrain. Liminality is the time of the journey encompassed by the leaving, arrival and return – exclusive of the moment of separation upon leaving and the moment of reaggregation upon return; departure and return mark not only points of time outside of the liminal but points in a cultural space where one's place or position is settled. Liminality is, 'a movement between fixed points [that] is essentially ambiguous, unsettled, and unsettling'; it is 'a no-place and no-time that resists classification' (Turner, 1974: 274). It may also be characterised as a state of 'disequilibrium,' 'in-between-ness,' or 'ambiguity' (Beech, 2011). The impact and outcomes of liminal experiences on reflection, critical thinking, futurizing and change/transformation are documented in the literature. For example, in management/marketing field, the concept of liminality is used to describe personal journeys and (life-changing) experiences in temporally and spatially bound extraordinary consumption contexts such as adventure tourism (e.g. sky diving and rafting), dark and spiritual tourism, and transformational festivals like Burning Man. People enter deliberately and with a purpose into such experiential contexts, because as research shows (Pastor & Kent, 2020), liminal experiences in such transitional and spatially–temporally limited settings may encourage productivity, creativity, (self)-reflection and transformational benefits.

COVID-19 disruptions have also created liminal experiences, as rules and routines of everyday life are suspended, social coordinates are lost and individuals find themselves in transition to an unknown future. However, although the positive outcomes of such time- and space-controlled environments are well documented in the literature, the impacts of liminal experiences elicited and created by COVID-19 are unknown and unpredictable. This is because the pandemic has created high-intensity liminal states that have been going on for long, while their duration is uncertain and while their impact is global disrupting every aspect of work, social, and personal life. COVID-19 liminal experiences are temporally uncertain, spatially unlimited, affecting every aspect of our mental, psychological, social, behavioural (e.g. economic), moral/ethical and spiritual remit of our lives. Thus, whether COVID-19 induced liminal experiences can result in positive tourism transformation through future thinking and imagination is still to be examined. To study and understand the implications of this question, science fiction simulations immersing people into such uncontrollable and unlimited catastrophic liminal experiences can be used as a method to examine the former question. So, within the context of Turner's (1998) writings, COVID-19 is a liminal world of passage, in which we do not know what the destination will be.

These concepts labelled in Figure 19.24 supports Turner's (1998) and Beech's (2011) findings, including 'reality and fiction', 'authentic….inauthentic' and 'untethered'. Turner's proposition of no-place, no-time, a feeling of ambiguity and feeling unsettled is portrayed by Picken in Chapter 3 in relation to the oceans. Picken sees oceans as deep, novel and unknowns. They are spaces where we only see the surface and are inaccessible, thus they are liminal, novel and awe-inspiring in the tourist's mind. Robinson in Chapter 4 positions tourism as a liminal experience as by its very nature tourism is 'away from home', untethered by the constraints of everyday life. In particular, Robinson takes us into a world of tourism places, which are dystopian in outlook. These places create a sense of the unnatural, of uncertainty and unfamiliarity to many and, as Turner (1998) would categorise, they are places of spatial change. Banaszkiewicz and Skinner in Chapter 17 explore apocalypse destinations such as Chernobyl and Montserrat. These places are empty spaces that are no-go areas, prohibited, unsafe spaces which are neither civilised nor entirely natural. They defy us as demilitarised No Man's Lands.

From another perspective, Guttentag in Chapter 12 introduces the concept of 'untethered' as technology becomes less cumbersome and

308 Part 5: Concluding Thoughts

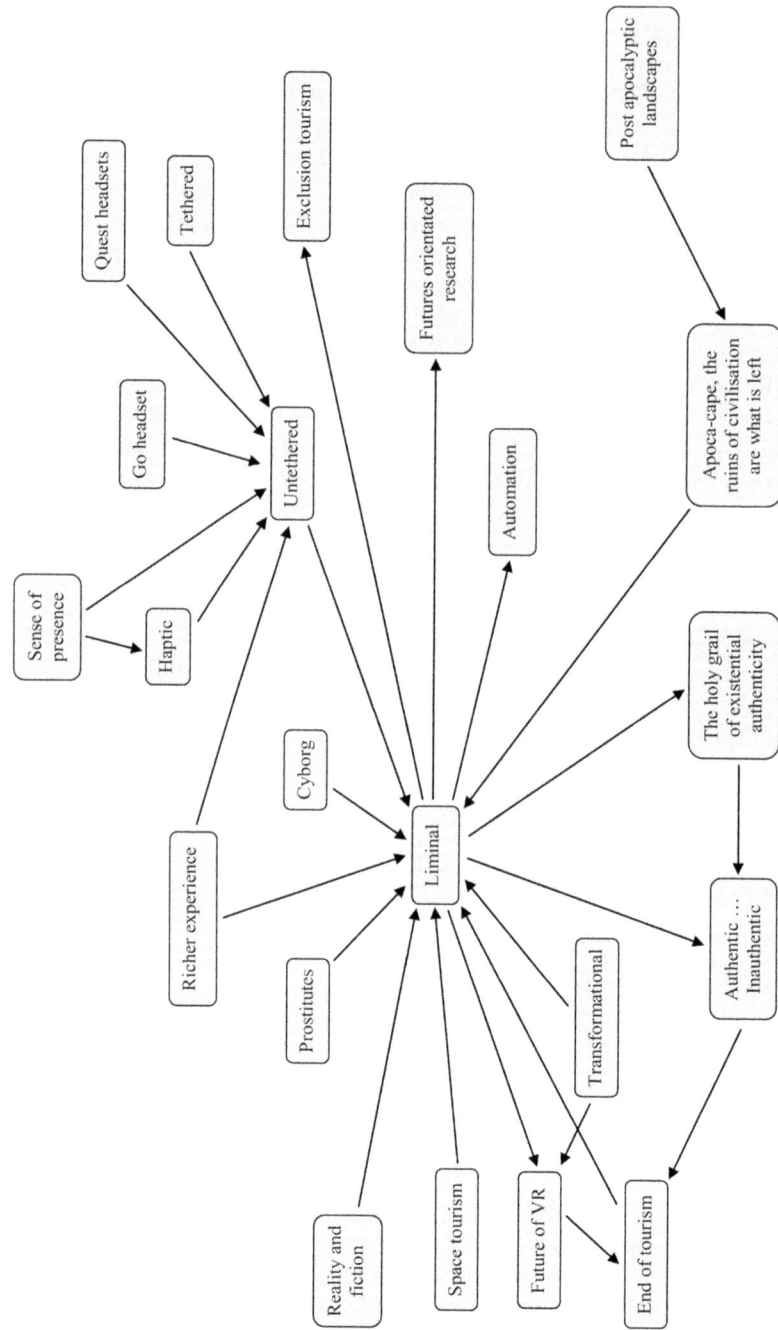

Figure 19.24 Liminality

freely available. VR and AR experiences can expose us to new, liminal worlds with new dimensions. In Chapter 2, Yeoman and McMahon-Beattie take us into a world of Amsterdam's red light district in 2050 with android prostitutes which are so real we can mistake them humans. This is a theme that Lovell and Hitchmough take further in Chapter 15. As androids' appearance becomes more human-like people's disposition toward them becomes more positive, until a point at which this human-likeness leads to androids being considered strange, unfamiliar and disconcerting.

Viewpoint 7: Tourism as portrayed as science fiction

As we see in Figure 19.25, every aspect of this book links to science fiction and tourism, hence the purposes of the book. As a starting point, consider H.G. Wells' (1895) *The Time Machine* which links, travel and tourism to scientific fiction. In the story, the leading character travels to the year AD 802,701 and encounters two posthuman societies: the Eloi are naïve inhabitants of an undemanding leisure utopia, while crude Morlocks toil with machinery underground, providing for the Eloi. Lashua (2018) notes that the book was the first science fiction novel written for an adult audience and that it encapsulates cautionary futurology, that is, it is an attempt to predict possible futures based on the present and the past. Roughly contemporary with Marshall's (1890) and Veblen's (1899) foundational texts on wealth and leisure, *The Time Machine* called critical attention to leisure and class inequities. Parallels can be drawn in Chapter 5, where Toivonen addresses the issue of space travel, a topic often written about by futurists (Toivonen, 2021). Here we envisage travelling through the cosmos at the speed of light visiting intergalactic planets. But fundamentally, space tourism is mainly for astronauts or the super-rich. The first space tourist was Dennis Tito, travelling on a Russian Soyuz rocket in 2001. His $20 million trip involved a five-day stay at the International Space Station and included six months of astronaut training and hours of physical exercise (Toivonen, 2021). Indeed, as the continued growth of tourism is based upon the wealth and the rise of the middle classes (Yeoman, 2013), there is an increasing divide between the haves and have-nots. Historically, as Bertella notes in Chapter 8, it was rich tourists who hunted animals on safari in Africa. Certainly, while zoo tourism or hunting tourism is controversial to some, Bertella argues that using science fiction to imagine futuristic scenarios which draw on technological advancements such as AR and VR, allows us to ponder solutions.

Travel is a central trope of horror and dystopian science fiction, a reality attributable to the core narrative tenets of the two genres (Power, 2014). Frequently these genres portray societies afflicted by unforeseen circumstances that force characters to flee their immediate environments to survive and/or strive for personal freedoms. Part of the fascination

310 Part 5: Concluding Thoughts

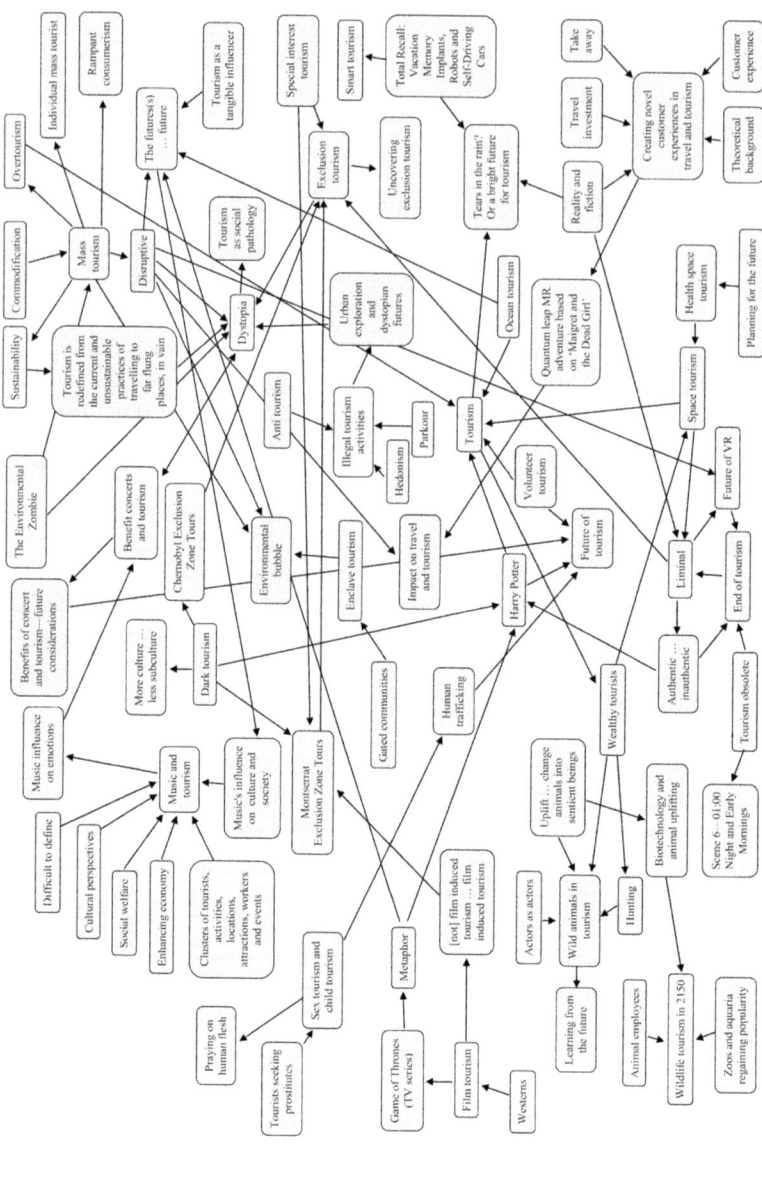

Figure 19.25 Tourism as portrayed as science fiction

attributed to the post-apocalyptic subgenre is the way in which it interacts with and portrays such devastated worlds and the attendant authenticity of any such depiction. Vicariously watching doomsday scenarios through the safety and comfort of the screen, the viewer can tour through alternate worlds. In Chapter 17, Banaszkiewicz and Skinner take us through the imagined endings of apocalypse, a sort of theatre of doom and gloom and it is the tourist who is right at the heart of the show. In this context tourists demonstrate their fascination with death and dark tourism. Banaszkiewicz and Skinner present Chernobyl and Montserrat as tourism experiences which evoke illusions of 'the end' but which are packaged so that visits to these destinations can be experienced safely. In a similar fashion, de Bernardi (Chapter 13) explores the concept of safety bubbles exemplified by tourism experiences in resorts and enclaves. The safety and familiarity of resorts and enclaves create a sense of security hence the tourists develop a personal environmental bubble. This sense of security extends to organised trips from the resort as they are still within the bubble and responsibility of the resort.

Climate change places major transformational demands on modern societies. For some, however, it is a notoriously difficult and elusive crisis to make sense of, particularly compared to other human-impact catastrophes. In the disaster movie *The Day After Tomorrow* a global warming induced shutdown of the North Atlantic thermohaline circulation system triggers extreme weather events worldwide and a new ice age. According to Leiserowitz (2004) and Hansen *et al.* (2004) the film heightened people's awareness of climate change but many are sceptical of the scenario portrayed in the film seeing it as mere fiction and big screen entertainment. As du Cros (2017) ponders, these dramatic Hollywood films exaggerate the future but stakeholders in tourism need to engage with climate change in the short to medium term. In Chapter 18, Gren and Höckert's story approach to the future using *Hotel Anthropocene*, is a metaphor from a science fiction genre about the ecological system of planet earth and the role of tourism. They emphasise that tourists are causing the eradication of other living things, therefore, there is an urgent need for alternative environmental and planetary ethics that decentres humans and enhances kinship among human and non-human nature. This is a call for tourism to lead the debate about change. In line with this, Reid and Ek in Chapter 14 tap into the fear of social and existential irrelevancy telling the story of tourism's selfishness through hedonism, consumption and continued growth. Tourists are portrayed as zombies who are simply unwilling to countenance any change to their increasingly unsustainable behaviour.

Although the future of tourism in science fiction can be portrayed as dystopian given the nature of the science fiction genre (Clute & Nicholl, 1999; Robertson *et al.*, 2015; Slaughter, 1998), there is another side to it. In Chapter 7 Reichenberger explores the universal appeal of Harry Potter through popular culture. While there are elements of the dark side and death in the story, Harry Potter, an orphaned boy who learns to be a wizard, goes on to defeat the villain Voldemort who murdered his parents (Kidd, 2007). Reichenberger draws parallels between tourism and Harry Potter by exploring issues such as power, authority, morality, prejudice, discrimination and how the actions of individuals can change the world for the better – as well as for the worse. Harry Potter can be seen as a metaphor of good overcoming evil, as an example of how indivdiauls can contribute to social change. Thus, tourism can and should be used for the good of society.

Viewpoint 8: Technological singularity as arrived: *Westworld*

The TV series *Westworld* is an emblem of disruption (Engels & South, 2018). Westworld is a fictional, technologically advanced Wild West themed amusement park populated by android 'hosts'. The park caters to high-paying 'guests' who may indulge their wildest fantasies within the park without fear of retaliation from the hosts, who are prevented by their programming from harming humans. Later, in the third season, the plot expands to the real world, in the mid-21st century, where people's lives are driven and controlled by a powerful artificial intelligence named Rehoboam. Westworld is fundamentally based upon the scientific concept of technological singularity. With this concept, the decision making processes, both rational and emotional, are superior to human capabilities (Callaghan *et al.*, 2017). Technological singularity is embedded in automation, artificial intelligence, voice recognition systems, robotics, facial recognition and speech patterns to name a just few technologies.

This technological singularity is not a science fiction, but a current reality. Mirrorworld is a technology project aiming to combine the power and functionality of many technologies into a single technology platform that will be a complete simulation of our real world (Kelly, 2019). In some way, the mirrorworld is the next unified technology (going beyond AR, VR and MR) that will combine all smart technologies and data to provide us a holistic technological representation of the real world. This mirrorworld will reflect not just what something looks like but its context, meaning and function. People will be able to interact with it, manipulate it, and experience it like what they do in the real world. To achieve this, mirrorworld will aggregate the affordances of many technological advances to create a digital landscape that will feel real and will convey the meaning of what landscape architects call *placeness* (Shim & Santos, 2014). For example, the Street View images in Google Maps are just facades, flat images hinged together. But in the mirrorworld: a virtual building will have volume and accommodate past, current and prospective occupants and visitors with virtual social profiles, behaviours and feelings, a virtual chair will exhibit chairness, a virtual street will have layers of textures, gaps and intrusions. All these brought together through the synergies and compatibility of various technologies will convey a sense of 'street'.

In Figure 19.26, the concept of singularity is at the heart of the cognitive map. Key concepts and linkages include 'Westworld', 'Transformational' and 'Liminal'. Two chapters in this book are devoted to Westworld, namely Chapter 6 by Gurevitch and Chapter 15 by Lovell and Hitchmough. Gurevitch focus on the role of design and technology in the future of tourism, with automation disrupting the labour of service industries. Westworld imagines an industry fulfilling one of the few services traditionally assumed to be impervious to automation: human interaction. Throughout Westworld runs a tension between the design of experiences and the disruptive potential of technology to deliver such experiences within a framework of human morality. From another perspective, Lovell and Hitchmough deconstruct the narrative of the series focusing on hyper-staged authenticity of the 'technological frontier', examining how period detail such as costumes is produced by android technology, in the form of nostalgic-futurism. The relevance of the chapter relates to how tourists in Westworld seek pseudo 19th-century dangers in a Western theme park frontier. From a reality perspective, Westworld is a metaphor for change, demonstrating how, as inventions become more sophisticated, future tourism trends will reflect the fear of a post-human and post-Anthropocentric universe, a dark tourism and dystopian perspective which runs throughout this book. Combined, both chapters represent how science fiction is a transformational ontology represented by the concept of technological singularity. It is a literary or cinematic genre in which fantasy, typically based on speculative scientific discoveries or developments, environmental changes, space travel, or life on other planets, form part of the plot or background (Plotz, 2018). Westworld is an extension of Kurzweil's (2005) prediction of technological singularity which is embedded in technological futurism. According to Gordon Moore, one of the pioneers of the modern computer industry, as long as electronics have been manufactured the processing power of integrated circuits, relative to the cost of their production, has doubled approximately every two years (Track et al., 2017). However, Kurzweil goes further, positing that technology develops on exponential curve thus being transformational and organic and where, at a point in the future, technological growth becomes uncontrollable and irreversible, resulting in unforeseeable changes to human civilisation. This is called the intelligence explosion, where an independent, autonomous agent or entity will enter a runaway reaction of self-improvement cycles. Each new and more intelligent generation or version appears more and more rapidly, causing an explosion in intelligence. This is the case in Westworld where the processing power of the androids becomes far superior to that of humans, with the robots developing the ability to make irrational and emotion decisions (Engels & South, 2018) just like humans but better. Combined, these developments would be transformational, so imagine a future world where robots are the superior species not humans. So, who is the tourist in this case? More existentially, one might ask what reality is, as everything is liminal.

314　Part 5: Concluding Thoughts

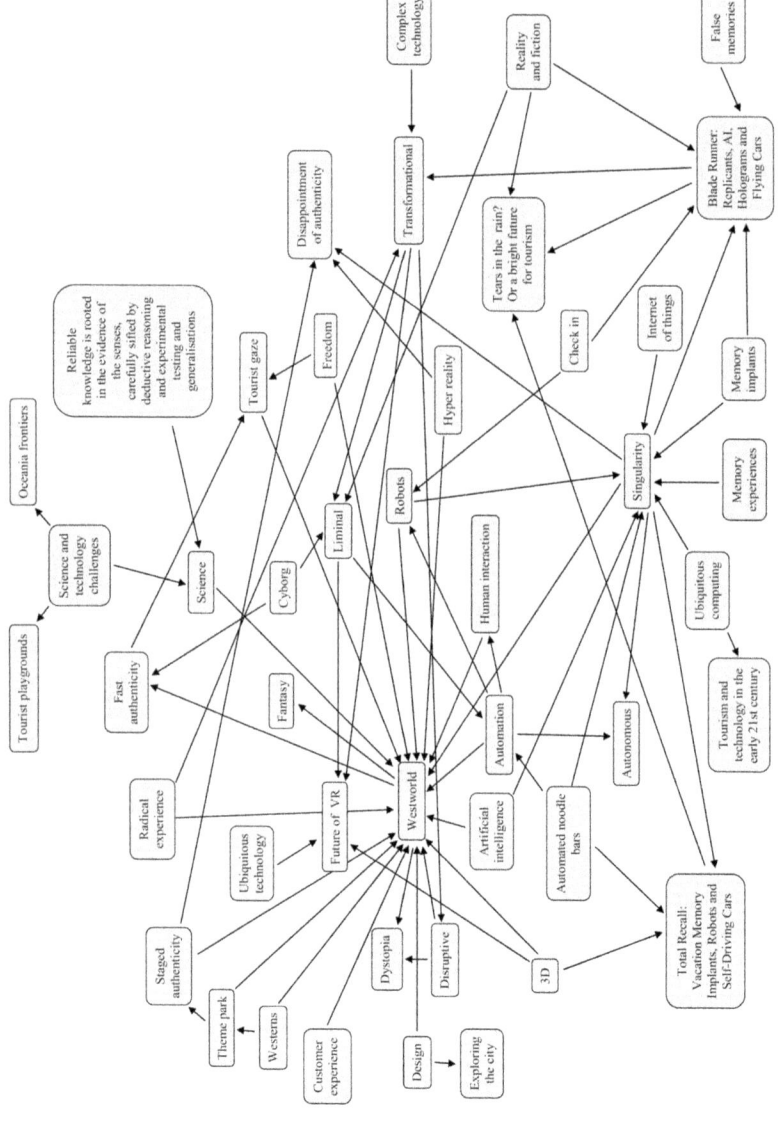

Figure 19.26 Technological singularity has arrived: *Westworld*

Viewpoint 9: Sustainability

One prime concern and powerful motive of sustainable tourism is to ensure its future by maintaining the economic and social benefits of tourism development while reducing or negating any adverse impacts on the natural, heritage, cultural or social environment (Edgell, 2006; Gössling et al., 2012). What will the future look like? Given the critical uncertainties surrounding climate change, disruption, overtourism and the inability of policy makers to make a decision about the future (Becken, 2012; Behrensmeyer, 2006; Gössling et al., 2010; Schott, 2010), there is a case for cultivating 'radical creativity' by opening a dialogue between science fiction literature – which also concerns itself with the future of humankind – and social research within the field of sustainability. As Gendron et al. (2017: 1544) state:

> Such a dialogue may help researchers to 'think without a banister', hence unlocking new theorizing potentials that would have remained invisible under the exclusive (and restrictive) light of institutional science. Such creativity may contribute to reinventing products and services or to reshaping business models.

Thus, the use of 'radical creativity' is better suited for recasting sustainability research based on an increased epistemological pluralism. To the extent that issues affect society as a whole, problematizing firm and management theories through 'radical creativity' indeed always partly amounts to problematising the relationships between researchers, research and society. This approach allows us to face the global challenges of sustainable tourism using new ontological and epistemologies to frame sustainable tourism (Yeoman & Postma, 2014). These concepts and arguments draw parallel with Figure 19.27, where the central concepts that dominate the cognitive map are 'Overtourism', 'Sustainability', 'Climate change' and 'The Environmental Zombie'. In Chapter 10, McEntee and colleagues discuss the destination of the dead and raise the concept of the environmental zombie. Using a dystopian scenario, the authors draw on the parallels with overtourism, in which the negative impacts of tourism outweigh the positive for the local community The chapter describes the invasion of destinations by 'herd like' tourists, who – to the local communities – are seen as a zombie horde. This metaphor calling tourists zombies has some mileage as tourism involves the separation of the individual from normal their 'instrumental' life. Being a tourist requires a person to something other and beyond themselves and often results in change. Just as zombism changes the person, tourists are also altered irrevocably by being on holiday. Reid and Ek in Chapter 14 extend this proposition as tourism exceeds the thresholds of overtourism and sustainability thus becoming unsustainable. Combined with the effects of social media and digital devices, a plague of blind tourism could arise in the future, with visitors auto-communicating rather than mindfully

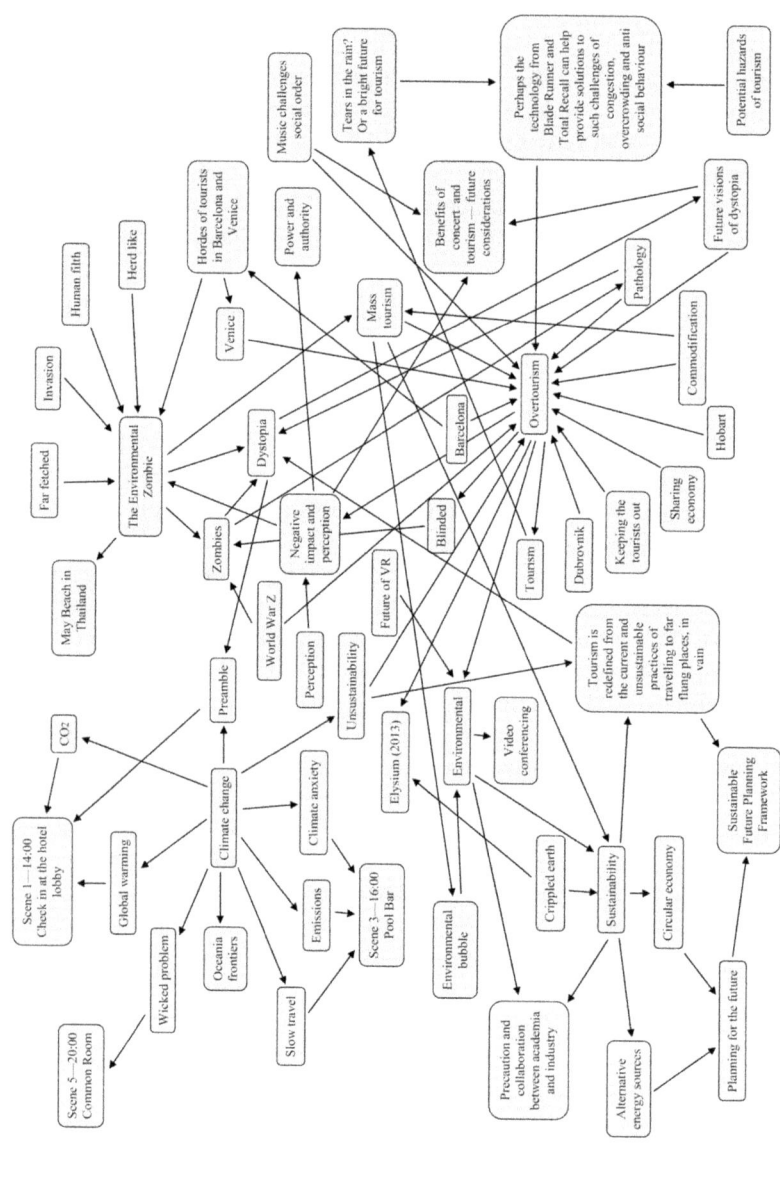

Figure 19.27 Sustainability

interacting with the places they visit. This is a tourist that is not engaging and is oblivious to the culture that surrounds the place that he or she is visiting. Reid and Ek conclude with an interesting question, that is, can a future tourism, even if it is less based on unsustainable travelling practices, be anything other than dystopic as long as it is embedded in the logic of a pathologic capitalism? One wonders if tourism evet change. In Chapter 18, Gren and Höckert present a day at the *Hotel Anthropocene,* during an emerging ecological catastrophe – thus capturing the arguments about the present decisions associated with climate change and tourism through and investigation of paradoxes and contradictions.

Concluding Arguments: Developing a Theoretical Framework – An Ontological and Epistemological Contribution

At the beginnings of the 20th century, H.G. Wells anticipated that futures research would become a scientific discipline. This thinking was fuelled by the post-World War II uncertainty and the need to deal with complex technologies in an uncertain world (Bell *et al.*, 2013). One hundred years later we have another cataclysmic event to deal with, COVID-19, and similarly we are talking about uncertainties, disruptions, liminal experiences and the urgency to reimagine 'better' tourism futures. For many, pre-pandemic, the future of tourism was simply an extension of the past, with the potential for linear growth. COVID-19 has disrupted not only the linear evolution of the past, but it has also challenged and questioned the fundamental taken for granted assumptions supporting this 'predicted evolution', i.e. the hypothesis that growth can happen only with numerical increases (Sigala, 2020). It has also been argued that COVID-19 will not also bring about the end of tourism as we know it (Duong, 2020; Grech, 2020; UNWTO, 2020). As the current uncertainties and challenges have emerged, academic and industry professionals and leaders have been using the lexicon of future studies to grapple with what tourism might look like post COVID-19. Words such as transformation, disruption, automation, dystopia, change and uncertainty have become prominent. These words are also commonly used in science fiction and embodied in many of the chapters in this edited collection. It is argued that, in order to make sense of the future, it is necessary to understand the concepts and theories of future thinking. Hence, this is the purpose of this edited collection that uses a science fiction genre to identify and develop theories associated with the future of tourism. Figure 19.18 represents a conceptualisation of tourism and science fiction as represented in this book.

However, no research is perfect and qualitative research is subjective in nature. The conceptualisation and writing of the book (chapters) started well before COVID-19 and without any suspicion of what was going to follow due to the pandemic. Given the emergence of COVID-19 and its resultant disruption, it become apparent that the relevance, value and role of this book in assisting scholars and industry alike to re-image and reset tourism futures in the post-pandemic period is more than warranted. To that end, we undertook a postscript analysis and relooked at the chapter outputs. This process has raised several questions: 'what if the contributors had written their chapter in a COVID-19 world?' or 'as experienced editors what do we think is missing?'. To address these two questions, we reflected back to our ontological framework (Figure 19.18) and incorporated in each of the seven concepts (determining tourism and science fiction) our personal reflections as well as current thinking and research about the role and the impact of COVID-19 on tourism transformation, the next normal and the tourism futures. In addition, we added three further tentative concepts (*Narrative, Scepticism, COVID-19 and Pandemics*), which are displayed in Figure 19.27 and explained in the next sections.

Why Tourism Futures Needs a Theoretical Framework: The Contribution of Science Fiction

Documenting the past is easy, as it has already happened and, indeed, this is the dominant perspective in tourism research (Yeoman & McMahon-Beattie, 2020). Why? It is empirical, factual and can be published (Bricker & Donohoe, 2015; Lohmann & Netto, 2017; Pearce & Butler, 2010; Tribe, 2010; Tribe & Xiao, 2011). As Tribe (2004; 2006) argues tourism management research is biased towards scientific positivism because of the nature of academic research, a focus on creation, measurement, consciousness, accountability and the context of business. On the other hand, imagining and forming tourism futures, can be seen as both a science and an art.

However, tourism is not as mature as other academic disciplines such as law, science, economics or geography. In 'Tourism Tribes, Territories and Networks', Tribe (2010) concedes tourism was a soft and relatively new field of study. The issue of contention is about the maturity of tourism and the development of appropriate underpinning theory. The study of tourism is not singular but draws on a multiplicity of disciplines. It has emerged as a consequence of the fluidity and changing patterns of consumption shaped by accelerated economic, social and technology drivers (Cohen & Cohen, 2012; Yeoman, 2006, 2012). These changes make the study of the future of tourism important, as the future is only place you can prepare for or move to. Tourism futures needs a theoretical framework to contribute towards the evolution of tourism research. But as Yeoman and Postma (2014) note, tourism futures is often presented without a foundation, is often misunderstood, and those that write about the future tend to emphasis presentism. What COVID-19 has taught us is the importance of moving beyond presentism and not thinking about the

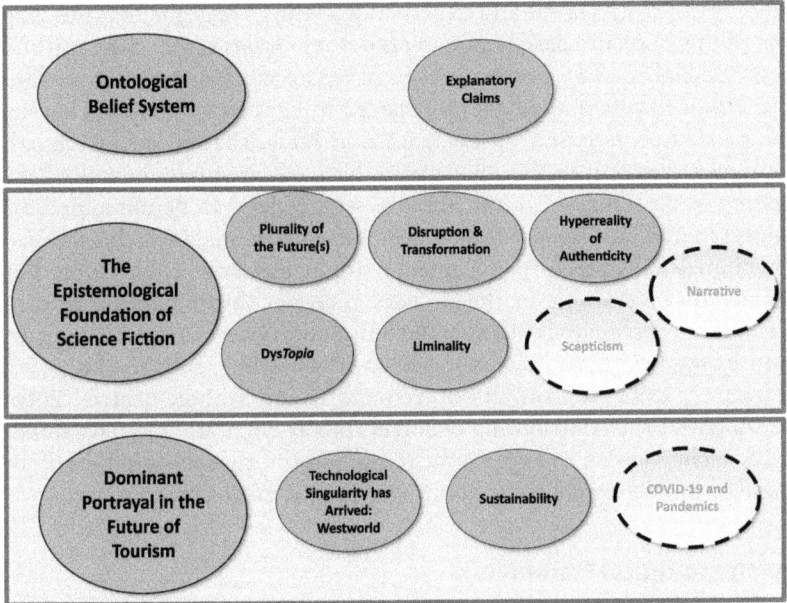

Figure 19.28 The future of tourism: a science fiction theoretical framework

future as a linear projection based upon previously studied inter-relations of known (economic) variables. Hence, the value of this edited collection as it encourages us to make a quantum leap in the terms of how we view and how we can afford to think about the future of tourism and tourism research. It takes us beyond the positivism to the non-linearity of interpretivism (Guba, 1990; Denzin & Lincoln, 2008, 2011) and a multiplicity of futures.

Figure 19.18 represents a conceptualisation of tourism and science fiction as represented in this edited collection. We propose a conceptual framework based upon an ontological and epistemological perspective, which is shown in Figure 19.28 and explained in the next section.

An Ontological Contribution: Explanatory Claims

An ontological perspective focuses on the belief system of those that execute tourism futures research. Ontology is concerned with the study of being and assumptions are concerned with what constitutes reality, in other words, what is (Scotland, 2012). Every paradigm is based upon its own ontological assumptions. Since all assumptions are conjecture, the empirical underpinnings can never be empirically proven or disproven, as from an ontological perspective, belief is personal and perceptual (Scotland, 2012). But what we can do is understand how that belief system is constructed. Drawing upon Bergman *et al.*'s (2010) ontological typology of truth claims and explanation claims, we argue that the chapters of this book are based upon *explanatory claims* rather than truthfulness (see Figure 19.1). Truthfulness is based upon a belief system in which the author beliefs that what is forecasted to happen will happen because the prediction is based on evidence and facts, that is, strong signals. However, the contributing chapters in this book are fundamentally based upon assumptions and weak signals, which need to be explained and interpreted. Thus, science fiction acts like a mechanism which explains a phenomenon. Therefore, the reader might ask how could something happen or what would be the circumstances for this to happen (or come true)? This explanatory claim is the foundation of scenario planning and futures studies; it is the classic 'What if' question (Schoemaker, 1995; Schwartz, 1993; Wilson, 2000; Yeoman & McMahon-Beattie, 2005). Throughout the contributing chapters, this is what authors have done – used science fiction to explain meaning, interpret weak signals and how these phenomena may come about and what it means for tourism.

Epistemological Framework

While science fiction might seem too far-fetched for some, we must acknowledge that change is necessary. Knowledge and change are intertwined, knowledge bringing change to society and organisations (Antonacopoulou, 1999). This takes us to the debate about what knowledge is and its relevance when studying the future (Johnson *et al.*, 2006; Johnson & Anthony, 2004). Knowledge is epistemology: the word is derived from two Greek words, *episteme* which means knowledge or science and *logos* which means knowledge, information, theory or account. This aetiology demonstrated how epistemology is usually understood as being concerned with the nature of knowledge and different methods of gaining knowledge. Epistemology leads to lots of questions about the forms of knowledge associated with science fiction and how it can be seen as a method of gaining knowledge. In Figure 19.27 we determine there are seven epistemological forms of knowledge associated with science fiction. Five are determined and adapted from Figure 19.18, namely *Plurality of Futures, Disruption & Transformation, Hyperreality of Authenticity,* Dys*Topia* and *Liminality*. In addition, two additional forms of knowledge have been added to Figure 19.27, namely *Scepticism* and *Narrative* based upon further analysis or lower ranked concepts from the cognitive mapping analysis.

First, it is not 'the future' but both 'futures' and 'future'. Thus, science fiction is based upon a *Plurality of the Future(s)* as the ontological belief system draws upon explanatory claims which support alternative perspectives or cosmic pluralism. Second, given the popularity of dystopian literature and the genre of dark tourism, science fiction can be seen to capture our fears, our criticisms of tourism and the dark side. This is clearly portrayed throughout this edited collection by the contributing authors. But dystopia is just one side of topia (Tavera, 2018). Topia is a place of specified characteristics which unifies dystopia and utopia (Latham, 2017). Indeed, Bergman et al. (2010) demonstrate that dystopia and utopia are a continuum. Given the present pandemic, we are seeing many destinations focusing on the reimagination of tourism, thereby adopting an utopian perspective in a dystopian world (Sigala, 2020). Thus, we have called this concept Dys*Topia* as the edited collection takes a predominately dysoptian perspective, with no contributor embracing a utopian perspective. Given different circumstances or different contributors, this may have been different.

Third, *Liminality* is the blurring of reality which can be seen as that space between fact and fictional or the passageway of in-between. Liminality is also a concept which can describe the transient, ambivalent inequilibrium state that COVID-19 has generated stressing every one of us to re-think and re-imagine the next normal. Fourth, is *Authenticity*: making sense of the future and creating a sense of reality is a central feature of scenario planning, hence the importance of authenticity. Authenticity in this context is about a narrative of 'how I see it' (Winter, 2002), but this is not about universal truth. It is an unfolding narrative, combining explanations, weak signals, possibilities and interpretations that are bound together. From a postmodernist perspective, we can describe this as the authenticity of hyperreality or simulacra in which the experience is simulated and imagined as a possibility (Rickly-Boyd, 2012). Therefore, science fiction is a narrative of signs of what could be. We label this concept as the *Hyperreality of Authenticity*. Fifth, given the present debate about the reimagination of tourism in a post-COVID-19 world (Gössling *et al*., 2020; Kock *et al*., 2020; UNWTO, 2020), *Disruption and Transformation* are prominent. Science fiction represents radical alternatives, the unthinkable, science which hasn't been invented yet and an opposite to the status quo.

We have added two further concepts based upon the analysis of the cognitive maps, namely, *Scepticism* and *Narrative*. Good futures research has a degree of scepticism or fallibility (Dator, 2014). As Nozick (2016) notes the future is a series of futures of unknowns. Scepticism is ripe within the genre of science fiction as it is seen as is a distant future, based upon weak signals, which asks the reader to suspend belief or is about technology that has not been invented (Polak, 1955; Slaughter, 1998).

Scenarios are stories. In the diverse field of scenario planning, this is perhaps the single point of universal agreement. Storytelling has long been a central ingredient in scenario planning, with key players, such as Shell, hiring professional storytellers like Betty Sue Flowers to work with management to better imagine scenario stories (Burnam-Fink, 2015). Therefore, we propose a final concept of *Narrative*. Stories are imaged and created, but when they are told they are interpreted differently by their receivers (i.e. the narrative of the story, what people understand from a story). In addition, when stories are shared and discussed, the interpretations of the stories are further shaped through co-creation. The narratives and the co-creation of the stories' meanings also reflect the multiplicity of (tourism) futures.

Theories of scenario planning have described scenarios as aids for reperceiving (Wack, 1985), decision wind-tunnels (Van der Heijden, 2005) and memories of the future (Weick, 1989), but in all cases they serve as a method for turning the unknown into a resource for strategic planning. By extrapolating the interplay of driving forces, predetermined elements and critical uncertainties, a set of scenarios spanning the space of plausible futures is developed and fleshed out (Schwartz, 1993). As Burnam-Fink (2015) points out that scenario planning narratives are often constrained by the politicisation of the organisation, thus bringing a degree of rationality to scenarios and squeezing out scepticism and fallibility, while science fiction narrative brings transformational thinking. It is an imaginative launch pad which is evocative and entertaining, without having to be strategic, useful or insightful like scenario planning narrative. Thus, the use of science fiction narrative in scenario planning is about imagination, thinking the impossible or how something could occur.

Overall, the science fiction narrative is a useful mechanism for imagining the future beyond normality (Gernsback & Westfahl, 1994). It provides a story to be interpreted through (subjective) scepticism, to enable individuals to generate their own story narratives that they can then share and discuss to co-create multiple futures.

Dominant Portrayal in the Future of Tourism

COVID-19 aside, technology and climate change are the prominent issues affecting tourism today. COVID-19 has magnified and intensified the urgency and the vitality to futurise a 'next tourism' that addresses technological advances and climate change. Thus, from a science fiction perspective they are the dominant issues in this book, for example, the concepts discussed in the section titled *Technological Singularity has Arrived: Westworld* and *Sustainability*. Futures studies has always been linked to technology, in fact futurism was the artistic movement that originated in Italy in the early 20th century which emphasised speed, technology, disruption and violence (Dator, 1986; Masini, 1989). Science fiction often portrays end of the world disasters. Such catastrophes might be brought about by climate change and/or technological advances such as robots and AI. By reading these stories and/or by living a simulated and controlled liminal experience of a hero/character featured in a science fiction movie, we generate our *narratives* of the story that trigger us to consider how things could fall apart or help create a picture of a future that we do not want. Thus, science fiction simulates a liminal stage that in turn allows us to think about the worst case scenario relating to climate change and an unsustainable future (Gendron *et al.*, 2017), and so, the inclusion of the concept *Sustainability* in Figure 19.27. The final concept, we have added is indeed *COVID-19 and Pandemics,* which we recognise would have had a stronger presence in this edited collection should COVID-19 had been emerging during the writing of the contributing chapters. Like the word *overtourism,* which was probably the tourism word of the year of 2019, COVID-19 is certainly the tourism word of the year in 2020 and beyond. As a transformational agent and stressor, COVID-19 has triggered and fueled many changes, some of which were already overdue. Hence, COVID-19 is also a major milestone event and concept to which many tourism transformations and futures will be attributed and related to, if only as a starting point.

What Next?

So, what will be next? Many science fiction movies and books come in sequels. So, COVID-20 we suppose but hope not! Indeed, that is not a science-fiction-inspired thought anymore but very much a possibility. We can only suggest you read a good science fiction novel and draw your own imaginings about the future of tourism. That's what we did, and Captain Kirk was our inspiration along with films such as *Soylent Green, Star Wars* and *Blade Runner.* Delve into those alternative, imaginative worlds and ask yourself, what if they were to come true?

References

Abbott, S. (2009) Arthouse SF Film (Part IV Subgenres). In M. Bould, A.M. Butler, A. Roberts and S. Vint (eds) *The Routledge Companion to Science Fiction*. Abingdon: Routledge, Taylor & Francis Group.

Ackermann, F. (2011) How OR can contribute to strategy making. *Journal of the Operational Research Society* 62 (5), 921–923. See doi:10.1057/jors.2010.128 (accessed 19 Janaury 2021).

Amara, R. (1981) The futures field: Searching for definitions and boundaries. *The Futurist* 15 (1), 25–29.

Antonacopoulou, E. (1999) Individuals' responses to change: The relationship between learning and knowledge. *Creativity and Innovation Management* 8 (2), 130–39.

Atwood, M. (2004) *Oryx and Crake: A Novel*. New York: Anchor Books.

Baudrillard, J. (2001) *Selected Writings* (2nd edn). Cambridge: Polity.

Becken, S. (2012) *Climate Change and Tourism from Policy to Practice*. New York: Routledge.

Beech, N. (2011) Liminality and the practices of identity reconstruction. *Human Relations* 64 (2), 285–302.

Behrensmeyer, A. (2006) Climate change and human evolution. *Science* (Washington), 311 (5760), 476–478.

Bell, F., Fletcher, G., Greenhill, A., Griffiths, M. and McLean, R. (2013) Science fiction prototypes: Visionary technology narratives between futures. *Futures* 50 (June), 10.

Bell, W. (1993) *Foundations of Futures Studies: History, Purposes, Knowledge: Human Science for a New Era*. London: Transaction Publishers.

Bergman, A., Karlsson, J.C. and Axelsson, J. (2010) Truth claims and explanatory claims – an ontological typology of futures studies. *Futures* 42 (8), 857–865. See doi:https://doi.org/10.1016/j.futures.2010.02.003 (accessed 21 January 2021).

Bhaskar, R. (1978) *A Realist Theory of Science*. Hemel Hamstead: Harvester Press.
Blanchot, M. (1995) *The Writing of the Disaster (L'écriture du désastre)*. Lincoln: University of Nebraska Press.
Braidotti, R. (2013) *The Posthuman*. Cambridge: Polity Press.
Brake, M. (2006) On the plurality of inhabited worlds: A brief history of extraterrestrialism. *International Journal of Astrobiology* 5 (2), 99–107.
Bricker, K.S. and Donohoe, H. (2015) *Demystifying Theories in Tourism Research*. Boston, MA: CABI International.
Burnam-Fink, M. (2015) Creating narrative scenarios: Science fiction prototyping at Emerge. *Futures*, 70 (June), 48–55. See doi:https://doi.org/10.1016/j.futures.2014.12.005
Callaghan, V., Miller, J., Yampolskiy, R. and Armstrong, S. (2017) *The Technological Singularity: Managing the Journey*. Berlin, Heidelberg: Springer.
Camus, A. (1948) *The Plague*. New York: A. A. Knopf.
Change, I.P.O.C. (2014) Climate Change 2014 Synthesis Report. See https://www.ipcc.ch/site/assets/uploads/2018/05/SYR_AR5_FINAL_full_wcover.pdf (accessed 21 January 2021).
Chaturvedi, S. and Doyle, T. (2010) Geopolitics of fear and the emergence of 'climate refugees': Imaginative geographies of climate change and displacements in Bangladesh. *Journal of the Indian Ocean Region* 6 (2), 206–222.
Clarke, A.C. (1972) *2001: A Space Odyssey*. London: Arrow Books.
Clute, J. and Nicholl, P. (1999) *Encyclopedia of Science Fictions*. London: Orbit.
Cohen, E. (1979) Rethinking the sociology of tourism. *Annals of Tourism Research* 6 (1), 18–35.
Cohen, E. and Cohen, S.A. (2012) Current sociological theories and issues in tourism. *Annals of Tourism Research* 39 (4), 2177–2202.
Corbin, J.M. (2015) *Basics of Qualitative Research: Techniques and Procedures for Developing Grounded Theory* (4th edn). Los Angeles: Sage.
Curry, A. and Hodgson, A. (2020) Seeing in multiple horizons: Connecting futures to vision and strategy. In R.A. Slaughter and A. Hines (eds) *Knowledge Base of Futures Studies* (pp. 66–85). Houston: Association of Professional Futurists and Foresight International.
Dator, J. (1986) Futures report: The futures of futures studies — the view from Hawaii. *Futures* 18 (3), 440–445.
Dator, J. (2009) Alternative futures at the Manoa School. *Journal of Futures Studies* 1 (2), 1–18.
Dator, J. (2014) Four images of the future. *Set: Research Information for Teachers* 1, 61–63. doi:10.18296/set.0319
Dator, J. and Yeoman, I. (2015) Tourism in Hawaii 1776-2076: Futurist Jim Dator talks with Ian Yeoman. *Journal of Tourism Futures* 1 (1), 36–45.
Denzin, N.K. and Lincoln, Y.S. (2008) *The Landscape of Qualitative Research*. Thousand Oaks, California, USA: Sage Publications Inc.
Denzin, N.K. and Lincoln, Y.S. (eds) (2011) *The Sage Handbook of Qualitative Research* (4th edn). Thousand Oaks, California, USA: Sage Publications Inc.
Dick, P.K. (1968) *Blade Runner – Do Androids Dream of Electric Sheep*. New York, N.Y: Ballantyne Books.
Dick, P.K. (2002) *We Can Remember it for you Wholesale*. New York, N.Y: Citadel Press Books.
Dredge, D. and Jenkins, J. (2011) *Stories of Practice Tourism Policy and Planning*. Farnham: Ashgate.
du Cros, H. (2017) A work of science fiction?: Engaging tourism stakeholders on climate change before it's too late. In *Proceedings of CAUTHE: Time for big ideas? Re-thinking the field for tomorrow* (pp. 644–648): Department of Tourism, University of Otago (February, 7–10).

Duedahl, E. (2020) Co-designing emergent opportunities for sustainable development on the verges of inertia, sustaining tourism and re-imagining tourism. *Tourism Recreation Research*, 47, 1–16. See https://doi.org/10.1080/02508281.2020.1814520 (accessed 24 January 2021).

Duong, M. (2020) COVID-19 and after: Impact and ways forward. *East Asia Forum Quarterly* 12 (2), 31–33.

Eco, U. (1986) *Faith in Fakes: Essays*. London: Secker and Warburg.

Eden, C. and Ackerman, F. (1998) *Making Strategy: The Journey of Strategic Management*. London: Sage Publications.

Edgell, D.L. (2006) *Managing Sustainable Tourism: A Legacy for the Future*. Binghamton, NY: Haworth Hospitality Press.

Ellis, A., Park, E., Kim, S. and Yeoman, I. (2018) What is food tourism? *Tourism Management* 68, 250–263.

Engels, K.S. and South, J.B. (2018) *Westworld and Philosophy: If you go Looking for the Truth, Get the Whole Thing*. Hoboken, NJ: Wiley Blackwell.

Farkic, J. (2020) Consuming dystopic places: What answers are we looking for? *Tourism Management Perspectives*, 1–33.

Fergnani, A. and Song, Z. (2020) The six scenario archetypes framework: A systematic investigation of science fiction films set in the future. *Futures*, 124, 102645. See doi:https://doi.org/10.1016/j.futures.2020.102645 (accessed 21 January 2021).

Gendron, C., Ivanaj, S., Girard, B. and Arpin, M.-L. (2017) Science-fiction literature as inspiration for social theorizing within sustainability research. *Journal of Cleaner Production* 164, 1553–1562. See doi:https://doi.org/10.1016/j.jclepro.2017.07.044 (20 January 2021).

Gernsback, H. and Westfahl, G. (1994) How to write 'science' stories: The editor of 'Scientific Detective Monthly' tells how to and how not to write them. *Science Fiction Studies* 21 (2), 268–272.

Gillespie, A. (2006) Tourist photography and the reverse gaze. *Ethos* 34 (3), 343–366.

Gössling, S., Hall, C.M., Ekström, F., Engeset, A.B. and Aall, C. (2012) Transition management: A tool for implementing sustainable tourism scenarios? *Journal of Sustainable Tourism* 20 (6), 899–916.

Gössling, S., Hall, C.M., Peeters, P. and Scott, D. (2010) The future of tourism: Can tourism growth and climate policy be reconciled? A climate change mitigation perspective. *Tourism Recreation Resarch* 35 (2), 119–130.

Gössling, S., Scott, D. and Hall, C.M. (2020) Pandemics, tourism and global change: A rapid assessment of COVID-19. *Journal of Sustainable Tourism* 29 (1), 1–20.

Grech, V. (2020) Unknown unknowns – COVID-19 and potential global mortality. *Early Human Development*, 144, 105026. See doi:https://doi.org/10.1016/j.earlhumdev.2020.105026 (accessed 19 January 2021).

Guba, E.G. (1990) *The Paradigm Dialog*. Newbury Park, Calif.: Sage.

Guttentag, D.A. (2010) Virtual reality: Applications and implications for tourism. *Tourism Management* 31 (5), 637–651. See doi:https://doi.org/10.1016/j.tourman.2009.07.003 (accessed 20 January 2021).

Hansen, B., Østerhus, S., Quadfasel, D. and Turrell, W. (2004) Already the day after tomorrow? *Science* 305 (5686), 953–954.

Huff, A.S. and Jenkins, M. (2002) *Mapping Strategic Knowledge*. London: Sage.

Johnson, P., Buehring, A., Cassell, C. and Symon, G. (2006) Evaluating qualitative management research: Towards a contingent criteriology. *International Journal of Management Reviews* 8 (3), 27.

Johnson, R.B. and Anthony, J.O. (2004) Mixed methods research: A research paradigm whose time has come. *Educational Researcher* 33 (7), 14–26.

Johnston, J. (Writer) (1995) *Jumanji*. USA: Tristar Pictures.

Jones, M. (1993) *Decision Explorer: Reference manual (Version 3.1)*. Kendal: Banxi Software Ltd.

Kelly, G. (1977) Personal construct theory and the psychotherapeutic interview. *Cognitive Therapy and Research* 1 (4), 355–362.

Kelly, G.A. (1955) *The Psychology of Personal Constructs*. New York: Norton.

Kelly, K. (2019) AR will spark the next big tech platform – call it Mirrorword. See https://www.wired.com/story/mirrorworld-ar-next-big-tech-platform/ (accessed 7 March 2021).

Keough, S.M. and Shanahan, K.J. (2008) Scenario planning: Toward a more complete model for practice. *Advances in Developing Human Resources* 10 (2), 166–178.

Kidd, D. (2007) Harry Potter and the functions of popular culture. *The Journal of Popular Culture* 40 (1), 69–89.

Kock, F., Nørfelt, A., Josiassen, A., Assaf, A.G. and Tsionas, M.G. (2020) Understanding the COVID-19 tourist psyche: The evolutionary tourism paradigm. *Annals of Tourism Research* 85, 103053. See https://doi.org/10.1016/j.annals.2020.103053 (accessed 31 January 2021).

Koontz, D. (1981) *The Eyes of Darkness*. New York: Pocket Books.

Kubrick, S. (Writer) (1968) *2001: A Space Odyssey*. In S. Kubrick (Producer). USA: Metro Goldwyn-Mayer.

Kurzweil, R. (2005) *The Singularity Is Near: When Humans Transcend Biology*. New York: Viking.

Lashua, B.D. (2018) The time machine: Leisure science (fiction) and futurology. *Leisure Sciences* 40 (1–2), 85–94.

Latham, R. (2017) *Science Fiction Criticism: An Anthology of Essential Writings*. London: Bloomsbury Publishing Plc.

Leiserowitz, A.A. (2004) Day after tomorrow: Study of climate change risk perception. *Environment: Science and Policy for Sustainable Development* 46 (9), 22–39.

Lennon, J.J. and Foley, M. (2000) *Dark Tourism: The Attraction of Death and Disaster*. New York: Cengage Learning EMEA.

Light, D. (2017) Progress in dark tourism and thanatourism research: an uneasy relationship with heritage tourism. *Tourism Management* 61, 275–301.

Lohmann, G. and Netto, A.P. (2017) *Tourism Theory: Concepts, Models and Systems*. Wallingford: CABI International.

MacCannell, D. (1973) Staged authenticity: Arrangements of social space in tourist settings. *American Journal of Sociology* 79 (3), 589–603.

MacCannell, D. (1976) *The Tourist: A New Theory of the Leisure Class*. New York: Schocken Books.

Manjikian, M. (2012) *Apocalypse and Post-politics: The Romance of the End*. New York: Lexington Books.

Mars, M.S., Yeoman, I.S. and McMahon-Beattie, U. (2017) Ping pong in Phuket: The intersections of tourism, porn and the future. *Journal of Tourism Futures* 3 (1), 39–55.

Marshall, A. (1890) *Principles of Economics*. London: Macmillian.

Masini, E.B. (1989) The future of futures studies: A European view. *Futures* 21 (2), 152–160.

Mayurika, C. (2020) Science fiction explores the interconnectedness revealed by the coronavirus pandemic. *The Conversation*. See https://search.proquest.com/docview/2425924871?rfr_id=info%3Axri%2Fsid%3Aprimo (accessed 26 January 2021).

Mingers, J. (2014) *Systems Thinking, Critical Realism and Philosophy: A Confluence of Ideas*. Florence: Routledge.

Mishan, E.J. (1969) *The Costs of Economic Growth*. London: Weidenfeld and Nicolson.

Moriarty, J. (2012) Theorising scenario analysis to improve future perspective planning in tourism. *Journal of Sustainable Tourism* 20 (6), 779–800.

National Research Council (2010) *Persistent Forecasting of Disruptive Technologies – Report 2*. Washington, D.C.: National Academies Press.

Nietzsche, F.W. (1986) *Human, All too Human: A Book for Free Spirits*. Cambridge: Cambridge University Press.

Nozick, R. (2016) Knowledge and SCEPTICISM. In H. Arló-Costa, V.F. Hendricks and J. van Benthem (eds) *Readings in Formal Epistemology: Sourcebook* (pp. 587–603). Cham: Springer International Publishing.

Ocident, B., Gilbert Gilibrays, O., Eric Oyondi, N., Alex, M. and Timothy, O. (2020) Exponential disruptive technologies and the required skills of industry 4.0. *Journal of Engineering* 2020. See https://downloads.hindawi.com/journals/je/2020/4280156.pdf (accessed 30 January 2021).

Page, S., Yeoman, I., Munro, C., Connell, J. and Walker, L. (2006) A case study of best practice – visit Scotland's prepared response to an influenza pandemic. *Tourism Management* 27 (3), 361–393. See doi:https://doi.org/10.1016/j.tourman.2006.01.001 (accessed 23 January 2021).

Pastor, D. and Kent, A.J. (2020)Transformative landscapes: liminality and visitors' emotional experiences at German memorial sites. *Tourism Geographies* 22 (2), 250–272.

Pearce, D. and Butler, R. (eds) (2010) *Tourism Research: A 20-20 Vision*. Oxford: Goodfellow Publishers.

Plotz, J. (2018) Science fiction. *Victorian Literature and Culture* 46 (3–4), 854–858.

Podoshen, J.S., Venkatesh, V., Wallin, J., Andrzejewski, S.A. and Jin, Z. (2015) Dystopian dark tourism: An exploratory examination. *Tourism Management* 51, 316–328.

Polak, F. (1955) *The Image of the Future*. Amsterdam: Elsevier.

Poser, H. (1990) Struktur und Dynamik wissenschaftlicher Theorien: Beitrage zur Wissenschaftsgeschichte und Wissenschaftstheorie aus der bulgarischen Forschung. *Isis* 81, 154–155.

Power, A. (2014) Panic on the streets of London: Tourism and British dystopian cinema. *Science Fiction Film and Television* 7 (1), 77–98, 158.

Preece, T. and Price, G. (2005) Motivations of participants in dark tourism: A case study of Port Arthur, Tasmania, Australia. In C. Ryan, S.J. Page and M. Aicken (eds) *Taking Tourism to the Limits: Issues, Concepts and Managerial Perspectives* (pp. 191–198). New York: Elsevier.

Rand, A. (1967) *Capitalism: The Unknown Ideal*. New York: Sgnet.

Rickly-Boyd, J.M. (2012) Authenticity & aura: A Benjaminian paproach to tourism. *Annals of Tourism Research* 39 (1), 269–289. See doi:https://doi.org/10.1016/j.annals.2011.05.003 (24 January 2021).

Rigby, C.E. (2015) *Dancing with Disaster: Environmental Histories, Narratives, and Ethics for Perilous Times*. Charlottesville: University of Virginia Press.

Robertson, M. and Yeoman, I. (2014) Signals and signposts of the future: Literary festival consumption in 2050. *Tourism Recreation Research* 39 (3), 321–342.

Robertson, M., Yeoman, I., Smith, K. and McMahon-Beattie, U. (2015) Technology, society, and visioning the future of music festivals. *Event Management* 19 (4), 567–587.

Romero, G.A. (2004) *Dawn of the Dead*. Troy, MI: Anchor Bay Entertainment.

Schänzel, H.A. and Yeoman, I. (2014) The future of family tourism. *Tourism Recreation Research* 39 (3), 343–360.

Schnaars, S.P. (1987) How to develop and use scenarios. *Long Range Planning* 20 (1), 105–114.

Schoemaker, P.J.H. (1995) Scenario planning: A tool for strategic thinking. *Sloan Management Review* 36 (2), 25.

Schott, C. (ed.) (2010) *Tourism and the Implications of Climate Change Issues and Actions*. Bingley: Emerald.

Schwartz, P. (1993) Composing a plot for your scenario. *Planning Review* 20 (3), 4–46.

Scotland, J. (2012) Exploring the philosophical underpinnings of research: Relating ontology and epistemology to the methodology and methods of the scientific, interpretive and critical research paradigms. *English Language Teaching* 5 (9), 9–16.

Scott, R. (1982) *Blade Runner*. Burbank, CA: Warner Home Video.

Searle, J.R. (1995) *The Construction of Social Reality*. New York: Free Press.

Shim, C. and Santos, C.A. (2014) Tourism, place and placelessness in the phenomenological experience of shopping malls in Seoul. *Tourism Management* 45, 106–114.

Shklar, J. (1965) The political theory of utopia: From melancholy to nostalgia. *Daedalus* 94 (2), 367–381.

Sigala, M. (2020) Tourism and COVID-19: Impacts and implications for advancing and resetting industry and research. *Journal of Business Research* 117 (June), 312–321. See doi:https://doi.org/10.1016/j.jbusres.2020.06.015 (accessed 22 January 2021).

Sigala, M. and Steriopoulos, E. (2021) Does emotional engagement matter in dark tourism? Implications drawn from a reflective approach. *Journal of Heritage Tourism*, 1–21.

Silver, C. and Lewins, A. (2014) *Using Software in Qualitative Research: A Step-by-Step Guide* (2nd edn). London: Sage.

Slaughter, R.A. (1998) Futures beyond dystopia. *Futures* 30 (10), 993–1002.

Stableford, B. (2003) Science fiction before the genre. In E. James and F. Mendlesohn (eds) *The Cambridge Companion to Science Fiction* (pp. 15–31). Cambridge: Cambridge University Press.

Stapledon, W.O. (1930) *Last and First Men*. London: Methuen.

Strugatsky, A. and Strugatsky, B. (1971) *Roadside Picnic*. Basingstoke: Macmillan.

Tarkovsky, A. (Writer) (1979) *Stalker*. USSR: Goskino.

Tavera, S.P. (2018) Her body, Herland: Reproductive health and dis/topian satire in Charlotte Perkins Gilman. *Utopian Studies* 29 (1), 1–20.

Toivonen, A. (2021) *Sustainable Space Tourism: An Introduction*. Bristol: Channel View Publications.

Track, E., Forbes, N. and Strawn, G. (2017) The end of Moore's law. *Computing in Science & Engineering* 19 (2), 4–6.

Tribe, J. (2004) Knowing about tourism. Epistemological issues. In J. Phillimore and L. Goodson (eds) *Qualitative Research in Tourism* (pp. 46–62). London: Routledge.

Tribe, J. (2006) The truth about tourism. *Annals of Tourism Research* 33 (2), 360–381. See doi:http://dx.doi.org/10.1016/j.annals.2005.11.001 (accessed 20 January 2021)

Tribe, J. (2010) Tribes, territories and networks in the tourism academy. *Annals of Tourism Research* 37 (1), 7–33.

Tribe, J. and Liburd, J.J. (2016) The tourism knowledge system. *Annals of Tourism Research* 57, 44–61.

Tribe, J. and Xiao, H. (2011) Developments in tourism social science. *Annals of Tourism Research* 38 (1), 7–26.

Turner, V.W. (1974) *Dramas, Fields, and Metaphors*. Ithaca, NY: Cornell University Press.

Turner, V. (1998) Betwixt and between: The Liminal period in rites de Passage. In L. Mahdi, S. Foster and M. Little (eds) *Betwixt and Between: Patterns of Masculine and Feminine Initiation* (pp. 4–20). New York: Open Court.

UNWTO (2020) Policy brief: COVID-19 and transforming tourism. See https://webunwto.s3.eu-west-1.amazonaws.com/s3fs-public/2020-08/SG-Policy-Brief-on-COVID-and-Tourism.pdf (accessed 22 January 2021).

Urry, J. and Larsen, J. (2011) *The Tourist Gaze 3.0*. Los Angeles; London: Sage.

Van der Heijden, K. (2005) *Scenarios: The Art of Strategic Conversation* (2nd edn). Chichester, West Sussex: John Wiley.

Van der Heijden, K., Bradfield, R., Burt, G., Cairns, G. and Wright, G. (2002) *Sixth Sense Accelerating Organizational Learning with Scenarios*. Chichester: Wiley.

Veblen, T. (1899) *The Theory of the Leisure Class: An Economic Study of Institutions*. London: Allen & Unwin.

Wack, P. (1985) Scenarios: uncharted waters ahead. *Harvard Business Review* 63 (5), 72.

Weick, K.E. (1989) Theory construction as disciplined imagination. *Academy of Management Review* 14 (4), 516–531.

Wells, H.G. (1895) *The Time Machine*. Minneapolis: Lerner Publishing Group.

Wells, H.G. (1902) The discovery of the future. *Nature* 65 (1684), 326–331.

Wilson, I. (2000) From scenario thinking to strategic action. *Technological Forecasting & Social Change* 65 (1), 23–29.

Winter, R. (2002) Truth or fiction: Problems of validity and authenticity in narratives of action research. *Educational Action Research* 10 (1), 143–154.

Yeoman, I. (2004) The development of a conceptual map of soft operational research practice. (PhD), Napier University, Edinburgh. See http://researchrepository.napier.ac.uk/id/eprint/3873 (accessed 20 January 2021).

Yeoman, I. (2006) How technologies are shaping the future? *Journal of Revenue and Pricing Management* 5 (1), 1.

Yeoman, I. (2012a) *2050: Tomorrow's Tourism*. Bristol: Channel View Publications.

Yeoman, I. (2012b) Authentic learning: My reflective journey with postgraduates. *Journal of Teaching in Travel & Tourism* 12 (3), 295–311.

Yeoman, I. (2013) Tomorrow's tourist and New Zealand. In J. Leigh, C. Webster and S. Ivanov (eds) *Future Tourism* (pp. 161–188). London: Routledge.

Yeoman, I. and Beeton, S. (2014) The state of tourism futures research. *Journal of Travel Research* 53 (6), 675–679.

Yeoman, I., Galt, M. and McMahon-Beattie, U. (2005) A case study of how VisitScotland prepared for war. *Journal of Travel Research* 44 (1), 6–20.

Yeoman, I., Lennon, J.J. and Black, L. (2005) Foot-and-mouth disease: A scenario of reoccurrence for Scotland's tourism industry. *Journal of Vacation Marketing* 11 (2), 179–

Yeoman, I. and McMahon-Beattie, U. (2020) Does the past shape the future of tourism? A cogntive map(s) perspective. In I. Yeoman and U. McMahon-Beattie (eds) *The Future Past of Tourism: Historical Perspectives and Future Evolutions* (pp. 243–307). Bristol: Channel View Publications.

Yeoman, I., McMahon-Beattie, Backer, E., Robertson, M. and Smith, K. (eds) (2014) *The Future of Events and Festivals*. Oxford:Routledge.

Yeoman, I. and McMahon-Beattie, U. (2005) Developing a scenario planning process using a blank piece of paper. *Tourism and Hospitality Research* 5 (3), 273–285.

Yeoman, I. and McMahon-Beattie, U. (2018) Framing tourism futures research: An ontological perspective. In C. Cooper, S. Volo, W. Gartner and N. Scott (eds) *The Sage Handbook of Tourism Management: Theories, Concepts and Disciplinary Approaches to Tourism* (Vol. 2, pp. 463–484). London: Sage.

Yeoman, I. and McMahon-Beattie, U. (2020) *The Future Past of Tourism: Historical Perspectives and Future Evolutions*. Bristol: Channel View Publications.

Yeoman, I., McMahon-Beattie, U. and Wheatley, C. (2015) The future of food tourism: A cognitive map(s) perspective. In I. Yeoman, U. McMahon-Beattie, K. Fields, J.N. Albrecht and K. Meethan (eds) *The Future of Food Tourism: Foodies, Experiences, Exclusivity, Visions and Political Capital* (pp. 237–278). Bristol: Channel View Publications.

Yeoman, I., Palomino-Schalscha, M. and McMahon-Beattie, U. (2015) Keeping it pure: Could New Zealand be an eco paradise? *Journal of Tourism Futures* 1 (1), 19–35.

Yeoman, I. and Postma, A. (2014) Developing an ontological framework for tourism futures. *Tourism Recreation Research* 39 (3), 299–304.

Yeoman, I. and Watson, S. (2011) Cognitive maps of tourism and demography: Contributions, themes and further research. In I. Yeoman, C.H.C. Hsu, K.A. Smith and S. Watson (eds) *Tourism and Demography* (pp. 209–236). Oxford: Goodfellow.

Zheng, C., Zhang, J., Qiu, M., Guo, Y. and Zhang, H. (2019) From mixed emotional experience to spiritual meaning: Learning in dark tourism places. *Tourism Geographies* 22 (1) 105–126.

Index

Abandoned site, 12, 45, 49–52, 126, 213–214, 219, 225–226
Activism, 88–93, 267
Adventure, 9, 11, 20, 33, 47, 52, 57–58, 62, 77, 133–143, 162–164, 170, 188, 191, 193, 218, 226, 230, 264, 274–275, 307
Algorithm, 79–82
Alien, 30, 32, 162–163, 177–178, 189–190, 218, 224–225, 298, 302
Alienation zone, 224–225
American Frontier, 188
Animal ethics in tourism, 98–99
Animal uplifting, 99–101, 103–105
Anthropocene, 33, 215, 234–251, 285–287, 300
Apocalypse, 122–123, 128, 215, 230, 285, 300, 307, 311
Artificial intelligence (AI), 67, 73–74, 77, 109, 148, 195–196, 264, 271, 312
Augmented reality (AR), 97, 100–101, 109, 147, 269
Authenticity, 9–10, 43, 46–47, 52–53, 77–78, 81, 111, 114, 148, 150, 152, 189–197, 262, 281
Authority, 10, 84–85, 87–88, 90–92, 267, 312
Automata, 194
Automation, 73–83, 257, 264, 306, 312

Biotechnology, 97, 99–100, 102
Blue planet, 31–32, 36

Capitalism, 33, 47, 178, 184, 215, 264, 278, 281, 317
Charity, 205
Chatbots, 109, 112–113, 271
Chernobyl Exclusion Zone, 12, 215–220, 229, 285

Circular economy, 60–61, 64, 67, 69
City tourism, 48–51
Climate change, 19, 33, 68, 116, 181, 238, 246–250, 256, 287, 300
Code, 73, 79–82, 156, 179, 256–257, 264
Collective responsibility, 87–88, 92
Communities, 44, 49, 52, 86, 89, 91, 93, 102–104, 120–123, 126, 147, 165–166, 169, 200–203, 207–210, 214, 283
Conflict, 11–12, 85–87, 100, 201, 207–208, 299
Concerts, 11, 49, 65, 90, 149, 154, 195, 201–210
Conspicuous Consumption, 177–184
Consumer behaviour, 44, 51
Consumption, 177–184
COVID-19, 4–6, 8, 13, 26–27, 166, 255–256, 260, 293–307
Crime, 125, 161, 164–170, 269, 278
Culture, 3, 10, 35, 45, 48, 51–52, 65, 76, 80, 85, 88–92, 104, 114–115, 120, 122, 125–128, 142, 147, 164–165, 178, 180, 184, 201–210, 215, 217, 220, 230, 267–269, 271, 283, 295, 298, 311
Cultural capital, 180, 183
Cultural heritage, 35, 189
Customer experience, 135, 142

3D environment, 146
Dark tourism, 215, 217, 219, 226, 299, 300, 311, 320
Design, 10–11, 32, 34–35, 48, 74–75, 79, 81–82, 106, 135, 177, 179, 191, 196, 244, 264–267, 298, 313
Digital tourism, 176, 179–180
Digital transformation, 110
Disasters, 201, 205, 207, 210, 230, 299, 322

Disease, 25–26, 46, 161, 163, 166, 169, 269, 299, 304
Discrimination, 10, 86–87, 90–92, 103, 267, 312
Diving, 32, 34, 36, 307
Dystopia, 4–6, 9, 11–12, 19, 22–23, 27, 31, 42–53, 116, 121, 126, 175–251

Earth, 20, 32–33, 38, 56–58, 62, 64–66, 68–69, 123, 162–167, 184, 205, 208, 218, 226, 236, 247–251, 262, 274, 281, 302
Earth system, 236, 247, 251
Ecological catastrophe, 244–245, 249, 317
Economic, 4–5, 10, 26, 46, 48, 51, 58, 61–62, 67, 69, 74, 125–126, 151, 161, 165–166, 177, 179, 183, 195, 203–204, 215–216, 242, 267
Emotions, 149–150, 167, 204–205, 208–209, 247
Enclave tourism, 161–170, 278
Environmental, 12, 37, 59, 63–64, 67–70, 81, 98, 101, 103, 105, 121–122, 126, 152, 154, 181, 183, 201, 215–217, 248–249, 251, 264
Environmental bubble, 163, 170, 269, 278, 311
Ethics, 5, 60, 97–101, 154, 248–249, 287, 311
Events, 3, 6, 23–24, 26, 45, 90, 110–111, 135, 138, 153, 188, 190, 192, 201–206, 209–210, 258, 293, 295, 311
Exclusion zone, 214, 215–230
Existential Authenticity, 42–53, 262
Experience, 9–10, 13, 19, 21, 26, 30, 32–39, 42–53, 57, 65–67, 73–83, 98, 100–105, 109–117, 120–124, 135–142

Fake news, 168
Fandom, 85, 88–92
Fans, 21, 88–94, 188, 202, 205–206, 267
Feelings, 43, 47, 64–65, 146, 149, 165, 169, 204–205, 247, 249
Fiction, 3–8
Fordism, 81, 196
Frontier, 148, 187–197
Frontier tourism, 189
Futures map, 59–61, 67

Gaming, 147, 191, 193, 197
Global warming, 208, 238, 250, 311
Grounded theory, 60, 67, 69, 257
Guests, 77, 81, 120, 192–197, 235–244, 247–251, 281, 287, 312

Harry Potter, 10, 84–94, 267–269, 311, 312
Heritage, 33–35, 42, 45–47, 49–51, 86, 135, 147, 154, 189, 191–192, 262, 300, 315
Hermeneutic Circle, 50
Holocene, 234, 236, 239, 243, 245, 247, 250, 285
Holograms, 10, 111–114, 116, 194, 271
Hosts, 43, 80–82, 120, 187, 189, 191, 193–197, 249, 283, 312
Hotel, 35, 38, 43, 58, 62, 66, 78, 109–110, 113, 115, 134, 140, 142, 147, 153, 194, 225–228, 234–251, 260, 271, 285–287
Hyper-consumerism, 43
Hyper-reality, 114, 179
Hyper-staged authenticity, 191–194, 281, 313
HTC Vive, 146, 156, 219

Immortality, 162, 193, 197, 300
Implantable Memories, 114, 149, 176, 182, 196
Information search, 43, 52, 79–80, 83, 110, 112, 115, 135–140
Information reliability, 60
Innovative, 10, 31, 33, 44, 64–65, 128, 135
Intellectual property, 156
Internet of things, 109, 271

Knowledge, 12–13, 19, 24, 30–35, 59–61, 64, 66, 68–69, 80, 89, 106, 111, 138, 142, 165, 167, 201, 230, 245, 256, 260, 287, 291

Liminality, 13, 33, 219, 229, 290, 295, 306, 320
Literature, 3, 7, 10, 12–13, 20–25, 43, 60, 85, 111, 134–135, 138, 195, 201, 257, 274, 306, 315

Maigret, 133–143, 274–275, 306
Marketing, 43, 45, 52, 128, 141–142, 147, 150, 179, 204, 215, 224
Melancholia, 205, 241, 244, 249

Mindfulness, 183
Mixed reality, 11, 133–143, 274–275
Montserrat, 12, 213–230, 285, 307, 311
Moral disengagement, 92–93, 269
Morality, 10, 85, 87–88, 90–92, 267, 312
Music, 51, 65, 76, 200–210, 225, 283–285
Myth, 27, 34, 188–192, 281–283, 302

Neural Implants, 101, 105, 148, 275
New Media, 79–81
Nuclear tourism, 63, 208

Oculus, 146, 150, 219
Overtourism, 10–11, 13, 52, 106, 116–117, 120–128, 166, 189–191, 177, 195, 197, 209, 271, 274, 278, 281, 315, 323

Perceived risk, 161–170, 269, 278
Planetary ethics, 248, 287, 311
Popular culture, 3, 10, 76, 85, 88–93, 128, 184, 215, 220, 267–269, 311
Post-apocalyptic site, 230
Post-human, 11, 36, 187, 197, 260, 264, 281–283, 313
Post Tourist, 44–45
Postmodern theory, 44
Power, 6, 10, 23, 37, 43–44, 51, 58, 63, 74, 81, 84–93
Precautionary principle in tourism, 102–104
Presence, 34, 37–38, 102, 146, 149, 166, 168, 206, 217, 260, 269, 278, 323
Product development, 90, 133
Prosumption, 43–44, 49, 262

Road mapping, 59
Robots, 67, 81, 109–110, 114–116, 155–157, 169, 191, 194–195, 271, 313, 322

Safety, 46–47, 57, 62, 64, 69, 101, 161–170, 278, 311
Safety bubble, 161–170, 278, 311
Scenario technique, 274
Science, 3–13, 19–27, 56–69, 73–119, 255–323
Sculpture museum, 35–36
Second Life, 67, 147, 151, 155
Selfie Pathology, 124–125

Sharing economy, 65, 183
Simenon, 134, 136
Simulation, 77, 81, 100, 114, 147, 155, 194, 307, 312
Smart tourism, 112, 117
Social change, 85, 88, 93, 209, 267, 312
Social media, 65, 83, 101, 110, 112, 117, 124, 168, 175, 178–179, 184, 217, 269, 278, 281, 315
Societal, 85–86, 91–92, 175–178, 281
Society, 6–7, 22–23, 34, 41–53, 60, 65, 85, 87, 98, 105, 111, 115–116, 120, 122–125, 177–178, 200–210, 214, 267, 271, 283, 299, 304, 312, 315, 320
Solutions, 11, 48, 63, 65, 104, 106, 116, 128
Soviet history, 56, 218
Space, 7, 30, 32–39, 42–53, 74, 77, 80, 88, 90, 92, 111, 126, 135, 151, 153, 162–163, 166, 169, 179, 189, 191–193, 206–207, 214, 216–225, 229, 250, 262
Space legislation, 62–63
Space tourism, 9, 56–69
Staged authenticity, 111, 187–192, 281, 298
Staycation, 176, 183, 249, 281
Submerged hotels, 35
Surveillance, 37, 156, 193, 196, 207, 283
Sustainable planning, 60, 63–64, 67–69, 315
Sustainability, 60, 63–64, 67–69, 315

Technological frontier, 11, 187–198
Technology, 4, 10–11, 32–38, 57–64, 73–77, 79, 83, 99–106, 109–117, 142–143, 146–157, 169, 178, 182–183, 188–197, 216, 258, 260, 262
Terrorism, 161, 165–169, 269
Theme parks, 11, 84, 90, 100, 189–197
Tourism connectivity, 37–38, 109–116, 190, 271
Tourism experience design, 10, 48, 52, 57, 67, 100, 102, 105, 146, 149–155, 260, 275, 296
Tourism products, 48, 67, 133, 262
Tourist gaze, 73, 75–77, 81–82, 189, 205, 238, 264, 298
Transformative, 30, 87–93, 260
Travel Photography, 35, 45, 49–53, 179–180
Travel trends, 162

Television, 75–76, 82, 100, 112, 115, 153, 187–189, 195, 206, 274, 281
Twenty-first century, 9, 30, 32, 35, 38, 52, 57, 110–112, 260, 262, 304, 312

Undersea tourism, 9, 31–33, 260
Underwater pods, 35
Urban Exploration, 44, 49–51, 262, 300
Urbex, 220
Ukraine, 12, 214–217, 285
Utopia, 4, 6, 9–12, 19, 22–23, 31, 42, 45, 48, 101, 183–184, 191, 206–210, 258, 260, 262, 298, 309, 321

Value drivers, 133–143, 274
Victimisation, 162, 164–167
Virtual, 11, 60–61, 66–69, 100–101, 104, 106, 109–110, 117, 133–134, 137, 142, 145–157, 182, 219, 269, 271, 275, 298, 312

Virtual reality (VR), 11, 100–101, 104, 106, 109–110, 117, 133–134, 137, 142, 145–157, 182, 219, 269, 271, 275, 298, 312
Virtual travel, 60–61, 66–69
Virtual world, 11, 156
Volcano, 220–230

Weak signals, 9, 12, 25, 59, 67, 69, 258, 260, 296, 320, 321
Web 2.0, 43, 49, 262
Westworld, 8, 10–11, 13, 73–83, 187–197, 264–267
Wildlife tourism, 97–106
Wildlife tourism experience, 104–105
Wildlife tourism challenges, 105

Zombie Trope, 119–128
Zombies, 10, 119–128

For Product Safety Concerns and Information please contact our EU Authorised Representative:

Easy Access System Europe

Mustamäe tee 50

10621 Tallinn

Estonia

gpsr.requests@easproject.com